IFIP Advances in Information and Communication Technology 780

IFIP Advances in Information and Communication Technology

The IFIP AICT series publishes state-of-the-art results in the sciences and technologies of information and communication. The scope of the series includes: foundations of computer science; software theory and practice; education; computer applications in technology; communication systems; systems modeling and optimization; information systems; ICT and society; computer systems technology; security and protection in information processing systems; artificial intelligence; and human-computer interaction.

Edited volumes and proceedings of refereed international conferences in computer science and interdisciplinary fields are featured. These results often precede journal publication and represent the most current research.

The principal aim of the IFIP AICT series is to encourage education and the dissemination and exchange of information about all aspects of computing.

More information about this series at https://link.springer.com/bookseries/6102

Eunika Mercier-Laurent · Gülgün Kayakutlu ·
Iwona Chomiak-Orsa · Lukasz Przysucha ·
Dominique Verdejo
Editors

Artificial Intelligence for Sustainability and Global Security

IFIP WG 12.11 and WG 12.13 International Workshops, IJCAI 2025
Montréal, QC, Canada, August 16–17, 2025
Revised Selected Papers

Springer

Editors
Eunika Mercier-Laurent
Université Reims Champagne-Ardenne
Reims, France

Gülgün Kayakutlu
Istanbul Technical University
Istanbul, Türkiye

Iwona Chomiak-Orsa
Wroclaw University of Economics
Wrocław, Poland

Lukasz Przysucha
Wroclaw University of Economics
Wrocław, Poland

Dominique Verdejo
Université Reims Champagne-Ardenne
Reims, France

ISSN 1868-4238 ISSN 1868-422X (electronic)
IFIP Advances in Information and Communication Technology
ISBN 978-3-032-21183-5 ISBN 978-3-032-21184-2 (eBook)
https://doi.org/10.1007/978-3-032-21184-2

This Springer imprint is published by the registered company Springer Nature Switzerland AG
The registered company address is: Gewerbestrasse 11, 6330 Cham, Switzerland

AI4S 2025 Preface

Since the initial IFIP TC12.6 AI4KM workshop at ECAI (European Conference on Artificial Intelligence) 2012 held in Montpellier, France, in 2025 due to growing interest in AI our workshop was split into three: AI4KM ran this year separately and AI4S and AI4GS were added. Actually, these three workshops brought together in Montreal, Canada, the IFIP TC12.6, TC12.11 and TC12.13 working groups.

The previous workshop at IJCAI 2021 in Montréal was held online. The virtual conference, exposition and social events were amazing; however, it still lacked human exchanges, energy and serendipity.

Combining Knowledge Management, Energy, Sustainability and Global Security leads to better understanding of the scope of each and of connections between the related topics. This year we included three papers on AI for Global Security. They are in the second part of this volume.

Sustainability covers many concerns often addressed separately. The answers of Chat GPT clearly demonstrate this separation. However, all elements influence each other and should be considered from various perspectives. The holistic and system approaches to sustainability provide a deeper understanding of the interconnections between the concerned domains and their interplay. It also includes security issues induced by technology, globalization and changing behaviors.

During these two days the participants improved their understanding of overall aspects of sustainability through various lenses and their connections with Climate Change.

Some information about the AI4S reviewing process:

AI4S received 35 submissions, of which 19 papers were accepted for presentation during these two days. The submitted papers received an average of two single-blind reviews each, with three in the case of a conflict; the acceptance rate was 54%. The invited speakers were selected based on current trends in the addressed fields, the interests of the audience and local proximity. Their presentations are available on the workshop website and the corresponding papers, extended after the exchanges with the audience, are included in this volume.

The accepted papers covered a large scope of topics related to sustainability in various contexts such as smart cities, agriculture, energy and gas production and distribution, industry, management and biodiversity. The authors covered most of the 17 UN Sustainable Development Goals.

Three invited talks at AI4S by Toktam Fabrizi, Lynne Schneider and Margaret Havey emphasized deep understanding of the impacts of AI. We had also an opportunity to meet Ioanna Roussaki representing the EU sponsored project DIVINE on Sustainable Agriculture, who we invited to discuss this promising project.

At the end of the second day the present PC members and organizers selected the best papers to be extended and included in this volume.

After the introduction Eunika Mercier-Laurent introduced the covered fields and their interconnections.

Our first invited speaker Toktam B. Fabrizi from Western Sydney University presented **"A Hybrid Knowledge-Based and Large Language Model Framework for Sustainable Design Education"**, which proposes a practical pedagogical framework that integrates artificial intelligence to enhance the teaching of circular design in Life-Cycle Assessment (LCA). The environmental footprint of generative AI is reviewed, addressing energy use, carbon emissions, water consumption, material resources and electronic waste. The study discusses how the educational benefits of AI can outweigh these environmental costs when sustainable AI practices are applied.

The following article **"The Future of Work: Exploring Expectations and Fears Toward Artificial Intelligence in the Labor Market"** by Jolanta Kowal, Dmitriy Yanchylik and Paweł Weichbroth addresses AI anxiety among Polish workers using a mixed-methods approach. Applying AI Anxiety Scale and thematic analysis, the authors compared anxiety levels between automation-prone occupations and resilient ones. The study emphasizes that effective AI adoption requires emotional, not only technical, readiness—especially in vulnerable job sectors.

The next paper entitled **Generative AI for sustainable problem solving in I5.0 Generative AI** by Cristina R. Monsone, Eunika Mercier-Laurent and Ádám B. Csapó discusses complex industrial design problems with consideration of sustainability. In line with the human-centric approach of Industry 5.0, generative AI enables personalized manufacturing, where products are tailored to individual customer preferences. This proactive approach minimizes resource use and waste, contributing to more sustainable manufacturing processes. This helps managers and technicians make informed decisions that balance economic, social and environmental considerations. This facilitates better collaboration between humans and machines, enhancing productivity and innovation.

The authors of **"Causal Analysis of Machine Deviations on Espresso Sensory Quality"**, Kouassi Kouakou Kan Maxime, Michal Ptaszynski, Fumito Masui, Tanjim Mahmud, Tsuyoshi Gotoda, Hirofumi Arai, Chao-Hui Feng and Jun Watanabe, investigate the causal effects of deviations from targets by espresso machines on consumer-rated sensory attributes as Aroma, Bitterness, Concentration and Richness. This work consists in understanding the correlation between the features and the final satisfaction of consumers through a causal inference framework to estimate the Average Treatment Effect of parameters on sensory outcomes. This research addressing well-being highlights the complexity of espresso extraction and the estimated influence of the deviation of the parameters on the features of this beverage, and the need for further research with larger datasets and consideration of other influential factors.

The next paper entitled "**Studying Neural Network Robustness to Weight Perturbations for Sustainable Incremental Learning**" by Parth Mehta and Xiaoyin Wang proposes incremental learning to avoid retraining of the model with the full dataset with the aim of saving a huge amount of computation resources and energy. However, this sustainable technique also faces additional threats such as adversarial attacks, system errors, and software bugs. They compare the performance of several architectures under various conditions, including different initial training durations, error percentages and error types with the aim to better understand robustness of neural network models.

This paper is followed by "**Time Series Augmentations with Unsupervised View-makers for Robust Disruption Prediction in Nuclear Fusion**" by Dhruva Chayapathy, Lucas Spangher, Tavis Siebert, Om Patil and Akshata Kishore Moharir. The authors study the problem of detecting disruptions - i.e. plasma instabilities that can cause significant damage, impairing the reliability and efficiency required for the real-world viability of nuclear fusion. Machine learning often struggles to generalize the diverse characteristics and dynamics of different machines. This is a critical barrier to scaling fusion technology. Given the success of data augmentation in improving model robustness and generalizability in other fields, this study explores the use of a novel time series view maker network to generate diverse augmentations or "views" of training data. This is essential for advancing fusion technology and ultimately addressing climate change through reliable and sustainable energy production.

The next article "**Self-Refining Segment Anything: A Perturbation-Driven Approach for Semi-Supervised Satellite Image Segmentation**" by Renzo Larenas, Luis Caro and Billy Peralta addresses the growing demand for accurate information in precision agriculture, which, driven by climate change and the need for resource optimization, has led to the development of advanced techniques for satellite image segmentation. They propose a lightweight approach based on the Segment Anything Model, enhanced through an iterative pseudo-labeling strategy that relies on segmentation consistency under salt-and-pepper noise perturbations. This methodology is validated through experiments on real satellite imagery.

Aleksander Kruczkowski and Paweł Weichbroth in their paper entitled **"A review of reinforcement learning methods and performance measures in algorithmic stock trading**" investigate the application of reinforcement learning methods in algorithmic stock trading. They determined sixteen performance evaluation measures useful for both the research community and novice stock traders.

This is followed by the review by Konstantin M. Golubev "**Artificial Intelligence Business and Global Sustainability Review"**, which focuses on the impact of the hype of Artificial Intelligence LLMs on global sustainability. The author studied Large Language Models, AI agents and AI reinforcement learning and discusses the limitations of current methods in solving real-life problems and proposes the need for a wise evaluation of its applicability.

The second invited talk by Lynne Schneider elaborated on "**Supply Chain Considerations for AI Implementation – Critical Minerals, Data Centers, and the Environment".** This chain is crucial for AI and data centers because they depend on a steady, secure supply of critical minerals such as copper, silicon, gallium, aluminum and rare earth elements, essential for manufacturing semiconductors, power distribution systems, cooling networks, high-speed processors and storage devices that underpin AI infrastructure and data center operations. Mining and processing these minerals have significant environmental footprints, including habitat destruction, water usage, waste generation and greenhouse gas emissions. She suggested balancing AI growth with environmental safeguards with the aim of fostering a more sustainable future.

Robert Tracey, M. Emre Sahin and Carol Mak in **"Accelerating LSTM model training through biologically inspired NeuroAI and investigating novel applications"** explore the impact of Spiking Neural Networks in LSTM via the NeuroAI toolkit. Compared with the traditional LSTM, the training time of the adapted LSTM significantly decreased by as much as 75% but with an increase in accuracy. They also demonstrate that using the SNN decreases the amount of CPU used by as much as 14% during training, thereby being more efficient and sustainable.

This is followed by **"Optimal Planning and Machine Learning for Responsive Tracking and Enhanced Forecasting of Wildfires using a Spacecraft Constellation".** Sreeja Roy-Singh, Vinay Ravindra, Richard Levinson, Mahta Moghaddam, Jan Mandel, Adam Kochanski, Angel Farguell Causo, Kurtis Nelson, Samira Alkaee Taleghan, Archana Kannan and Amer Melebari propose a novel concept of operations using optimal planning methods and machine learning to collect spaceborne data for monitoring wildfires, process it to create new or enhanced products in the context of wildfire danger or spread monitoring, and assimilate them to improve existing, wildfire decision support tools delivered to firefighters within latency appropriate for time-critical applications.

The next article **"Classification Framework for Stream Type Identification"** by Golnaz Mesbahi, Shadan Golestan, Buddy Brown, Joey Ton and Jerome Cranston proposes a framework that leverages diverse geospatial features and domain knowledge to curate representative datasets and design hierarchical classification models for stream type identification. Empirical results demonstrate that the proposed framework outperforms standard multi-class models, offering great promise for exploring data representation and structure in environmental classification tasks.

This is followed **by "Explainable AI for Wildfire Management and Ecological Sustainability: A Short Survey of Techniques, Case Studies, and Operational Pathways"** by Akshata Kishore Moharir, Sameera S Kashyap, Jay Prakash Thakur and Divya Nayak. They tackle the very timely topic of the growing destructiveness of wildfires, amplified by climate change, which demands transparent predictive tools. AI models have advanced wildfire prediction, but their "black-box" nature hinders operational adoption. This short survey investigates explainable AI (XAI) methods for wildfire management and ecological sustainability, focusing on bridging the experimental to operational gap. They systematize XAI techniques (model-agnostic, model specific, hybrid/emerging) into a unified taxonomy, evaluating their interpretability, computational efficiency and fidelity. Deployed case studies show that XAI integration yields substantial benefits but also reveals critical implementation barriers such as resource constraints and legacy system integration. The paper provides actionable guidelines for selecting the right XAI methods that balance technical stability with operational practicality. It also highlights research priorities to align the advances of XAI with the needs of firefighters, policymakers and ecosystem sustainability managers to promote climate resilience and biodiversity conservation.

The next article **"RAG4MuniESG: ESG Assessment for Municipal Sustainability through Retrieval-Augmented Generation"** by Dorsaf Sallami and Juste Rajaonson presents an AI-driven framework combining Large Language Models (LLMs) with Retrieval-Augmented Generation (RAG) to facilitate ESG (Environmental, Social and Governance) assessments specific to municipalities. Using a public sector-oriented ESG scheme, the system semantically merges structured indicators with unstructured documents. The authors propose a hybrid AI-human approach wherein automated extraction and scoring are augmented with transparent and traceable justifications to support human-in-the-loop verification and auditability.

Marek Szajt, Klaudia Smoląg, Katarzyna Kozioł-Nadolna, Zuzana Hajduová and Iwona Chomiak Orsa in **"Changing the perception of digital competences in 2018 and 2025 among Polish and Slovak students"** present a comparative analysis of the self-esteem among Polish and Slovak students and researchers related to their competencies in the use of technology. This study points out the subjectivity in self-evaluation. This aspect may influence their evaluation of the environmental impact of technology.

"The current intensity of AI use and the perception of AI applications in strategic decision-making" by Katarzyna Grzesik, Renata Brajer-Marczak, Michał Nadolny, Katarzyna Liwarska-Fulczyk, Iwona Chomiak-Orsa and Anna Marciszewska presents a study of the impact that the frequency of AI use has on managers' perceptions of AI applications in strategic decision-making. The obtained results indicate the need for further education in the use of AI and building trust in AI. This is particularly important for those people who rarely use AI solutions. The research results confirmed the irreversibility of the process of adapting AI solutions in the approach to strategic decision-making.

"A Simulation-Based Approach to Processes in the Municipal Waste Management Sector" by Aleksander Binsztok, Beata Butryn, Krzysztof Hauke and Maciej Pondel discusses waste management, one of the major challenges faced by contemporary cities, requiring the optimization of collection, transport and processing operations while taking into account environmental, economic and social aspects. They propose to use computer simulation as a tool to support efficient municipal decision making and planning of waste management at the level of a residential housing estate. Their simulation model reflects residents' behaviors, waste collection schedules, and the spatial distribution of waste containers and it allows the analysis of various scenarios, such as changes in collection frequency, reorganization of waste collection routes, or ecological education initiatives targeted at residents. The results indicate that computer simulation can serve as a valuable decision-making tool for local governments and estate managers, supporting the planning of sustainable municipal waste management strategies.

Our third invited speaker Margaret Havey elaborated on the **Canadian policy of sustainable development** in the context of the 17 UN Sustainable Development Goals and the AI era. The accelerating deployment of artificial intelligence (AI) is unfolding in parallel with the world's triple planetary crisis of climate change, biodiversity loss and pollution. M. Havey explains how AI intersects with sustainable development, with a specific focus on Canada's national context, governance model and emerging AI ecosystem. She traces the evolution of global sustainability efforts from the UN's early environmental initiatives to the adoption of the 2030 Agenda and its 17 Sustainable Development Goals (SDGs), and outlines how Canada has translated these commitments into a whole-of-government approach through its Federal Sustainable Development Strategy (FSDS). She also explains how Canada is leveraging AI to accelerate clean energy, critical minerals development, industrial decarbonization and ecosystem protection, while simultaneously navigating the trade-offs that accompany sustainable development—particularly those related to mining, AI data center infrastructure, water consumption and energy demand. The societal implications of AI adoption, including labour market disruption, risks associated with generative AI, and impacts on creative industries and education, are explored as key considerations for responsible technology governance. Ultimately, this paper argues that AI can be a powerful enabler of sustainability if deployed within strong ethical, environmental and institutional frameworks, but warns that without careful oversight, AI's resource intensity and social impacts could undermine the very goals it seeks to advance.

Next, Lukasz Przysucha and Adam Sulich in their paper entitled "**Data-Driven Sustainability: The Role of Urban Platforms in Participatory Smart-City Governance**" examine how urban digital platforms that integrate sensing, data governance, analytics and participatory tools translate citizen inputs into sustainability outcomes and greenjob formation. A comparative multiplecase study is conducted across Amsterdam, Athens, Seville and Osaka. The results indicate that platforms with transparent pipelines from citizen input to implementation show stronger links to SD indicators and ecosystem activity; robust datagovernance mechanisms broaden participation and enhance the durability of results; and close alignment between platform analytics and procurement/operations fosters local greenanddigital employment. Implications for platform design and governance are outlined, together with research needs on participation quality, impact measurement and transferability across institutional contexts.

In the article that follows by Michał Petri, Iwona Chomiak-Orsa and Łukasz Przysucha **"Algorithmic Governance in ESG Reporting: A Comparative Study of AI Tools' Effectiveness in Mitigating Greenwashing"** investigates the dual nature of AI as both a facilitator and potential disruptor of transparency in ESG reporting, a phenomenon conceptualized herein as the "transparency paradox." While AI has the potential to enhance data accuracy, automate reporting processes, and enable more comprehensive sustainability assessments, it also introduces challenges related to algorithmic bias, interpretability, and unintended opacity in decision-making. The study critically evaluates how AI systems may both mitigate and exacerbate issues such as greenwashing and unreliable disclosures.

The next paper **"Technological challenges, energy poverty and other barriers to the expansion of energy communities in Europe"** by Evaldo Costa and Eunika Mercier-Laurent investigates how renewable energy communities adopt cutting-edge technologies in Europe, what barriers they face to their expansion and how these aspects are linked to energy poverty. The results indicate that technologies such as smart grids and digital platforms are the most frequently adopted, while newer solutions such as blockchain, IoT, VPPs and battery energy storage remain underutilized, requiring targeted policy support. The study highlights the need for greater attention to supporting research and innovation, which can be achieved by harmonizing specific policies to promote inclusive and low-carbon energy transitions.

We had also the opportunity to include **"Towards sustainable, AI-powered agriculture: Emerging technologies, benefits and barriers"** by Marios Paraskevopoulos, George Routis and Ioanna Roussaki.

Currently, agriculture faces pressures from climate change, resource scarcity, biodiversity loss and increasing food demand, intensifying the need for sustainable and resilient transformation. Artificial intelligence has emerged as a key part of this transition, offering advanced capabilities for precision agriculture, resource management, supply-chain optimization and data-driven decision-making. The authors examine the role of AI in supporting sustainability across the agri-food sector, thoroughly explaining its main benefits, including increased efficiency and productivity, environmental and socio-economic gains and improved risk assessment and resilience. They assess the challenges and barriers that hinder broader AI adoption, including infrastructural, regulatory, economic and cultural constraints that restrict equitable deployment.

We hope you enjoying reading these papers.

December 2025

Eunika Mercier-Laurent
Gülgün Kayakutlu
Iwona Chomiak-Orsa
Lukasz Przysucha
Dominique Verdejo

AI4S Organization

Program Chairs

Eunika Mercier-Laurent	Prof. Dr-Eng, University of Reims Champagne Ardenne, France and Chair IFIP TC12 (AI)
Iwona Chomiak-Orsa	Dr Hab University of Economics Wroclaw, Poland

Programme Committee AI4S

Weronika Adrian	AGH University of Science and Technology, Poland
Joel Alleyne	University of Toronto, Canada
Cecilia Coelho	University of Minho, Portugal
Noël Conruyt	University of Reunion Island, France
Iwona Chomiak-Orsa	Wroclaw University of Economics and Business, Poland
Anne Dourgnon	Research Center EDF, France
Konstantin M. Golubev	Independent Researcher, Ukraine
Krzysztof Hauke	Wroclaw University of Economics and Business, Poland
Gulgun Kayakutlu	Instanbul Technical University, Turkey
Krzysztof Kluza	AGH University of Science and Technology, Poland
Antoni Ligeza	AGH University of Science and Technology, Poland
Nada Matta	University of Technology of Troyes, France
Eunika Mercier-Laurent	University of Reims Champagne-Ardenne, France
Cristina R. Monsone	Italy
Simeon Simoff	Western Sydney University, Australia
Michael Stankosky	George Washington University, USA
Abdul Sattar	Griffith University, Australia
Frederique Segond	Inria, France
Guillermo Simari	Universidad Nacional del Sur, Argentina
Eric Thivant	University Lyon 3, France
Dominique Verdejo	University of Reims Champagne-Ardenne, France
Paweł Weichbroth	Gdańsk University of Technology, Poland

Janusz Wojtusiak	George Mason University, Fairfax, USA

Local Organizing Committee

Joel Alleyne	University of Toronto, Canada
Margaret Havey	Shared Services, Canada

AI4GS 2025 Preface

Considering the growing integration of Artificial Intelligence into critical infrastructures and defense systems, the need to address the multiple facets of global security has become imperative. In a digitized world facing increased geopolitical tensions, transversal technologies like AI offer unprecedented opportunities but also raise major challenges regarding sovereignty, trust, and tactical coordination.

AI4GS 2025 was organized in continuation of the first conference AI4GS 2024, held in Paris. The main themes that emerged this year were **Autonomous Defense**, **Digital Sovereignty**, and the **Ethics of Trust**.

The presentations covered theoretical and practical aspects, ranging from data governance to collaborative air combat simulation.

Articles reviewing process: From the submissions received, the program committee made a rigorous selection. The selected articles underwent a thorough review process to guarantee the relevance and scientific quality of the contributions. The three papers presented in this volume perfectly illustrate the diversity and complementarity of the approaches necessary to secure the future of AI: a strategic approach through sovereignty analysis, a tactical approach through multi-agent coordination, and an ethical approach through explainability.

After the introduction by Eunika Mercier-Laurent the session was structured as follows:

Sovereignty and Technological Strategy

The first article, titled **"A SWOT analysis of AI technology sovereignty landscape assessment"** addressed the critical issue of digital sovereignty in the defense domain, particularly for military aircraft. Facing challenges of national security and strategic autonomy, the authors Mariya Evtimova-Gardair, Frédérique Segond, Patrick Fabiani, Thibault Lahire, and Michel Rottmaier propose a classification of sovereign AI and Machine Learning technologies. The presented assessment based on a SWOT analysis (Strengths, Weaknesses, Opportunities, Threats) is applied to sovereign software. This approach helps to identify technology priorities and measure the "sovereignty degree" of an AI pipeline using fuzzy logic. The study highlights determining criteria such as source code ownership, data origin, and interoperability. Through a concrete use case on the *Scikit-learn* library, the authors demonstrate how the auditability and transparency of open-source tools can strengthen national control over critical functions, such as target recognition, against risks of vendor lock-in or hidden vulnerabilities.

Defense and Multi-Agent Systems

In the operational domain, the article **"Rule-Based Coordination Strategies for Protection in Multi-Agent Defense Scenarios"** by Thibault Lahire explores the challenges of protecting High-Value Assets (HVA) with autonomous systems in contested environments. The study focuses on coordination among allied drones tasked with escorting a strike drone through hostile territory. The author introduces a discrete grid simulation environment, inspired by tactical military settings, to evaluate various rule-based coordination strategies. Unlike Deep Reinforcement Learning (DRL) approaches, which can be opaque, this work prioritizes rule-based systems for their interpretability and robustness—qualities essential for certification in aerospace contexts. Results show that when outnumbered by enemy drones, hybrid strategies—balancing close protection of the asset with support for engaged allies—yield higher success rates. This work contributes to the standardization of coordination protocols for collaborative defense missions.

Trust, Ethics, and Governance

The article **"Trust and Explainability in Ethical AI for Global Security"** by Brian Challita and Pierre Parrend offers a systematic literature review examining the intersection of explainability (XAI), trust, and ethical governance in high-stakes security contexts. While AI is becoming integrated into critical infrastructures like surveillance and threat detection, the authors emphasize that explainability must not be treated merely as a technical feature, but as a fundamental pillar of accountability. The analysis of 23 selected studies reveals a critical gap: over 70% of existing XAI approaches fall short of fully incorporating human-centric design principles suitable for sensitive operational environments. The paper highlights that current models (such as LIME or SHAP) often struggle to meet stakeholder requirements regarding security. The authors advocate a multidisciplinary approach bridging law, ethics, and systems engineering to ensure that AI technologies deployed for global security are not only effective but also transparent and socially acceptable.

At the end of the session, we had the pleasure to exchange with Hans G. Bathija on the rising topic of Cognitive Security, referring to the practice of protecting individuals and societies from manipulative information and psychological influences that can distort perceptions and decision-making. We discussed the use of AI to detect misinformation, deepfakes, and other forms of cognitive manipulation that target human thinking. However, AI is also used to massive generation of fakes and nudges. While AI can help to detect consistency in information, humans have to practice more and more critical thinking.

We hope that you enjoy reading these papers as much as we had pleasure in collecting and reviewing them.

Dominique Verdejo
Eunika Mercier-Laurent

AI4GS Organization

Dominique Verdejo	University of Reims Champagne Ardenne, France
Eunika Mercier-Laurent	University of Reims Champagne Ardenne, France, Chair IFIP TC12 (AI)

Program Committee

Joël Courtois	P4S, France
Antoine Guilmoto	Independent Researcher, France
Ammar Mohammed	Modern Sciences and Arts University and Cairo University, Egypt
Didier Mascarelli	KADLOG, France
Brett Van Niekerk	Durban University of Technology, South Africa
Albrecht Schmidt	Ludwig Maximilian University of Munich, Germany
Sanjeet Kumar Nayat	Indian Institute of Information Technology, Design and Manufacturing, Kancheepuram, India
Perumal Thinagaran	Universiti Putra Malaysia, Malaysia
Renumol V. G.	Cochin University, India
Patrick Hebrard	RHEA Group, France
Reda Yaich	IRT SYSTEM-X, France
Kave Salamatian	Université de Savoie, France
Guillaume Acca	Cyber Defence Command, France
Paul Askell Dowland	Edith Cowan University, Australia
William Puech	LIRMM, Université Montpellier, CNRS, France
Ysens de France	Aivancity, France
Simeon Simoff	Western Sydney University, Australia
Solange Ghernaouti	University of Lausanne, Switzerland

Contents

Artificial Intelligence for Sustainability

A Hybrid Knowledge-Based and Large Language Model Framework for Sustainable Design Education

Toktam B. Fabrizi(✉)

Western Sydney University, Sydney, Australia
toktam.tabrizi@westernsydney.edu.au

Abstract. Sustainable design principles, including circular design and life cycle assessment (LCA), are increasingly vital in architecture and product design education, yet they remain challenging to teach with traditional methods. This paper proposes a practical pedagogical framework that integrates artificial intelligence (AI) tools – specifically a knowledge-based system (KBS) combined with a large language model (LLM) – to enhance the teaching of circular design and LCA. It justifies the use of AI in sustainability-focused design curricula, highlighting how AI-driven tools can provide interactive, real-time feedback and expert knowledge to students. The environmental footprint of generative AI is reviewed, addressing energy use, carbon emissions, water consumption, material resources and electronic waste, with up-to-date data from 2019–2025. The study discusses how the educational benefits of AI can outweigh these environmental costs when sustainable AI practices are applied. The proposed AI-KBS + LLM framework is presented as a conceptual prototype: the KBS component supplies domain-specific LCA data and design heuristics, while the LLM offers natural language interaction and contextual reasoning, enabling students to receive tailored sustainability advice during the design process. The study outlines the framework's components and workflow. Potential pedagogical benefits – from democratising access to sustainability expertise to fostering iterative, reflection-driven design – are analysed alongside ethical considerations, such as managing AI's biases and environmental impact. The paper concludes that, with mindful implementation, AI-augmented learning can empower future designers to create more sustainable, circular solutions while the AI's own footprint is mitigated through responsible practices.

Keywords: Sustainable Design Education · Circular Economy · Life Cycle Assessment · Generative AI · Knowledge-based Systems · Large Language Models

E. Mercier-Laurent et al. (Eds.): AI4S 2025, IFIP AICT 780, pp. 3–15, 2026.
https://doi.org/10.1007/978-3-032-21184-2_1

1 Introduction

In response to accelerating climate change and the depletion of natural resources, architecture and design programs face mounting demands to cultivate students' competencies in sustainability [1, 2]. Circular design, designing products and buildings for reuse, disassembly, and material recovery, and life cycle assessment (LCA), the systematic evaluation of environmental impacts from cradle to grave, have become essential components of sustainable design education. Together, they equip future designers with strategies to reduce waste and optimise resource efficiency across the life cycle of products and buildings [3, 4]. These strategies encourage designers to minimise waste and optimise resource efficiency across the life span of a building or product [1, 5].

Despite their theoretical importance, circular design and LCA are often underrepresented in standard curricula [6, 7]. Many students in architecture and product design complete their studies with minimal exposure to hands-on LCA workflows or circular economy strategies such as design for disassembly [8, 9]. This deficiency partly stems from the inherent complexity and resource intensiveness of these methods: rigorous LCA demands detailed data, for instance: material breakdowns, manufacturing processes, end-of-life assumptions and domain expertise that can intimidate both learners and instructors [8, 10]. Consequently, students may regard LCA as the domain of specialists rather than an integral part of the design process [11, 12].

Recent advances in artificial intelligence (AI) present a promising avenue to close this pedagogical gap. AI systems can provide on-demand computational support and interpretive guidance, thus demystifying sustainability assessments [13]. In particular, hybrid systems combining knowledge-based rule engines and large language models (LLMs) can serve as intelligent tutors for sustainable design, enabling real-time simulation of design impacts and iterative feedback [1, 14, 15]. For instance, an AI assistant might alert a student that selecting reinforced concrete significantly increases embodied carbon, while also suggesting alternative materials. Such instantaneous, context-sensitive feedback is difficult to replicate in conventional instruction but may be feasible via AI "co-pilots." This can help students internalise trade-offs in circular design and life cycle thinking [16, 17].

However, integrating generative AI into design education introduces critical considerations. Training and executing large AI models incur environmental costs, including energy consumption, carbon emissions, water use, and electronic waste [18, 19]. Therefore, any educational deployment must ensure that the pedagogical gains outweigh the AI's own footprint. Encouragingly, research into sustainable AI practices, such as smaller specialised models, prompt engineering to reduce compute, and deployment on renewable energy infrastructure, indicates significant reductions in resource use are possible [20, 21]. Moreover, if AI tools can train a generation of architects and designers to create lower-impact buildings and products, the resultant lifetime carbon savings may substantially exceed the emissions associated with the AI systems themselves.

This paper explores these issues in detail. First, Sect. 2 reviews the environmental footprint of generative AI, grounding the discussion in empirical evidence. Section 3 argues that through sustainable computing strategies, the pedagogic benefits of AI in circular design education can justify its deployment. Section 4 proposes a hybrid AI architecture integrating a knowledge base with an LLM, including a conceptual model and prototype workflow for tutoring in LCA and circular design. Section 5 examines pedagogical advantages such as enhanced student engagement and improved decision-making and ethical considerations such as bias, transparency, and model reliability. Finally, Sect. 6 concludes by reflecting on future directions for AI-assisted sustainability education and suggesting priorities for implementation and research.

2 Environmental Impact of Generative AI

Before describing the AI-based teaching framework, it is essential to examine the environmental footprint of generative AI systems themselves, spanning energy use and carbon emissions, water consumption, material extraction, and electronic waste, and to ground later design decisions in this context.

2.1 Energy Use and Carbon Emissions

Training and inference operations for large AI models are highly energy intensive. Data centres already consume approximately 1–2% of global electricity, a proportion that is projected to rise significantly with increased AI workloads [22]. Training state-of-the-art models can demand enormous power: for instance, GPT-3's training has been estimated to consume energy equivalent to that required for powering about 120 US homes for one year, and newer models (e.g. GPT-4) likely require magnitudes more [23]. In inference mode, the cumulative energy cost of millions of queries can be substantial. A recent benchmarking study reports that LLM inferences in large-scale production can consume nontrivial watt-hours per prompt and lead to significant aggregate emissions when scaled globally [24]. Unless powered by low-carbon electricity, this energy usage translates into considerable greenhouse gas (GHG) emissions, particularly in regions reliant on fossil-fuel-based grids [22].

2.2 Water Consumption

Data centres and power infrastructure also impose a water burden, primarily for cooling. In many facilities, for every kWh of electricity consumed on site, an estimated 1–2 L of water are used in cooling processes. A UK government report warns that AI's water demand could reach billions of cubic metres annually, placing strain on regional water security [25]. IEEE Spectrum analysis emphasises that indirect water use, i.e. water embedded in electricity generation, often dominates total consumption, sometimes exceeding 80% of the data centre water footprint. Regional concentrations of AI infrastructure in water-stressed areas heighten risks to local supply systems [26].

2.3 Materials, Raw Resources, and Manufacturing

The production of high-performance AI hardware (e.g. GPUs, TPUs, specialized accelerators) relies on extraction and processing of critical minerals (such as lithium, cobalt, tantalum, rare earth elements), which carry environmental and social costs (deforestation, soil erosion, toxic waste, biodiversity degradation). The material input demands are large: for instance, manufacturing an electronic device often entails many times its mass in raw materials when accounting for upstream processes. Each new generation of chips tends to intensify resource demands and complexity, reinforcing pressure on supply chains and raising concerns about long-term sustainability [25, 27].

2.4 Electronic Waste (E-Waste)

As hardware is upgraded or decommissioned, obsolete units contribute to e-waste. Many components (e.g. PCBs, capacitors) contain hazardous materials such as lead, mercury, and arsenic, posing risks if improperly disposed [28]. Accelerated replacement cycles in the AI sector amplify this problem: shipments of GPUs to support AI infrastructure increased sharply in recent years, reflecting rapid turnover [29]. Without robust circular-economy strategies (reuse, refurbishment, modular design), e-waste from AI systems may exacerbate environmental harm [28].

In summary, generative AI systems impose significant environmental burdens across multiple axes, energy and carbon, water, raw materials, and e-waste. Any educational architecture that utilises AI must therefore contend with these trade-offs. In the following chapter, this study explores how sustainable AI practices and design decisions can help mitigate these impacts, and whether the pedagogical benefits of AI in sustainable design justify its use under constrained environmental conditions.

Table 1 summarises key impact areas and recent statistics, highlighting the paradox that proposals to use AI in education for sustainability must also address AI's own environmental footprint. Accordingly, the adoption of such technologies requires parallel strategies to mitigate these impacts.

Table 1. Key Environmental Impact Areas of Generative AI

Impact Area	Details and Recent Findings
Energy & Carbon	• Data centres supporting AI workloads account for about 1–2% of global electricity use [30], with projections suggesting demand could reach 20–21% by 2030 if unchecked [22] • Training OpenAI's GPT-3 consumed energy equivalent to powering about 120 US homes for a year [31] • Each ChatGPT query may use about 10 × more energy than a standard web search [23] • In regions reliant on fossil-fuel grids, this demand translates into significant CO_2 emissions
Water Usage	• AI data centres rely heavily on water for cooling. On average about1.5–2 L of water are used per kWh of electricity consumed [25] • By 2027, global data centre water use may equal ~ 50% of the UK's annual consumption [25] • AI infrastructure could consume six times Denmark's total water use by the late 2020s [26] • Concentrated withdrawals risk exacerbating water scarcity and ecological degradation

(continued)

Table 1. (*continued*)

Impact Area	Details and Recent Findings
Materials & Mining	• Manufacturing AI hardware requires extensive raw materials. Producing a 2 kg computer can demand about 800 kg of raw inputs across the supply chain [27] • Critical minerals such as cobalt, lithium, and tantalum are linked to deforestation, water pollution, and hazardous working conditions [28] • Demand surged in 2023, with 3.85 million GPUs shipped to data centres, up from 2.67 million in 2022, further intensifying extraction pressures. [29]
Electronic Waste	• AI accelerators and GPUs typically have about 4-year operational lifespans [28] • With rapid upgrade cycles, discarded components contribute to the about 50 Mt of e-waste generated globally each year [28, 32] • Many contain toxic materials (lead, mercury, arsenic) that can leach into ecosystems if not responsibly recycled. Without robust circular economy practices, AI-driven hardware turnover will exacerbate e-waste challenges

3 Sustainable AI Strategies and Educational Trade-Offs

While deploying AI to teach sustainability may appear paradoxical, this chapter argues that deliberate adoption of sustainable AI strategies can tilt the balance in favour of pedagogical benefit. By minimising the environmental cost of AI and maximising its educational "handprint," a net gain is achievable. Key technical, infrastructural, and pedagogical strategies were reviewed, and their effectiveness was assessed how educational multiplier effects can justify AI use in design pedagogy.

3.1 Efficiency Optimisation of AI Models

A central mechanism to reduce AI's environmental burden lies in software-level optimisations. Techniques such as quantisation (reducing numerical precision), pruning (removing redundant parameters), distillation (training a simpler model to mimic a larger one), and early stopping have shown substantial energy savings with minimal performance losses [33]. Benchmarking studies of quantization methods report large variation in energy profiles, underlining the need to evaluate them by energy metrics, not just compression ratio [34]. More broadly, a Green AI agenda emphasises energy and compute efficiency alongside accuracy as first-class metrics [35]. A recent LLM optimisation study reported up to 45% reduction in energy consumption via quantization and efficient inference strategies, without significant accuracy loss [36]. These optimisations suggest that an educational AI assistant can be configured to operate with a distilled or quantised model, constrained response lengths, and prompt engineering to avoid unnecessary computation. Such design aligns with educational best practice: students generally benefit more from crisp, targeted feedback than from verbose outputs.

3.2 Green Infrastructure and Lifespan Extension

Efficiency alone is insufficient; sustainability also depends on the energy source and hardware practices. AI workloads should be hosted on cloud or data infrastructures powered by renewables (solar, wind, hydro) to lower marginal carbon intensity. Some providers already advertise carbon-aware operations and net-zero commitments. The selection of platforms with sustainability credentials should become a criterion in institutional adoption decisions [37, 38].

Additionally, scheduling compute-intensive tasks during periods of surplus renewable generation or off-peak hours can reduce carbon footprints. Extending the operational lifespan of AI servers via component upgrades or reassigning older hardware to lighter tasks prevents premature e-waste. In educational contexts, models do not always need to be the latest or largest; a fine-tuned model of modest size may suffice for design tutoring tasks if properly engineered [39, 40].

3.3 Usage Policies and Pedagogical Norms

Technical and infrastructural mitigation strategies must be complemented by responsible usage policies. Educators should teach students the environmental implications of AI and cultivate habits of efficient query formulation (precise prompts, minimal reiteration). Use of the AI assistant can be constrained to defined phases of project work, rather than permitting continuous access, thereby controlling server uptime. These norms mirror sustainable design ethics, just as waste avoidance is taught in material design, so should "compute frugality" be taught in AI usage.

Many institutions are beginning to adopt "AI ethics and sustainability" policies, integrating considerations of energy and resource use alongside data and equity [41]. Embedding such norms into curriculum signals that sustainability extends beyond building design into all technological systems.

3.4 Educational Multiplier Effects and Trade-Offs

The principal justification for employing AI in sustainability education arises from its multiplier impact. If students, aided by AI, internalise life cycle thinking and circular design, they may apply these insights throughout their careers, averting far more emissions and waste than the AI consumed in the classroom. For example, a 30% reduction in embodied carbon achieved across multiple projects compounds to large savings over a designer's lifetime.

Moreover, AI allows scalable dissemination of expertise. Many design programs lack faculty with specialist LCA or circular economy expertise; an AI tutor encapsulating curated knowledge (e.g. LCA databases, rules for disassembly) can democratise access. This scaling effect means educational returns may outweigh centralised resource costs.

The balance is justified only if mitigation is effective. If AI is deployed without sustainability safeguards, with large unoptimised models, fossil-powered servers, and indiscriminate usage, its footprint could dominate. But with conscientious design, the pedagogical and environmental returns can dominate.

4 AI–KBS + LLM Framework for Teaching Circular Design and LCA

To address gaps in circular design and LCA education, this study proposes a hybrid AI-driven tutoring framework that integrates a knowledge-based system (KBS) with a large language model (LLM). The objective is to provide students with instant, context-aware sustainability feedback during the design process, transforming abstract environmental metrics into actionable insights. This system is designed with both pedagogical effectiveness and sustainable AI practices in mind, following efficiency and resource-awareness principles outlined earlier [32, 39, 40].

4.1 Framework Overview

The architecture combines the interpretive and conversational capabilities of an LLM with the factual reliability of a KBS. The LLM enables natural language interaction and contextualises the outputs, while the KBS ensures grounding in validated data, including LCA datasets, building codes, and passive design rules. This hybrid approach mitigates common LLM limitations such as "hallucination" by constraining generative reasoning to curated sustainability knowledge [27, 28].

4.2 System Components

(See Table 2).

Table 2. AI–KBS + LLM Framework Components

Component	Function	Sustainability/Pedagogical Considerations
Student Interface	Provides a front-end for student input and feedback; can be embedded in CAD/BIM software or accessed via a web tool. Students input design parameters (materials, building type, climate zone) or ask questions in natural language	Interface designed to be intuitive & non-intrusive, prompting sustainability checks at key design stages. Aligns with educational best practice of timely, focused feedback
Large Language Model (LLM)	Interprets student inputs, formulates structured queries for the KBS, and translates results into natural language. Fine-tuned for sustainability & design contexts	Constrained to vetted KBS data to minimise errors; provides concise, pedagogically effective explanations rather than verbose outputs

(continued)

Table 2. (*continued*)

Component	Function	Sustainability/Pedagogical Considerations
Knowledge-Based System (KBS)	Curated repository of LCA databases, climate files, codes, & sustainability rules (e.g. embodied carbon benchmarks, passive design heuristics) Performs calculations and rule-based reasoning	Simplified LCA models focus on key high-impact parameters (materials, WWR, insulation, lifespan), balancing accuracy with usability for students. Enables actionable & context-specific recommendations

4.3 Workflow

The typical interaction follows a five-step loop:

Student input – via natural language (e.g. "What is the impact of using concrete walls?") or by making a design choice (e.g. selecting material).
LLM processing – interprets intent, formulates a query to the KBS.
KBS response – retrieves or computes relevant data (e.g. embodied carbon of concrete walls, regulatory benchmarks, design heuristics).
LLM output – converts KBS results into contextualised advice (e.g. "Concrete contributes around 300 kg CO_2e/m3; timber alternatives could cut this by around 50%").
Student feedback loop – student refines the design or queries alternatives, reinforcing learning through iteration.

This cycle operates in near real time by using cached calculations, a lightweight KBS, and a mid-sized LLM (e.g. 6–8B parameters) running on energy-efficient infrastructure. Such design choices balance responsiveness, pedagogical quality, and sustainability.

4.4 Prototype Scenario

Consider a case where a student designs a detached house in Sydney's temperate climate.

- Inputs: building type = detached house; lifespan = 50 years; wall material = double brick; window area = 50% of façade.
- The KBS flags excessive glazing relative to optimal window-to-wall ratio (WWR) for Sydney, drawing on passive design guidelines.
- It retrieves embodied carbon data for double brick versus lower-impact alternatives (e.g. timber framing, low-carbon concrete).
- The LLM generates context-specific advice: recommending shading devices for west façades, high-SHGC glazing on north-facing windows (southern hemisphere), and suggesting timber cladding to reduce embodied emissions.

This response combines quantitative metrics (e.g. 300 kg CO_2e/m^3 for concrete) with qualitative reasoning, illustrating both environmental consequences and alternative strategies. Students not only improve their design but also learn why these adjustments matter.

4.5 Contribution to Pedagogy and Sustainability

The hybrid AI tutor acts as a virtual studio mentor, continuously available to provide feedback grounded in validated sustainability data. It enables learning by doing, encouraging systems thinking and reinforcing concepts of circularity and LCA in real-time design contexts. Importantly, by deploying lean models on renewable-powered infrastructure and extending hardware life, the framework adheres to sustainable AI strategies. This balance of reduced footprint (through efficiency and green compute) and increased handprint (via improved student sustainability literacy) justifies the adoption of AI in design education.

5 Discussion: Pedagogical Benefits and Ethical Considerations

5.1 Educational Benefits

Integrating the proposed AI–KBS + LLM tutor into design studios offers several clear pedagogical advantages. First, it democratises access to expertise. Many students lack regular access to sustainability consultants or LCA specialists; the AI tutor fills this gap by providing reliable, evidence-based guidance. This is particularly important for institutions where structured LCA or circular design instruction remains limited. Early exposure to such tools helps embed sustainability as a default design parameter rather than an afterthought, addressing the curricular gaps identified in Sect. 2.

Second, the system enhances experiential learning. Architecture and industrial design are fundamentally project-based, and iterative "learning by doing" is more effective than abstract instruction. The AI tutor reinforces this approach by providing immediate, context-specific feedback during design iterations. Students not only see quantitative improvements (e.g. reduced embodied carbon or improved thermal comfort) but also gain a tangible sense of agency in sustainable design decision-making. This iterative loop fosters systems thinking and supports the principle that "waste is a design flaw rather than an inevitability" [3, 16]. Such a mindset shift reframes sustainability from being a constraint to being a driver of creativity and innovation.

Third, it broadens the scope of accessible knowledge without overloading curricula. Traditional courses often restrict coverage to selected sustainability topics due to time and staffing limitations. By contrast, the AI assistant can address a wider range of queries—from green roofs to passive design heuristics—on demand. This enables personalised and self-directed learning, catering to varied student interests while reinforcing the curriculum. Moreover, by delivering rationales with its advice, the system exposes students to professional frameworks (e.g. ISO 14040, LEED, BASIX), bridging academic instruction and industry practice.

Preliminary workshop trials reinforce these benefits: students reported that the AI tutor transformed sustainability from an abstract concept into a series of concrete, actionable design strategies. Several described the interaction as "game-like," where iterative improvements motivated them to continually refine their designs. Such engagement can counteract the perception that sustainable design is restrictive or tedious, reframing it as an iterative, innovative challenge.

5.2 Ethical and Practical Considerations

Despite its promise, deploying AI in education raises ethical and practical challenges that must be carefully managed. Accuracy and trustworthiness. Large language models can generate incorrect or misleading responses. In an educational setting, this risk is amplified as students may uncritically accept outputs. The AI–KBS + LLM Framework addresses this by grounding responses in the vetted KBS, drawing on validated datasets (e.g. LCA databases, regulatory standards) to ensure factual reliability. Incorporating transparent citation features can further build trust and simultaneously train students to justify design decisions with evidence. Periodic audits by instructors or sustainability experts remain essential during deployment.

Bias and Contextual Limitations. Training corpora and rule bases risk embedding cultural or technical biases, such as privileging Western building practices. To be globally relevant, the AI tutor must adapt recommendations to local climates, economies, and construction practices. This requires continuously updating the KBS with region-specific data and ensuring exposure to innovative practices, such as bio-based materials or circular business models. Without such vigilance, the AI risks reinforcing the status quo rather than fostering transformative design strategies.

Learning Dependency. A central concern is whether reliance on AI could undermine independent learning. To mitigate this, the tutor is positioned explicitly as a support tool, not a design autopilot. Its role is to provide critique, alternatives, and reasoning, while leaving decision-making to the student. Assignments can require reflective components where students explain how AI feedback was used and what trade-offs were considered, ensuring critical engagement rather than passive reliance. This mirrors established precedents in education, where tools like calculators accelerate computation but do not replace conceptual understanding.

Privacy and Integrity. If student interactions are logged for system improvement, strict protocols for anonymisation and secure handling are required. Academic integrity must also be addressed. Since the AI tutor functions as a universally available teaching aid, it does not constitute an unfair advantage. Nonetheless, clear usage policies are needed, for example, restricting AI use during examinations but encouraging it during open-ended project work.

Environmental Accountability. Finally, the environmental cost of the AI itself cannot be ignored (see Sect. 2). Running the system on renewable-powered infrastructure and using optimised, mid-sized models reduces its footprint. One promising pedagogical extension is to display the AI's own energy or carbon footprint to students, transforming the tool into a live case study of ICT sustainability. Such transparency reinforces the core lesson: every tool, even one built to support sustainability education, carries impacts that must be managed.

5.3 Summary

The discussion highlights both the pedagogical opportunities and ethical challenges of integrating an AI–KBS + LLM tutor into sustainable design education. With careful governance, ensuring accuracy, inclusivity, critical engagement, privacy, and environmental responsibility, the benefits outweigh the risks. Moreover, teaching with AI provides an opportunity to teach about AI: its limitations, its responsible use, and its potential to amplify climate solutions if deployed wisely. This dual focus prepares graduates not only as sustainable designers, but also as critical, responsible users of emerging technologies.

6 Conclusion

This paper has examined the integration of AI into sustainable design education, with a particular focus on teaching circular design and LCA. Despite the acknowledged environmental costs of large-scale AI models, the study has argued that carefully designed AI systems can play a pivotal role in embedding sustainability competencies within architectural and industrial design curricula. By combining a knowledge-based system (KBS) with a large language model (LLM), the proposed hybrid framework provides instant, contextual feedback that transforms sustainability from an abstract principle into a practical, iterative design consideration.

The framework offers several pedagogical advantages: it democratises access to specialist knowledge, supports experiential learning through real-time feedback, and broadens opportunities for self-directed exploration. Early testing suggests that students become more confident in applying sustainability principles when the AI tutor is embedded in studio projects, perceiving sustainability as a creative driver rather than a constraint. Crucially, the system is designed not to replace expertise, but to scaffold critical engagement by providing explanations, alternatives, and evidence-based reasoning.

At the same time, the ethical and environmental implications of educational AI cannot be overlooked. This study has addressed these by outlining sustainable AI practices, optimised models, renewable-powered infrastructure, and responsible usage guidelines, that reduce the system's footprint.When balanced against the potential impact of equipping graduates with the ability to deliver substantially lower-carbon and circular designs, the net effect is strongly positive. In essence, the modest resource costs of the AI tutor are outweighed by the long-term benefits of better design decisions in practice.

Future work should involve structured pilot studies to assess the tutor's effectiveness across diverse educational contexts. Mixed-method evaluations, combining quantitative measures (e.g. quality of student LCA outputs) with qualitative insights (e.g. reflective learning narratives), will be critical for refinement. Expanding the framework to integrate with design software, BIM platforms, and emerging immersive technologies such as VR could further enhance its pedagogical reach. Equally, ethical governance and standard-setting, potentially by international bodies such as UNESCO, will be required to ensure accuracy, fairness, and transparency in AI-enabled education.

The central question posed in this study—*is sustainable AI possible?* can be answered provisionally in the affirmative, provided intentional choices are made. AI tools must be developed and deployed efficiently, powered by renewable energy, and embedded within teaching strategies that foreground sustainability. In this way, what may appear as a contradiction becomes a mutually reinforcing cycle: AI systems that teach sustainability while themselves operating within sustainable boundaries.

This study therefore envisions a future where every design student has access to an intelligent sustainability tutor, and where graduates carry forward both digital literacy and ecological responsibility. By cultivating such practices, AI can become not merely a "smart" tool but a wise partner in the transition toward a regenerative built environment and a more sustainable society.

References

1. Altomonte, S., Rutherford, P., Wilson, R.: Mapping the way forward: education for sustainability in architecture and urban design. Corp. Soc. Responsib. Environ. Manag. **21**(3), 143–154 (2014)
2. Taylor, A., Linking Architecture and Education: Sustainable Design for Learning Environments. UNM Press (2009)
3. Falsafi, A., et al.: Life cycle assessment in circular design process: a systematic literature review. J. Clean. Prod. **521**, 146188 (2025)
4. Altomonte, S.: Environmental education for sustainable architecture. Rev. Eur. Stud. **1**, 12 (2009)
5. Attia, S., Al-Obaidy, M.: Design criteria for circular buildings (2021)
6. Kirchherr, J., Reike, D., Hekkert, M.: Conceptualizing the circular economy: an analysis of 114 definitions. Res. Conserv. Recycl. **127**, 221–232 (2017)
7. Potts, E.: Teaching the circle. In: Proceedings of 112th ACSA Annual Meeting (2024)
8. Scott, J., Seddiki, M.: The adoption of a circular economy approach to pedagogical learning in architectural technology (2025)
9. Hegli, T., et al.: Circular prototyping: detect-conduct-disrupt. In: Structures and Architecture, pp. 1517–1524. CRC Press (2025)
10. Tabrizi, T.B., Brambilla, A.: Toward LCA-lite: a simplified tool to easily apply LCA logic at the early design stage of building in Australia. Eur. J. Sustain. Dev. **8**(5), 383 (2019)
11. Piekarski, C.M., et al.: LCA and ecodesign teaching via university-industry cooperation. Int. J. Sustain. High. Educ. **20**(6), 1061–1079 (2019)
12. Cosme, N., et al.: Learning-by-doing: experience from 20 years of teaching LCA to future engineers. Int. J. Life Cycle Assess. **24**(3), 553–565 (2019)
13. Abioye, S.O., et al.: Artificial intelligence in the construction industry: a review of present status, opportunities and future challenges. J. Build. Eng. **44**, 103299 (2021)
14. Gilner, E., Galuszka, A., Grychowski, T.: Application of Artificial Intelligence Methods in Sustainable Building Design, pp. 408–417. Springer, Cham (2019)
15. Gilner, E., Galuszka, A., Grychowski, T.: Application of artificial intelligence in sustainable building design - optimisation methods. In: 2019 24th International Conference on Methods and Models in Automation and Robotics (MMAR) (2019)
16. Tabrizi, T.B., Gocer, O., Sadrieh, A., Globa, A.: Leveraging AI to instruct architecture students on circular design techniques and life cycle assessment. In: 9th International Conference on Higher Education Advances (HEAd'23), Valencia, Spain (2023)
17. Costa, D., Quinteiro, P., Dias, A.C.: A systematic review of life cycle sustainability assessment: current state, methodological challenges, and implementation issues. Sci. Total Environ. **686**, 774–787 (2019)
18. Wu, C.-J., et al.: Sustainable AI: environmental implications, challenges and opportunities. Proc. Mach. Learn. Syst. **4**, 795–813 (2022)

19. Chauhan, D., Bahad, P., Jain, J.K.: Sustainable AI: environmental implications, challenges, and opportunities. In: Explainable AI (XAI) for Sustainable Development, pp. 1–15 (2024)
20. Oyewole, O.O., Joseph, J.F.: Sustainable AI and green computing: reducing the environmental impact of large-scale models with energy-efficient techniques. Int. J. Sci. Res. Netw. Secur. Commun. **13**(3), 19–26 (2025)
21. Miller, T., et al.: The role of lightweight AI models in supporting a sustainable transition to renewable energy: a systematic review. Energies **18**(5), 1192 (2025)
22. (FAS), F.o.A.S., Measuring and standardizing AI's energy footprint. Federation of American Scientists (FAS) (2025)
23. Zewe, A.: Generative AI's environmental impact (2025). https://news.mit.edu/2025/explained-generative-ai-environmental-impact-0117?utm_source
24. Yang, Z., Meng, L., Chung, J., Chowdhury, M.: Chasing low-carbon electricity for practical and sustainable DNN training. arXiv preprint (2023)
25. R., K., Water use in data centres and artificial intelligence, U. Government, Editor. Department for Digital, Culture, Media and Sport (2023)
26. Ren, S., Luers, A.: The real story on AI's water use–and how to tackle it. IEEE Spect. Maga. (2025)
27. OECD.AI. Measuring the environmental impacts of artificial intelligence compute and applications (2025). https://oecd.ai/en/wonk/documents/measuring-the-environmental-impacts-of-artificial-intelligence-compute-and-applications
28. (UNEP), U.N.E.P., AI has an environmental problem—here's what the world can do about it, U.N. Story, Editor (2024)
29. Knowledge, D.C.: Ticking time bomb: AI, data centers, and the looming e-waste crisis (2024). https://www.datacenterknowledge.com/green-materials/ticking-time-bomb-ai-data-centers-and-the-looming-e-waste-crisis
30. (IEA), I.E.A., Electricity 2023: Analysis and forecast to 2025, IEA, Editor. Paris (2023)
31. Jegham, N., et al., How hungry is AI? benchmarking energy, water, and carbon footprint of llm inference. arXiv preprint arXiv:2505.09598 (2025)
32. Schwartz, R., et al.: Green AI. Commun. ACM **63**(12), 54–63 (2020)
33. Tabbakh, A., et al.: Towards sustainable AI: a comprehensive framework for Green AI. Disc. Sustain. **5**(1), 408 (2024)
34. Rajput, S., Sharma, T.: Benchmarking emerging deep learning quantization methods for energy efficiency. In: 2024 IEEE 21st International Conference on Software Architecture Companion (ICSA-C), pp. 238–242. IEEE Computer Society (2024)
35. Rajput, S., Saad, M., Sharma, T.: Tu (r) ning AI green: exploring energy efficiency cascading with orthogonal optimizations. arXiv preprint arXiv:2506.18289 (2025)
36. Khan, T., et al.: Optimizing large language models: metrics, energy efficiency, and case study insights (2025). arXiv preprint arXiv:2504.06307
37. Hewage, T.B., et al.: A framework for carbon-aware real-time workload management in clouds using renewables-driven cores. IEEE Trans. Comput. (2025)
38. Hanafy, W.A., et al.: Carbonscaler: leveraging cloud workload elasticity for optimizing carbon-efficiency. Proc. ACM Meas. Anal. Comput. Syst. **7**(3), 1–28 (2023)
39. Services, A.W.: How Amazon is making its data centers even more sustainable (2024). https://www.aboutamazon.com/news/aws/aws-data-center-ai-circularity
40. Azure, M.: Circular hardware strategy: Four pillars of a successful circular datacenter hardware program (2025). https://datacenters.microsoft.com/wp-content/uploads/2025/04/CircularityWhitePaperApril2025.pdf
41. AlSagri, H.S., Sohail, S.S.: Evaluating the role of Artificial Intelligence in sustainable development goals with an emphasis on "quality education." Disc. Sustain. **5**(1), 458 (2024)

The Future of Work: Exploring Expectations and Fears Toward Artificial Intelligence in the Labor Market

Jolanta Kowal[1], Dmitriy Yanchylik[1], and Paweł Weichbroth[2(✉)]

[1] Institute of Psychology, University of Wroclaw, Wroclaw, Poland
[2] Faculty of Electronics, Telecommunications, and Informatics, Department of Software Engineering, Gdansk University of Technology, Gdansk, Poland
pawel.weichbroth@pg.edu.pl

Abstract. This study investigates AI anxiety among 62 Polish workers using a mixed-methods approach. Applying Wang and Wang's (2022) AI Anxiety Scale (AIAS) and thematic analysis, we compared anxiety levels between automation-prone occupations (e.g., retail) and resilient ones (e.g., teaching). Key results include: (1) the significant difference between the means (72.0 vs. 54.3, $p < .05$) suggests that high-risk workers report 32% greater anxiety, especially concerning job loss; (2) anxiety was driven by perceived lack of control, not technical complexity; and (3) digital literacy moderated anxiety levels. This research contributes by (a) mapping occupation-specific AI anxiety, (b) revealing psychological barriers to AI acceptance, and (c) proposing anxiety-aware design principles, including transparent skill mapping and adaptive interfaces. Findings confirm both research questions (RQ1) and the hypothesis (H1), indicating that occupational risk correlates with elevated AI anxiety. The study emphasizes that effective AI adoption requires emotional, not only technical, readiness, especially in vulnerable job sectors. These results inform workforce policy, education, and AI design. Future use of the AIAS framework may support cross-national comparisons of AI anxiety in European labor markets.

Keywords: AI anxiety · Human-AI collaboration · Explainable Artificial Intelligence · workforce adaptation · occupational psychology

1 Introduction

The rapid expansion of artificial intelligence (AI) is transforming work [1], promising efficiency and innovation [2] while intensifying fears around job security [3], displacement [4], and inequality [5]. This paradox makes understanding how AI is experienced, especially by workers in roles vulnerable to automation is essential [6]. This study explores how employees perceive and emotionally navigate AI's growing presence in their work lives, particularly in high-risk professions. We examine expectations, fears, and personal adaptation strategies amid

Published by Springer Nature Switzerland AG 2026
E. Mercier-Laurent et al. (Eds.): AI4S 2025, IFIP AICT 780, pp. 16–26, 2026.
https://doi.org/10.1007/978-3-032-21184-2_2

this technological shift. While prior research [7] has framed AI as both a threat and an opportunity [8], most focus on global trends. Few address how specific groups internalize their impact, particularly in Central and Eastern Europe. This study seeks to fill that gap.

In Poland's rapidly restructuring labor market [9], understanding how individuals emotionally respond to automation is crucial for developing inclusive policies and education [10]. Although occupational risk is well mapped [11], its psychological toll remains understudied, especially in local contexts. Thus, we put forth the following research question (RQ1): Do workers in automation-susceptible roles report higher AI anxiety than those in stable professions? Drawing on the findings of Wang &Wang [12], we hypothesize that **high-risk workers exhibit significantly greater AI-related anxiety** (H1).

We examined 62 participants using a mixed-methods design. The participants were divided into two groups: at-risk (cashiers, postal workers, and copywriters) and non-at-risk (teachers, nurses, and entrepreneurs). All completed the validated Artificial Intelligence Anxiety Scale [12], measuring anxiety across four dimensions: job replacement, learning, socio-technical blindness, and discomfort with embodied AI. Open-ended responses provided additional insight into fears, coping strategies, and the meaning-making process.

This study theoretically extends AIAS to a new cultural context. Empirically, it offers original insight into how vocational identities shape emotional responses to automation. It also highlights early signs of digital exclusion and the urgent need for psychological and policy support. Bridging psychological theory, labor sociology, and technology acceptance, the study contributes to broader debates on human adaptation in the age of intelligent systems. To our knowledge, this is the first study to psychometrically examine AI anxiety across occupational risk profiles in the Polish labor market using the validated AIAS scale.

2 Theoretical Background

As AI becomes increasingly embedded in workplace routines [13], discourse around its impact has intensified. This section outlines a conceptual and empirical framework for understanding how workers perceive AI's rise, particularly in automation-prone roles. It examines core definitions of intelligence and artificial intelligence, psychological reactions such as technophobia and AI anxiety, and validated tools for measuring these responses. These insights are contextualized within broader labor market shifts and the sociocultural dynamics of the Fourth Industrial Revolution (4IR) [14]. This theoretical extension is particularly novel given the absence of similar data from Central and Eastern Europe, where labor-AI dynamics remain underexplored.

2.1 Defining Intelligence and Artificial Intelligence

Intelligence remains a debated construct, seen as adaptability [15], the ability to perceive relationships tacitly [16], or to learn from experience [17]. Dolinski and Strelau [18] emphasized its internal psychological architecture, while modern views include emotional and social dimensions. McCarthy [19] first defined artificial intelligence as the science of creating intelligent machines. Since then, it has evolved into a complex field encompassing language processing, image recognition, and machine learning [20]. In addition, Owoc [21] frames AI as a system simulating human reasoning, planning, and decision-making through knowledge-based learning.

2.2 The Psychological Construct of Technophobia

Initially understood as computer phobia, technophobia now encompasses generalized anxiety toward new technologies. Defined by Rosen and Maguire [22] as a mix of fear, negative attitudes, and self-defeating thoughts, it impedes technological engagement. Kanfer and Heggestad [23] note its impact on cognitive function and adaptability. Studies confirm higher levels among women, older adults, and less-educated individuals [7,24], underscoring its roots in psychology and structural social vulnerability.

2.3 AI Anxiety: From Attitudes to Behavioral Intentions

Artificial Intelligence Anxiety (AIA), defined by Wang and Wang (2022), captures both emotional and cognitive responses to AI. Unlike general technophobia [24], AIA is more specific, comprising four interrelated dimensions. Learning anxiety reflects fear of not grasping AI tools, especially among older or digitally excluded individuals [25]. Job replacement anxiety involves fears of automation in routine or low-autonomy roles [26]. Sociotechnical blindness refers to limited awareness of AI's broader impact, while AI configuration anxiety captures discomfort with humanoid systems and the "uncanny valley" effect [27]. This validated model enables nuanced analysis of AI attitudes across professional and cultural contexts.

2.4 Measuring AI Anxiety: The AIAS Tool I Anxiety: From Attitudes to Behavioral Intentions

Wang & Wang's Artificial Intelligence Anxiety Scale (AIAS) [12] is a psychometrically validated instrument consisting of 21 items mapped across the four dimensions mentioned above. The scale demonstrated high internal reliability (Cronbach's $\alpha > 0.96$), content validity, and predictive power about learning motivation. Notably, AIA correlated positively with the willingness to upskill or adapt professionally, especially among individuals who view AI as a legitimate threat to their job role. The instrument has primarily been used in East Asian contexts. The current study's attempt to adapt it to the Polish labor landscape represents a novel cultural validation of the scale and broadens its theoretical generalizability.

2.5 AI and Labor Market Transformation

AI-driven job loss primarily affects routine cognitive roles, while complex, emotionally demanding professions remain more resilient [28]. This shift contributes to a growing divide between secure, high-paid cognitive jobs and precarious low-skill work [29]. Psychologically, workers face a dual pressure to upskill or risk obsolescence, often accompanied by fear, stress, and diminished agency. The Fourth Industrial Revolution and "hyper-automation" further amplify this uncertainty, especially in regions where technological change is uneven and top-down [30]. Recent studies have continued to emphasize the psychological burden posed by automation, urging the development of emotionally responsive AI strategies [31].

2.6 Research Gap and Theoretical Relevance

Despite the richness of global research, there is an apparent lack of qualitative and psychometric studies on AI anxiety in Poland, particularly among workers in high-risk occupations. This research addresses that void by investigating how individuals perceive and internalize the threat of AI across occupational lines. The theoretical contribution lies in the intersectional application of psychological theory and labor market analysis, enriched by the AIAS scale and structured qualitative feedback. This approach enables a more nuanced understanding of how different groups psychologically navigate technological disruption.

3 Methodology

This study employed a mixed-methods research design to investigate levels of artificial intelligence anxiety (AIA) among workers in different occupational categories. By combining psychometric and qualitative methods, we aimed to capture both the intensity of AI-related fears and the underlying cognitive-emotional frameworks through which individuals interpret the rise of intelligent technologies in the labor market.

3.1 Research Framework

This study adopts a socio-psychological lens to examine how individuals across professions emotionally and cognitively adapt to the rise of AI in the workplace. The framework integrates structural vulnerability, demographics, and psychological traits to offer a multidimensional view of AI-related anxiety. At its core is the distinction between automation-prone and automation-resistant roles, forming the primary axis of comparison. This is shaped further by demographic factors such as age, gender, and education, which influence technological responses [7].

AI anxiety is central to the analysis, assessed using the validated AIAS scale [12], which captures fear across four domains: learning discomfort, job insecurity, sociotechnical blindness, and unease with embodied AI. Open-ended responses were included to deepen the quantitative findings, allowing participants to express their expectations, concerns, and coping strategies. This qualitative layer enables a thematic synthesis that moves beyond metrics to explore how people interpret AI's role in their work lives. Ultimately, this framework reveals socio-emotional fault lines between occupational groups and highlights the importance of human-centered strategies in navigating technological change.

3.2 Participants and Data Collection via Tool

The study involved 62 participants from diverse sectors with varying exposure to AI technologies. Most were women (71%) aged 18–29, reflecting a digitally engaged demographic. The majority held or pursued higher education and lived in urban are-as, primarily cities with over 100,000 inhabitants. Participants represented 61 different vocations, including education, trade, and creative fields. Income levels ranged from below PLN 3,000 to over PLN 10,000. Data were collected using an online questionnaire.

The first section covered demographics and professional background. The second employed the Polish version of the Artificial Intelligence Anxiety Scale (AIAS) [12], a 21-item tool on a 7-point Likert scale assessing four anxiety dimensions: learning discomfort, fear of replacement, sociotechnical blindness, and unease with embodied AI. The final section included open-ended questions on AI awareness, perceived threats, coping strategies, and future expectations, enabling quantitative analysis and rich thematic insight.

3.3 Procedure and Ethics

Data collection was conducted via Microsoft Forms, then exported to Excel for analysis. All responses were anonymized and prepared for both statistical and the-matic evaluation. Participants took roughly 8–10 min to complete the instrument, which featured built-in logic to compute AIAS subscale scores [12].

Quantitative analysis included descriptive statistics and independent t-tests to assess AI anxiety differences across occupational groups, with Cohen's d reported for effect size. Reliability of the Polish AIAS version was confirmed with Cronbach's alpha ($\alpha > .70$), deemed suitable for exploratory research. Simultaneously, qualitative responses underwent thematic analysis using Braun and Clarke's six-phase framework [32]. Codes were organized into broader themes: automation inevitability, emotional ambivalence, lack of readiness, and the imperative for upskilling. The study received institutional ethical approval. All participants gave informed consent, and their data were managed with strict confidentiality and research integrity throughout.

4 Results

This section presents the empirical findings of our mixed-method study on AI anxiety among Polish professionals, classified by occupational vulnerability to automation. Both quantitative and qualitative results are discussed.

4.1 Quantitative Findings: AIAS Scores

Analyzing the AI Anxiety Scale (AIAS) scores across occupational categories reveals a discernible trend. Participants in at-risk roles such as warehouse work, brand ambassadorship, and retail, exhibited the highest anxiety ($M = 72.00$, $SD = 14.73$). Those in non-risk professions, including education and psychological services, showed the lowest levels ($M = 54.25$, $SD = 7.59$). Interestingly, the largest group, unclassified occupations not mapped to automation risk, displayed moderately elevated anxiety ($M = 63.46$, $SD = 14.90$), bridging the other two (see Table 1 for details).

These findings suggest that proximity to automation is linked with heightened stress, yet ambiguity itself may fuel anxiety just as powerfully. Notably, many "unclassified" workers reported similar discomfort to those in overtly vulnerable roles, implying that uncertainty, not merely risk, could be a critical psychological trigger. Despite limitations in sample balance, the pattern indicates an emerging socio-emotional fault line shaped less by technical threat and more by perceptions of occupational permanence.

Table 1. Descriptive statistics for AIAS scores by occupational group ($n = 62$).

Group	n	%	Mean AI Anxiety	SD	Min	Max
At-Risk	10	16.13	72.00	14.73	49.0	97.0
Non-Risk	4	6.45	54.25	7.59	44.0	62.0
Unclassified	48	77.42	63.46	14.90	34.0	94.0

Although the sample remains limited for robust inferential statistics (e.g., t-test), the expanded data reinforces a meaningful trend: individuals in roles vulnerable to AI automation report notably higher levels of anxiety. Importantly, comparable levels among unclassified workers suggest that uncertainty itself, rather than explicit risk, may be a key psychological driver of discomfort, fear, and unease regarding intelligent technologies. The sample is insufficient for robust inferential statistics (e.g., t-test), the pattern suggests a meaningful trend: individuals working in occupations vulnerable to AI automation report higher levels of discomfort, fear, and uncertainty regarding intelligent technologies.

4.2 Qualitative Analysis Results

Thematic analysis of open-ended responses revealed a nuanced emotional and cognitive landscape surrounding AI in the workplace. Six core themes emerged. First, fear of displacement was prevalent, especially among participants in sales, logistics, and routine service roles. Many described AI as a direct threat to job security. Conversely, augmentation optimism dominated responses from educators and analysts, who envisioned AI as a tool for personalization, planning, and information synthesis. Secondly, participants exhibited cognitive ambiguity, a simultaneous curiosity and confusion about AI's capabilities. This often fostered a cautious stance, rooted in technological mistrust, particularly toward autonomous systems. A fourth theme, value erosion, highlighted concerns that AI may diminish human judgment and expertise to the point of algorithmic dominance. Despite apprehensions, a fifth theme, conditional acceptance, emerged, contingent upon transparency, oversight, and proper training. Lastly, the theme of role transformation suggested that work is not vanishing but evolving, demanding new skills and mindsets.

- "*I use AI for correcting emails, but I would not trust it to make decisions for me.*" – Administrative worker.
- "*AI might improve service delivery, but it cannot replace empathy or judgment.*" – Psychologist (Table 2).

Table 2. Thematical analysis based on qualitative answers ($n = 62$).

Theme	n	%	Description	Example Quote
Fear of Displacement	22	35,5	Concern over being replaced by AI, especially in repetitive or low-autonomy jobs	My job as a cashier could easily be automated
Augmentation Optimism	18	29	Belief that AI can support or enhance professional tasks, especially in education	AI helps me plan lessons faster and more effectively
Cognitive Ambiguity	25	40,3	Mixed emotions, curiosity combined with confusion or uncertainty about AI	I don't fully understand it, that's what makes it scary
Technological Mistrust	20	32,3	Skepticism toward AI, particularly around decision-making and autonomy	I don't trust AI to make decisions about people
Value Erosion	15	24,2	Fear that reliance on AI may diminish human expertise and judgment	It might make us lose touch with our skills and instincts
Conditional Acceptance	28	45,2	Willingness to use AI if supported by transparency, training, and oversight	If I had training, I'd be more comfortable using it
Role Transformation	17	27,4	Recognition that AI is reshaping work roles, requiring adaptation and new skills	Jobs won't disappear, but they will definitely change

The emotional landscape of artificial intelligence in the workplace emerges not as a uniform reaction but as a layered spectrum, shaped by occupational context, identity, and technological fluency. A thematic analysis of open-ended responses revealed striking disparities between participants in automation-prone professions and those in more secure roles.

For instance, teachers and small business owners tended to frame AI as an ally. "*AI could help personalize student tasks. It saves me time in lesson planning,*" re-marked one respondent, illustrating a broader narrative of optimism, strategic integration, and task enhancement. By contrast, participants from high-risk fields—postal workers, cashiers, copywriters, expressed disorientation, loss, and even existential threat. "*I do not see how I could compete with AI. I feel lost,*" said a retail assistant, echoing a widespread fear of irrelevance.

Three themes were consistent. First, fear of job replacement and dependency on AI dominated the at-risk group. Second, non-risk participants spoke of AI in hopeful, even empowering terms. But most critically, digital literacy, or its absence, cut across both groups. Anxiety stemmed less from AI itself and more from an inability to comprehend or control it.

5 Discussion

This study reinforces what scholars have long proposed: integrating artificial intelligence into professional life is not merely technical or economic—it is fundamentally emotional. Participants in automation-prone roles expressed identity erosion and helplessness, aligning with Wang and Wang's [12] concept of sociotechnical blind-ness, where individuals feel subordinated to systems beyond their control (RQ1). These in-sights expand the quantitative patterns observed [28], grounding them in narrative form through direct occupational experiences.

Emotional responses depend on how directly AI intersects with one's role finds strong support. Educators and self-employed professionals often described AI as a collaborator. "*AI helps personalize student tasks and reduces planning time*", one teacher noted. By contrast, cashiers, warehouse staff, and clerical workers framed AI as a competitor, expressing distress and fatalism. "*I don't see how I can compete with it*", a retail assistant shared. These contrasts underscore how automation reshapes not only labor but the emotional contours of work (H1).

Quantitative data reinforce these patterns: the at-risk group reported the highest mean AI anxiety ($M = 72.00$), while those in non-risk roles scored much lower ($M = 54.25$) (H1). Unclassified roles, those ambiguously positioned on the risk spectrum, displayed intermediate levels ($M = 63.46$), suggesting that uncertainty itself may be a key driver of AI-related unease (RQ1). Thematic results such as conditional acceptance and value erosion point to complex emotional negotiations, not merely resistance or enthusiasm (RQ1).

These findings align with recent work on explainable AI (XAI), suggesting that transparency requirements may need to extend beyond technical explainability to include role clarity in human-AI teams. Our anxiety framework could inform the development of new evaluation metrics for collaborative AI systems.

Theoretically, the study advances a socio-emotional model of AI adaptation, linking anxiety to occupational risk, perceived control, and AI literacy.

Practically, digital fluency proved a robust predictor of affective response. Echoing Rosen and Weil's [24] technophobia theory, those with limited technological understanding, especially in high-risk jobs, expressed more significant anxiety (H1). Training was desired not only for skill acquisition but also for emotional safety. This aligns with Jeong et al. [33] and Wang and Wang's [12] emphasis on motivated learning as a path to agency.

Our findings complement recent advances in explainable AI (XAI) by suggesting that role clarity (not just technical transparency) may be critical for reducing anxiety. This aligns with proposals for human-centered AI evaluation metrics that incorporate psychological factors alongside technical performance. Our findings on role clarity complement recent advances in interactive XAI (e.g., 'why did AI suggest this?' interfaces), suggesting that both technical and occupational transparency are needed to reduce anxiety.

Furthermore, our findings align with recent workplace coaching research [34], which highlights how psychological resilience can be strengthened through personalized support mechanisms. These insights reinforce the idea that emotional adaptation is critical to sustainable labor transitions in AI-intensive environments. While this study provides valuable insights, several limitations should be noted. The exploratory nature of our research, combined with a modest sample size ($n = 62$) and reliance on self-reported data, suggests the need for verification through more extensive studies.

Future work could employ longitudinal designs to track anxiety trajectories and stratified sampling to better capture occupational and cultural variations, particularly in Central and Eastern Europe where institutional trust dynamics may uniquely shape AI perceptions. We propose three key directions for future research:

1. Design and evaluation of anxiety-aware interfaces incorporating real-time competency visualization (e.g., dynamic human-AI task allocation maps).
2. Development of occupation-specific AI onboarding protocols that address identified risk profiles.
3. Experimental studies examining causal relationships between anxiety levels and actual workplace AI adoption metrics.

These extensions would bridge our psychological findings with practical AI system design while providing more robust evidence for policy recommendations. Finally, the study carries applied significance. Policymakers can design emotionally responsive reskilling programs. Educators can scaffold AI literacy with emotional support. Employers can foster transitions that honor human dignity. Above all, this research restores agency to workers, not as passive adopters of innovation, but as interpreters of its meaning. These directions underscore the need for international adaptation of the model, especially in diverse labor markets undergoing rapid AI transitions [35] (RQ1).

Additionally, aligning emotionally adaptive technologies with SDG 8 (Decent Work and Economic Growth) enhances sustainability by supporting inclusive, resilient work transitions (H1). By addressing sustainability in labor through emotional adaptation strategies, the study contributes to designing AI transitions that promote long-term workforce well-being.

References

1. Pawełoszek, I., Bhambri, P.: Call to action: driving digital sustainability. In: Digital Sustainability, pp. 264–281. CRC Press (2025)
2. Majkutewicz, J., Szymański, J.: Aligning large language models with human preferences using historical text edits. Knowl.-Based Syst. 113566 (2025)
3. Chymis, A.: Should we fear artificial intelligence. In: Artificial Intelligence and its Impact on Business, pp. 39–53 (2020)
4. Weichbroth, P.: Ai and the law: evaluating chatgpt's performance in legal classification. arXiv preprint arXiv:2502.12193 (2025)
5. Bhargava, A., Bester, M., Bolton, L.: Employees' perceptions of the implementation of robotics, artificial intelligence, and automation (raia) on job satisfaction, job security, and employability. J. Technol. Behav. Sci. **6**(1), 106–113 (2021)
6. Jelonek, D., Walentek, D.: Impact of artificial intelligence innovations on the labor market of the European union. In: Scientific Papers of Silesian University of Technology. Organization & Management/Zeszyty Naukowe Politechniki Slaskiej. Seria Organizacji i Zarzadzanie, no. 214 (2024)
7. McClure, P.K.: "you're fired", says the robot: the rise of automation in the workplace, technophobes, and fears of unemployment. Soc. Sci. Comput. Rev. **36**(2), 139–156 (2018)
8. Zema, T., Wójcik, F., Sulich, A., Hernes, M.: Forecasting fuel retail sales volume using machine learning for sustainable. In: Emerging Challenges in Intelligent Management Information Systems: Proceedings of 26th European Conference on Artificial Intelligence ECAI 2023-IMIS 2023 Workshop, vol. 1079, p. 109. Springer, Heidelberg (2024)
9. Kowal, J., Jasińska-Biliczak, A., Weichbroth, P.: Organizational ethics and position relationship moderators among knowledge workers: a regional study of Poland. In: Information Technology for Development, pp. 1–23 (2024)
10. Gawlik-Kobylińska, M.: Harnessing artificial intelligence for enhanced scientific collaboration: insights from students and educational implications. Educ. Sci. **14**(10), 1132 (2024)
11. Zurada, J.: Classifying the risk of work related low back disorders due to manual material handling tasks. Expert Syst. Appl. **39**(12), 11125–11134 (2012)
12. Wang, Y.Y., Wang, Y.S.: Development and validation of an artificial intelligence anxiety scale: an initial application in predicting motivated learning behavior. Interact. Learn. Environ. **30**(4), 619–634 (2022)
13. Owoc, M.L., Sawicka, A., Weichbroth, P.: Artificial intelligence technologies in education: benefits, challenges and strategies of implementation. In: IFIP International Workshop on Artificial Intelligence for Knowledge Management, pp. 37–58. Springer, Heidelberg (2019)
14. Weichbroth, P., Jandy, K., Zurada, J.: Toward sustainable development: exploring the value and benefits of digital twins. Telecom **5**(3) (2024)
15. Sternberg, R.J.: A theory of adaptive intelligence and its relation to general intelligence. J. Intell. **7**(4), 23 (2019)
16. Shaw, G.A.: Hyperactivity and creativity: the tacit dimension. Bull. Psychon. Soc. **30**(2), 157–160 (1992)

17. Mumford, A.: Four approaches to learning from experience. Learn. Organ. **1**(1), 4–10 (1994)
18. Doliński, D., Strelau, J.: Psychologia Akademicka. Tom 1. Podręcznik akademicki. Gdańskie Wydawnictwo Psychologiczne, Gdansk (2024)
19. McCarthy, J., Minsky, M.L., Rochester, N., Shannon, C.E.: A proposal for the dartmouth summer research project on artificial intelligence, august 31, 1955. AI Mag. **27**(4), 12–12 (2006)
20. Kaplan, A., Haenlein, M.: Siri, siri, in my hand: who's the fairest in the land? On the interpretations, illustrations, and implications of artificial intelligence. Bus. Horiz. **62**(1), 15–25 (2019)
21. Owoc, M.L.: Wartościowanie wiedzy w inteligentnych systemach wspomagających zarządzanie. Prace Naukowe Akademii Ekonomicznej we Wrocławiu. Seria: Monografie i Opracowania (nr 100) (1047) (2004)
22. Rosen, L.D., Maguire, P.: Myths and realities of computerphobia: a meta-analysis. Anxiety Res. **3**(3), 175–191 (1990)
23. Kanfer, R., Heggestad, E.D.: Motivational traits and skills: a person-centered approach to work motivation. Res. Organ. Behav. **19**, 1–56 (1997)
24. Rosen, L.D., Weil, M.M.: Computer availability, computer experience and technophobia among public school teachers. Comput. Hum. Behav. **11**(1), 9–31 (1995)
25. Bulut, H., Kınoğlu, N.G., Karaduman, B.: The fear of artificial intelligence: dentists and the anxiety of the unknown. Sağlık Bilimlerinde İleri Araştırmalar Dergisi **7**(1), 55–60 (2024)
26. Damioli, G., Van Roy, V., Vertesy, D.: The impact of artificial intelligence on labor productivity. Eurasian Bus. Rev. **11**, 1–25 (2021)
27. Di Natale, A.F., Simonetti, M.E., La Rocca, S., Bricolo, E.: Uncanny valley effect: a qualitative synthesis of empirical research to assess the suitability of using virtual faces in psychological research. Comput. Human Behav. Rep. **10**, 100288 (2023)
28. Gerlich, M.: Public anxieties about AI: implications for corporate strategy and societal impact. Adm. Sci. **14**(11), 288 (2024)
29. Dottori, D.: Robots and employment: evidence from Italy. Economia Politica **38**(2), 739–795 (2021)
30. Birkbeck, A., Rowe, L.: Navigating towards hyperautomation and the empowerment of human capital in family businesses: a perspective article. J. Family Bus. Manag. **14**(4), 727–734 (2024)
31. Duong, C.D., Ngo, T.V.N., Khuc, T.A., Tran, N.M., Nguyen, T.P.T.: Unraveling the dark side of chatgpt: a moderated mediation model of technology anxiety and technostress. Inf. Technol. People **38**(4), 2015–2040 (2025)
32. Clarke, V., Braun, V.: Teaching thematic analysis: overcoming challenges and developing strategies for effective learning. Psychologist **26**(2) (2013)
33. Jeong, J., Kim, B.J., Lee, J.: Navigating ai transitions: how coaching leadership buffers against job stress and protects employee physical health. Front. Public Health **12**, 1343932 (2024)
34. Hubscher-Davidson, S.: Coaching for wellbeing: a study of the benefits and drawbacks for translators in the workplace (2025)
35. Dignam, C., Smith, C.M., Kelly, A.L.: The heart and art of robotics: from AI to artificial emotional intelligence in stem education. J. Educ. Sci. Environ. Health **11**(2), 151–169 (2025)

Generative AI for Sustainable Problem Solving in I5.0

Cristina R. Monsone[1](✉), Eunika Mercier-Laurent[2], and Ádám B. Csapó[3]

[1] Budapest, Hungary
ing.crismonsone@gmail.com
[2] CReSTIC EA 3804, University of Reims, Champagne Ardenne, France
eunika.mercier-laurent@univ-reims.fr
[3] Corvinus University of Budapest Corvinus Institute for Advanced Studies/Institute of Data Analytics and Information Systems, Budapest, Hungary
adambalazs.csapo@uni-corvinus.hu

Abstract. Generative AI (genAI) has a transformative role in Industry 5.0 (I5.0), particularly in addressing complex design problems and promoting sustainability, which emphasises partnerships for achieving sustainable development. By leveraging data from machinery and equipment, generative AI can predict maintenance needs, thereby reducing downtime and extending the lifespan of equipment. In line with the human-centric approach of Industry 5.0, generative AI enables personalized manufacturing, where products are tailored to individual customer preferences. This proactive approach minimizes resource use and waste, contributing to more sustainable manufacturing processes, aligning with UN17 sustainability goals. In particular, Generative AI can assist in sustainable decision-making processes by providing data-driven insights and recommendations also addressing data scarcity and imbalance. This helps managers and technicians make informed decisions that balance economic, social, and environmental considerations. It facilitates better collaboration between humans and machines, enhancing productivity and innovation. This synergy is crucial for achieving sustainable development goals, as it combines human creativity with machine efficiency. By integrating these capabilities, generative AI can enhance the efficiency, sustainability, and human-centricity of Industry 5.0 operations, driving progress towards a more sustainable future by creating synthetic data that mimics real-world data to create domain-specific datasets tailored to particular industrial applications. This is particularly useful in industrial settings where collecting real data can be expensive or time-consuming. By generating synthetic datasets, AI can provide a rich source of data for training machine learning models. However, datasets created with genAI come with several limitations and challenges, such as bias (body type/shape biases), concerns related to the creation of synthetic data that could be used to deceive or manipulate, lack of genuine creativity and deep contextual understanding. This lack of transparency can be problematic, especially in critical applications where understanding the decision-making

C.R. Monsone—Independent researcher.

Published by Springer Nature Switzerland AG 2026
E. Mercier-Laurent et al. (Eds.): AI4S 2025, IFIP AICT 780, pp. 27–39, 2026.
https://doi.org/10.1007/978-3-032-21184-2_3

process is essential. While Generative AI technologies offer unprecedented capabilities, they also raise pressing concerns regarding sustainability. The energy consumption of large-scale AI models is particularly critical. Training state-of-the-art models such as GPT-4 requires several megawatt-hours of electricity, contributing significantly to carbon emissions. This work investigates how these biases can affect a training model based on deep learning techniques, as well as defines potential standardized benchmarks and metrics to assess the effectiveness of generative AI in creating high-quality datasets for these models, focused on real-time detection of objects both in static and dynamic images, such as videos. To mitigate these environmental costs, researchers are actively developing energy-efficient architectures and training strategies. Techniques such as model pruning, quantization, and knowledge distillation aim to reduce computational overhead without significantly sacrificing performance. Moreover, our approach in using transfer learning—reusing pretrained models on new tasks—minimizes redundant training, improves both energy efficiency and accessibility, and reports the energy footprint of our models for a sustainable development to reduce the carbon footprint. These measures reflect a shift toward a more responsible AI ecosystem, balancing innovation with environmental stewardship.

Keywords: Sustainable GenAI · deep learning · I5.0

1 Introduction

The increasing complexity and volume of industrial data have necessitated the development of advanced methods for data generation and simulation. Traditional data collection processes in industrial settings, while essential, often face challenges such as high costs, time consumption, and operational interruptions. Consequently, the industry is progressively turning towards synthetic data generation techniques to supplement and enhance real-world datasets [1]. Among these techniques, diffusion models have emerged as a powerful tool due to their ability to produce high-fidelity, realistic data [2]. Diffusion models, originally rooted in the field of statistical physics, have seen significant adaptation and advancement within machine learning and artificial intelligence domains. These models operate by simulating the gradual diffusion of information through a medium, allowing for the creation of data that closely mimics the underlying statistical properties of the target dataset. By iteratively refining data points through a series of probabilistic transitions, diffusion models can generate complex, high-dimensional data that is often indistinguishable from real-world observations [3]. More specifically, a creation influenced by thermodynamics, diffusion models have seen a sharp increase in use in recent years [4]. By reversing a slow-noising process, diffusion models acquire the ability to produce data. The noising process is seen as a parameterized Markov chain, that corrupts an image over time by adding noise, until (asymptotically) pure Gaussian noise is produced. By gradually eliminating the noise over a number of time-steps, the Diffusion Model reverses this process by learning to traverse backwards along this chain [5]. A random sample of Gaussian noise followed by de-noising to produce a photorealistic image can be achieved by "cutting in half" the Diffusion Model after training. This paper explores the application of diffusion models in generating industrial datasets, addressing both theoretical underpinnings and practical implementations, defining a potential standardized benchmark and metrics to assess the effectiveness of generative AI in creating high-quality datasets and how the potential biases affect the performance of these models applied in an industrial environment. We investigate how these models can be tailored to meet the specific requirements of various industrial sectors, such as manufacturing, logistics, and energy. The primary objective is to demonstrate that diffusion models can not only augment existing datasets but also facilitate the development of robust predictive models, enhance anomaly detection systems, and improve operational efficiencies. Furthermore, we present a comprehensive analysis of the diffusion model's performance in generating synthetic industrial data, evaluating key metrics such as accuracy, scalability, and computational efficiency. Through a series of case studies and experimental results, we illustrate the model's potential to transform industrial data management and analytics. In summary, the integration of diffusion models into industrial data generation processes holds promising prospects for advancing the field of industrial analytics [6]. By leveraging the capabilities of these models, industries can achieve greater data availability, improved data quality, and enhanced analytical outcomes, ultimately driving innovation and operational excellence in a sustainable way.

2 Methods

2.1 Generative Models Applied

Our approach considers open-source models for generative artificial intelligence (genAI), such as "stable diffusion", [7] and Dall-E3 [8]. These generative models are developed for creating high-quality images from textual descriptions (prompts). CompVis/stable-diffusion-v-1-4-original [7]is a state-of-the-art diffusion model for text-to-image generation, known for its ability to produce high-quality, detailed images from textual descriptions. Its open-source nature and robust architecture make it a valuable tool for a variety of applications in creative and industrial domains. It uses a diffusion process for image generation, making it one of the prominent models in the field of generative AI. The generation process involves starting from a noise vector and reversing the diffusion process to produce coherent images, making them suitable for various applications inc including industrial uses. The model architecture is Transformer-based for both the text encoder and the image generation process. DALL-E3 [8]is a generative model developed by OpenAI that creates images from textual descriptions. It is built on a variant of the GPT-3 model and leverages the power of transformer architectures for both text and image generation. DALL-E represents a significant advancement in multimodal AI, capable of generating high-quality images from textual descriptions using a transformer-based architecture. Key technical components include a powerful text encoder, autoregressive image generation, and the use of a discrete VAE for latent representation. Its ability to understand and generate diverse and complex images makes it a versatile tool for various creative and industrial applications. It uses a transformer architecture similar to GPT-3. Transformers are highly effective for handling sequential data and capturing long-range dependencies, which is crucial for understanding complex textual descriptions and generating coherent images. The approach is multimodal, combining both text and image processing within a single model. The architecture is designed to handle multimodal input (text) and output (image). To encode the text, it utilizes a transformer-based text encoder to convert textual descriptions into a sequence of tokens. This encoding captures the semantic meaning of the text, allowing the model to understand and generate relevant images. The text is tokenized into discrete units (tokens), which the model processes to generate an image. The image generation process is autoregressive, meaning the model generates one part of the image at a time, conditioning each part on the previously generated parts. As well, this model uses a discrete variational autoencoder (Discrete VAE (dVAE) to encode images into a discrete latent space. This latent space representation is used during the generation process to produce high-quality images. Images are represented in a compressed latent space. The model learns to map textual descriptions to these latent representations. The latent representations are decoded back into pixel space to generate the final image. The decoder is typically another neural network, trained to convert latent vectors into detailed and realistic images. The sampling techniques are "Top-k Sampling", a technique used during image generation to select the top k most probable tokens at each step, ensuring diversity while maintaining quality, and "Temperature Scaling" controls the randomness of the sampling process. A lower temperature makes the model's output more deterministic, while a higher temperature introduces more variability. Both of the methods use "Text-to-Image Synthesis", transforming natural language descriptions into corresponding images, leveraging powerful text encoders to understand and interpret the input text, based on diverse and complex textual prompts. However, the model limitations are the potential Bias and Ethical Concerns, in particular, it may reflect biases present in the training data and can generate inappropriate or biased content and the model's output quality can be highly dependent on the specificity and clarity of the input prompts [9,10].

2.2 Cobots Images Created with GenAI

We have devised a method for generating the suggested small industrial dataset using GenAI images by assigning targeted prompts for image production, considering also the spatial need to guarantee safety and sustainable human-machine interaction, for example, in VR environments for training purposes and for Digital Twin Implementation in a smart factory. Examples of the images obtained using the aforementioned generative techniques are in the following figures Figs. 1, 2, 3 and 4):

- Prompt 1: "Create Cobots"
- Prompt 2: "Create Cobots in an industrial environment"

a b

Fig. 1. Cobot images created with DALL-E (a), COMpvis (b) - Prompt 1.

c d

Fig. 2. Cobot images created with DALL-E (c), COMpvis (d) - Prompt 2.

– Prompt 3: " Create Cobots detailing arms, joints and links in an industrial environment "

e f

Fig. 3. Cobot images created with DALL-E (e), COMpvis (f) - Prompt 3.

- prompt 4: " Create Cobots interacting with humans in an industrial environment."

e f

Fig. 4. Cobot images created with DALL-E (g), COMpvis (h) - Prompt 4.

The number of total images obtained using genAI was 120, with a size of 1024×1024.

2.3 Small Industrial Dataset Based on GenAI Cobots Images

Following the creation of Cobots images in various positions and interacting with humans, we used the DTnet, [11–13]approach to create a small dataset that was appropriate for training convolutional neural networks (CNNs) with a limited number of images. We used the Roboflow Annotate tool [14], along with the bounding box system and COCO format (Common Objects in Context) record generation, to label the images. Labelling parameters and settings were found to be effective in achieving satisfactory training metrics. This COCO format is commonly used as a dataset and benchmark for computer vision applications. Built on top of JSON, its annotation format provides bounding boxes, segmentation masks, and other useful metadata together with structured data about object instances. The primary I5.0 criterion pertaining to human-machine interaction was taken into consideration when we created the labelling classes [15,16]. Specifically, we looked at the classes "person" and "cobot" in order to highlight cobots collaborating with humans in shared workspaces. By accurately recognizing and differentiating between people and cobots, we can implement safety precautions in a sustainable way. For example, in order to avoid unintentional contact, a cobot may slow down or stop working if a human gets too close. In addition, we also applied augmentation techniques to the images created with genAI, more specifically: Auto-Orient, Grayscale, Shear: $\pm 32°$ Horizontal, $\pm 16°$ Vertical. In the end, 286 images in total were obtained by the dataset: 248 for train, 23 for valid, and 14 for test. An additional Test folder containing 35 images was made, of which 12 were produced using genAI and 24 were actual real Cobot images (Fig. 5).

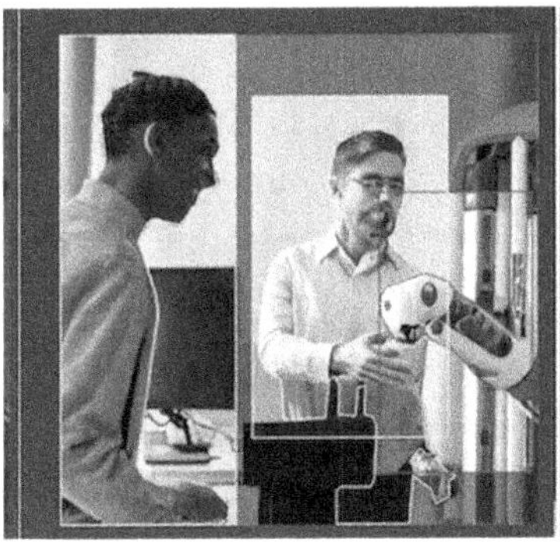

Fig. 5. Labeled image of DatasetGenAI120.

2.4 CNN: Yolov8 and Transfer Learning for Sustainable AI Model

We used Yolov8 [17] to train the model since our goal is to provide a standard for small datasets generated using genAI that can be used for real-time item detection. On a Tesla T4 GPU with 15102MiB of RAM, the model ran Torch version 12 (torch-2.0.1+cu118). We concentrated on examining the effect of hyperparameters on the final model during the first training phase. In this phase, the patience parameter was set to 5, and each model was trained for 10 epochs. Because of the GPU's capabilities, we were able to increase the batch size to 16. We changed the image dimensions to 640×640 pixels with 3 channels for training and validation in order to speed up the training process. Additionally, YOLO automatically applied further augmentation through Albumentations, including Blur (p = 0.01, blur limit = (3, 7)), MedianBlur (p = 0.01, blur limit = (3, 7)), ToGray (p = 0.01), and CLAHE (p = 0.01, clip limit = (1, 4.0), tile grid size = (8, 8)). Every period has five hundred steps. With this configuration, we were able to train 261 layers with 11,790,870 parameters and 11,790,854 gradients, yielding 42.7 GFLOPs. This model produces a global average CO_2 emission factor $\approx$ 0.4 kg CO_2/kWh and Carbon emissions = 0.0078 kWh $\times$ 0.4 kg CO_2/kWh = 0.0031 kg CO_2.

2.5 Metrics Considered to Evaluate the Performance of Our Training Model with GenAI Cobot Custom Dataset

In order to assess the Yolov8 training results using the GenAI-created and BBox-labelled dataset, we take into account the following metrics: accuracy, precision, recall, F1 score, confusion matrix, IoU (for detection), average precision, and mAP [13,18]. These metrics help in assessing the model's effectiveness and identifying areas for improvement, in particular, by ooking at the dataset suitability:

- Accuracy: high accuracy indicates that the CNN is generally performing well on the dataset. If the dataset is balanced and accuracy is high, it suggests that the dataset is suitable for the task. However, if the dataset is imbalanced, accuracy alone may not provide a complete picture.
- Precision: high precision means the CNN has a low false positive rate. As well as high precision on an imbalanced dataset, where certain classes are under-represented, indicates that the model is correctly identifying true positives without being misled by the imbalances.
- Recall (Sensitivity): high recall indicates that the CNN has a low false negative rate and it is crucial in applications where missing a true positive is costly. If recall is low, it might suggest that the dataset lacks sufficient examples of some classes, necessitating more data collection for those classes.
- F1 score balances precision and recall, providing a single metric that considers both false positives and false negatives. A high F1 score suggests that the dataset allows the CNN to perform well in terms of both precision and recall. A low F1 score might indicate issues with dataset quality or distribution.
- Confusion Matrix provides detailed insights into the model's performance across all classes, showing true positives, true negatives, false positives, and false negatives for each class. It can highlight specific classes where the model is under-performing, suggesting the need for more balanced data or additional examples of those classes. It also helps identify if the model is consistently confusing certain classes, indicating potential issues with the data labelling or features.

Understanding CNN performance metrics is crucial for assessing the suitability of a dataset for a given task and determining how well the CNN is learning from the data or if further data collection and pre-processing are needed.

3 Experimental Results

3.1 Labels

Multiple representations of the occurrences and spatial distributions of two categories in our dataset, "Cobot" and "Person," are shown in the following graph. This indicates that the tagged photos are unbalanced, with a majority of Cobot labels (Fig. 6).

3.2 Confusion Matrix

The confusion matrix shows the number of correct and incorrect predictions made by our classifier per category type. In our case, the proposed dataset works adequately. With the datasetGenAI120 we have (Fig. 7):

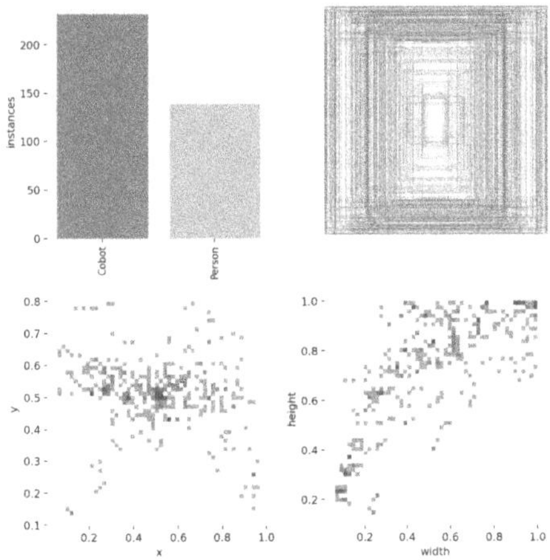

Fig. 6. Labels correlogram.

- **Cobot:** True Positive (TP): 0.79 (correctly identified as Cobot). False Negative (FN): 0,21 (actually a Cobot but predicted as Person or Background).
- **Person**: TP: 1 (correctly identified as Person). FN: 0 (actually a Person but predicted as a Cobot or Background).

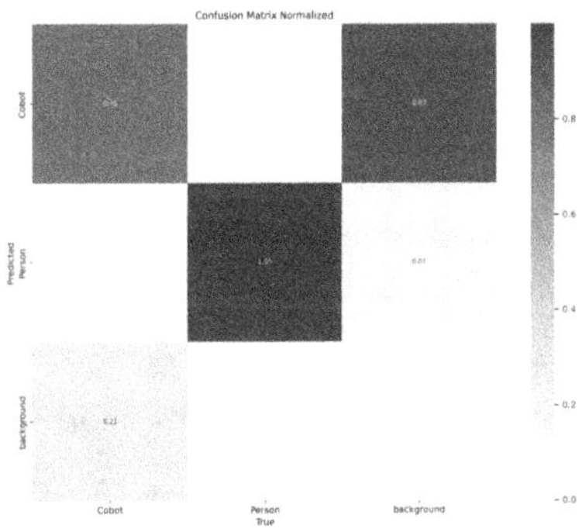

Fig. 7. Confusion Matrix - DatasetGenAI120.

3.3 F1-Confidence and Precision Curves

In the following Figure, the blue line represents the Cobot class, the orange line represents the person class, and the grey line represents all classes combined. The F1 curve (Fig. 8) shows values for different confidence levels of box prediction.

For our model, be obtained that:

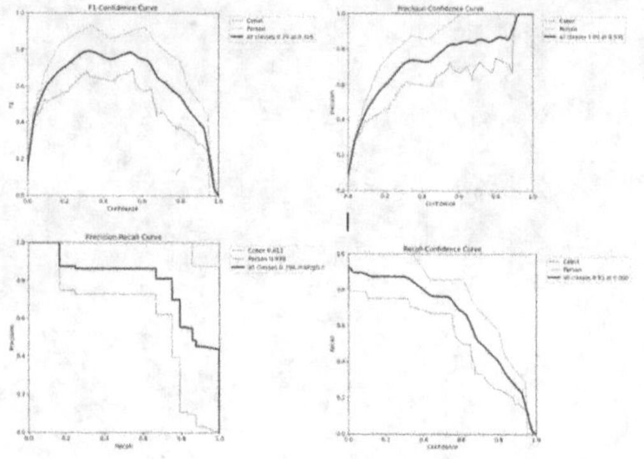

Fig. 8. F1, precision, recall curves - DatasetGenAI120.

- the average F1 score for all classes is 0.79, with a standard deviation of 0.325 in the case of box predictions.

These curves are useful for assessing the trade-off between a model's confidence (its certainty in predictions) and the F1 score (a metric that balances precision and recall). A higher F1 score indicates better overall performance of the model.

3.4 Precision-Recall and Recall-Confidence Curves

In Fig. 8, the PR curves show precision values for different confidence levels of box prediction. It is evident that the model performs better at increasing confidence levels in terms of precision. This is to be expected since the model gets less likely to report false positives as it gets more selective. On the other hand, as Fig. 8 illustrates, the more restrictive model is more likely to categorize genuine positive samples as negative, increasing the amount of false negatives. For this reason, the recall decreases with increasing confidence level. Figure 8 further illustrates this inverse link by showing that for each category, precision rises in proportion to recall. We conclude as follows, keeping in mind that an effective classifier will maintain both a high precision and a high recall:

- **Cobot:** Mean Average Precision (mAP) score: 0.611. As recall increases, the precision decreases gradually.
- **Person**: mAP score: 0.978. Precision remains high even at higher recall levels.
- **All Classes:**(combined): mAP score: 0.794. Precision declines as recall increases

These findings reveal that there are trade-offs between recall and precision in both approaches.

3.5 Object Detection in a Video in Real-Time

The implementation of real-time object detection using our dataset - DatasetGenAI120- generated by a diffusion model has demonstrated promising results in distinguishing between Cobots and humans. This approach not only ensures robust performance in varied environments but also highlights the potential of synthetic data in improving machine learning models in an industrial environment where sustainable safety conditions are a primary goal. Otherwise, the problem is linked to the fact: what happens if we use a dataset created with genAI, including the well-known biases, to detect objects/persons in real-time? There is an ethics problem. Future work could focus on further refining the model and exploring its application in other object detection tasks, considering biases at different levels (Fig. 9).

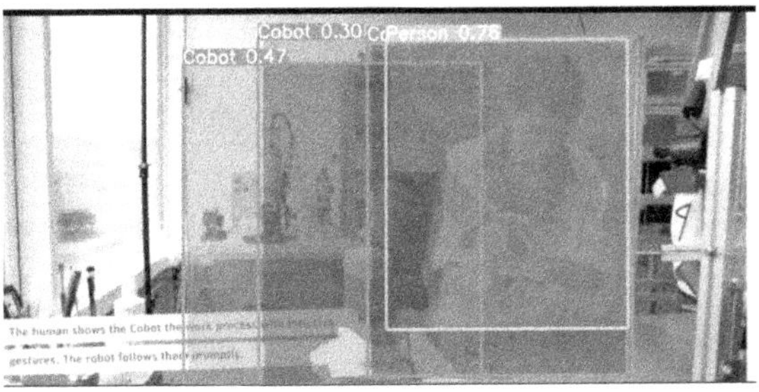

Fig. 9. Object detection in real-time using DatasetGenAI120.

3.6 Sustainability of Our AI Model in Terms of Carbon Footprint

Comparing the carbon footprint of our model based on transfer learning, and the same model trained from scratch, we have respectively 0,0031 kg CO_2 0,0934 kg CO_2 c

4 Conclusion

AI image generation's body type bias is a problem where some bodies are portrayed favourably, upholding particular aesthetic standards while obfuscating others. In the datasets used to train AI models, body forms that conform to prevalent cultural norms are preferred. This bias is evident in these selections. prejudicial and stereotyped characteristics pertaining to nationalities, gender, skin tone, professions, and more. Deepfake technology has the potential to introduce or increase biases, visible and quiet, in a number of ways. These biases can be inherited from the training data used to develop deep fake algorithms, or they can be intentionally introduced by creators [67]. Such biases can be seen across a variety of characteristics, including race, gender, and age. Ethical standards state that technology, including deepfake technology, should not continue or cause biases. A decent and equitable society must aim for equal treatment and respect for all people, regardless of background.

We have demonstrated that, considering appropriate prompts and image parameters, as well as model hyperparameters, these potential biases don't affect the object detection in real time when a small dataset, based on DTnet workflow, is created using images of genAI. As well, the use of transfer learning produce less carbon footprint compared to a model trained from scratch. Future work could focus on further refining the model and exploring its application in other object detection tasks where is required to identify technical details.

References

1. Ramdurai, B., Adhithya, P.: The impact, advancements and applications of generative AI. Int. J. Comput. Sci. Eng. **10**(6), 1–8 (2023)
2. Bengesi, S., El-Sayed, H., Sarker, M. K., Houkpati, Y., Irungu, J., Oladunni, T.: Advancements in generative AI: a comprehensive review of GANs, GPT, autoencoders, diffusion model, and transformers. IEEE Access (2024)
3. Li, Z., et al.: Diffusion model for data-driven black-box optimization. arXiv preprint arXiv:2403.13219 (2024)
4. Fuhr, A.S., Sumpter, B.G.: Deep generative models for materials discovery and machine learning-accelerated innovation. Front. Mater. **9**, 865270 (2022)
5. Kulkarni, A., Shivananda, A., Kulkarni, A., Gudivada, D.: Diffusion model and generative AI for images. In: Applied Generative AI for Beginners: Practical Knowledge on Diffusion Models, ChatGPT, and Other LLMs, pp. 155–177. Apress, Berkeley (2023)
6. Islam, S., Javeed, D., Saeed, M.S., Kumar, P., Jolfaei, A., Islam, A.N.: Generative AI and cognitive computing-driven intrusion detection system in industrial CPS. Cogn. Comput. 1–15 (2024)
7. Rombach, R., Blattmann, A., Lorenz, D., Esser, P., Ommer, B.: High-resolution image synthesis with latent diffusion models. In Proceedings of the IEEE/CVF Conference on Computer Vision and Pattern Recognition, pp. 10684–10695 (2022)
8. OpenAI. ChatGPT (2024)
9. Vázquez, A.F.D.C., Garrido-Merchán, E.C.: A taxonomy of the biases of the images created by generative artificial intelligence. arXiv preprint arXiv:2407.01556 (2024)
10. Zhou, M., Abhishek, V., Derdenger, T., Kim, J., Srinivasan, K.: Bias in generative AI. arXiv preprint arXiv:2403.02726 (2024)
11. Monsone, C., Csapó, Á.B.: Towards rapid prototyping of digital twins based on hand-held video. In: Accentuated Innovations in Cognitive Info-Communication, pp. 167–181. Springer, Cham (2022)
12. Monsone, C. R., Csapo, A.: A cobot image dataset for the creation of digital twins in industrial robot applications. In 12th IEEE International Conference on Cognitive Infocommunications: CogInfoCom 2021, vol. 12, p. 847 (2021)
13. Monsone, C. R., Csapo, A.: Instance segmentation in Industry 5.0 applications based on the automated generation of projected point clouds for VR. In: Acta Polytechnica Hungarica (2024). (in press)
14. Dwyer, B., Nelson, J., Hansen, T., et al.:Roboflow (Version 1.0) [Software] (2024). https://roboflow.com
15. Gorecky, D., Schmitt, M., Loskyll, M., Zühlke, D.: Human-machine-interaction in the Industry 4.0 era. In: 2014 12th IEEE International Conference on Industrial Informatics (INDIN), pp. 289–294. IEEE (2014)

16. Kalateh, S., Estrada-Jimenez, L.A., Pulikottil, T., Hojjati, S.N., Barata, J.: The human role in human-centric industry. In: IECON 2022–48th Annual Conference of the IEEE Industrial Electronics Society, pp. 1–6. IEEE (2022)
17. Jocher, G., Chaurasia, A., Qiu, J.: Ultralytics YOLOv8 (2023). https://github.com/ultralytics/ultralytics
18. Habibi Aghdam, H., Jahani Heravi, E., Habibi Aghdam, H., Jahani Heravi, E.: Convolutional neural networks (2017)

Causal Analysis of Machine Deviations on Espresso Sensory Quality

Kouassi Kouakou Kan Maxime[1(✉)], Michal Ptaszynski[1], Fumito Masui[1], Tanjim Mahmud[3], Tsuyoshi Gotoda[4], Hirofumi Arai[1], Chao-Hui Feng[1], and Jun Watanabe[2]

[1] Kitami Institute of Technology, 165 Koen-cho, Kitami 090-8507, Japan
maxkouassi14@gmail.com, michal@mail.kitami-it.ac.jp
[2] Obihiro University of Agriculture and Veterinary Medicine, Obihiro 080-8555, Japan
[3] Department of Computer Science and Engineering, Rangamati Science and Technology University, Rangamati 4500, Bangladesh
[4] 1stTAG, Workshop Cafe, Kitami, Japan

Abstract. Achieving consistent sensory quality in espresso is crucial for the coffee industry. So far, several work have been addressing this challenge, but remain some unexplored aspects such as deepening analysis by combining machines parameters information and consumers preferences. This study investigates the causal effects of deviations from target espresso machines parameters (water temperature, flow and pressure) on consumer-rated sensory attributes: Aroma, Bitterness, Concentration and Richness. Firstly, we tried to understand the correlation between the features and the final satisfaction of consumers, and secondly, combining datasets of consumer evaluations and machine parameters. We employed causal inference framework to estimate the Average Treatment Effect (ATE) of parameters on sensory outcomes using linear regression with backdoor adjustment. While the statistical analysis did not show significance regarding the conventional 0.05 alpha level, some trends were observed. Temperature deviation was the most influential parameter, positively affecting Aroma, Concentration, and Richness, while negatively affecting Bitterness. Robustness checks using placebo treatment, random common cause, and data subset refuters generally supported the findings. This research highlights the complexity of espresso extraction and the estimated influence of the deviation of the parameters on the features of this beverage, and the need for further research with larger datasets and consideration of other influential factors to fully elucidate the impact of brewing inconsistencies.

Keywords: Coffee · Espresso · Data Science · Causal Inference

1 Introduction

Known as a beverage obtained from roasted coffee bean, coffee swiftly spread worldwide and it has became nowadays a highly popular drink in modern society [7]. Samoggia et al. [11] further highlighted that it is one of the most drank

Published by Springer Nature Switzerland AG 2026
E. Mercier-Laurent et al. (Eds.): AI4S 2025, IFIP AICT 780, pp. 40–50, 2026.
https://doi.org/10.1007/978-3-032-21184-2_4

beverage in the world. In addition of being one of the most widely consumed beverage, coffee plays a crucial role in the economies, as Davis et al. [6] pointed it out, it is a significant beverage that fuels a multi-billion-dollar global industry and provides economics benefits and livelihoods for over 100 million coffee producers in numerous tropical countries. Coffee is a beverage that is extracted from coffee roasted seed grounded at a certain level, depending on the beverage type one would like to get from. The process involve forcing hot water over grounded coffee to obtain the beverage. Several extraction method of this beverage can be highlighted, such as Espresso, Moka, French Press, Cold Brew [3]. On the other hand, those extraction methods can be categorized in three categories such as decoction methods comprising boiled, turkish, percolator and vacuum coffee; infusion methods comprising filter coffee and Napoletana; and Pressure methods comprising Plugger, Moka, and Espresso coffee [5]. Among all of those methods, Espresso has been the mostly studied by researchers these last decades [5]. Originating in Italy, Espresso coffee is a rich, concentrated coffee made by pushing hot water through finely ground coffee beans under high pressure[1]. The process of Espresso involve several parameters and each of them has a certain influence throughout the extraction process [5], with the water temperature being one of those that are influencing the most, the final outcome of espresso, as it facilitate the extraction of several compounds from the ground [1]. So far, several studies have been carried out on the purpose of improving the quality of this beverage, while finding the right parameters that can achieve this goal [1,2,8–10]. Following the recent studies on this field and given the understanding of how parameters are playing a crucial role on the final quality of espresso coffee, this study aims to delve deeper into the impact of those parameters, specially how their deviations from the target settings are influencing the final sensory quality of the beverage. The objective of this research is to provide valuable insights into the importance of parameters consistency in achieving optimal espresso quality and contribute to a more data-driven approach for baristas and the broader coffee industry.

The key contributions of this study are as follows:

1. Explore and quantify the impact of machine parameter deviations on consumer sensory perception
2. Apply a causal inference framework to espresso quality analysis,
3. Provide data-driven insights for best practices and machine parameters management.

2 Materials and Methods

The Fig. 1 describes the overall workflow of this study, outlining the key steps of our process. The espresso machine records the subjective information (brewing parameters) of each single shot and store them in the cloud. Simultaneously, a questionnaire was conducted to collect feedback from consumers on the

[1] https://en.wikipedia.org/wiki/Espresso.

shots - those are the subjective information (sensory evaluation). Both types of data were collected, then analyzed independently and in combination, to extract insights that would gie the development of our model.

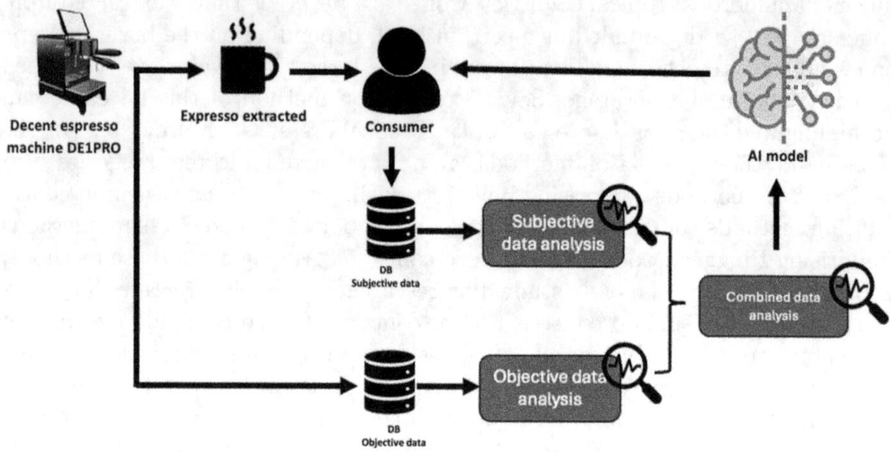

Fig. 1. Overview of the research process.

2.1 Data Collection

The dataset used in this research comprises two main parts: objective and subjective data. In collaboration with the Workshop Café[2]we are working with on this research, The objective data was collected from two coffee machines of the same model used in the research, The Decent Espresso Pro[3], which is an intelligent machine capable of saving all relevant information for each espresso shot made. Originally, before starting each extraction, we manually set on the machine, the target parameters such as water temperature, pressure, and flow rate. These targets represent the values we expect the machine to reach during the shot.

In addition to these targets, the machine automatically records the actual values reached throughout the entire brewing process. This includes a second-by-second log of the actual temperature, pressure, and flow, etc., allowing us to determine and monitor any deviations from the set targets.

Simultaneously, we collected feedback (subjective information) through a consumer survey. The shots were immediately served to participants who was asked to complete a short questionnaire.

The goal of this survey was to capture the participant's perception of the coffee they were tasting. Specifically, they were asked to rate coffee features—aroma, richness, sourness, concentration, as well as their overall satisfaction on

[2] Workshop Café: https://www.instagram.com/work_shop_cafe/.

[3] https://decentespresso.com.

the coffee on a scale from 1 to 5, where 1 represents the lowest and 5 the highest score.

Additionally, the questionnaire included questions about the perceived aftertaste, the participant's daily coffee consumption, age range, sex, and nationality. All responses were anonymous. The timing of the questionnaire was synchronized with the coffee preparation process to ensure that each set of survey responses was directly linked to a specific espresso shot. This approach allowed us to build a dataset of the machine-recorded brewing parameters and the subjective impressions of consumers. We were able to collect a total of 830 recorded shots from the coffee machines and 151 completed survey responses from participants for the first collection phase. The difference between the number of objective data entries and subjective responses is due to the fact the machines had been in use prior to the start of the research, which means the size of 830 includes both historical (before the research) and research-specific data.

The 151 survey responses were collected over a period of one year. After this period, we proceeded to link each survey response with its corresponding espresso shot information based on the timestamp and shot ID. Through this matching process, we were able to successfully pair 134 survey responses with their corresponding shot data. However, some of these responses corresponded to shots that were either double espressos or longer extractions, leading to a total of 164 matching shots. To ensure consistency and simplify the analysis, we decided to focus only on the subset of data that could be matched in a one-to-one relationship. From this filtering step, we identified 109 unique survey responses that could be matched directly with 109 individual espresso shots. This sample size formed the basis for our initial analysis, allowing us to study the relationship between machine-recorded brewing parameters and consumer feedback with clarity and precision. The summary of the dataset used in this work is presented on the Table 1

Table 1. Shot (Objective) & Survey (Subjective) data collected.

Type of information	No. of samples
Objective (ALL)	830
Subjective (ALL)	151
Combined (Matched)	109

2.2 Data Analysis

The analysis was carried out in two main parts with separate evaluations conducted for the subjective and objective information. We began by analyzing the subjective data, which consists of participants' sensory evaluations collected through the questionnaire. We performed statistical Analysis to explore how each sensory feature relates to overall consumer satisfaction by using Pearson correlation. The results in Table 2 indicate that aroma, concentration, and richness are positively correlated with overall satisfaction. On the other hand, bitterness shows a negative correlation with consumers overall satisfactions. These findings suggest that the more aromatic, rich, and concentrated a coffee is, the more likely it is to be appreciated. Conversely, higher levels of bitterness tend to be associated with lower satisfaction. By identifying these correlations, we gained valuable insights into how sensory attributes influence the overall perception of espresso quality, helping us better understand which features contribute most to consumer preference.

Secondly, we performed the analysis of objective data. The information recorded for each shot includes several brewing parameters where the most critical we used are water temperature, flow, and pressure. For each of them, the espresso machine captured both the target values that we manually set before extraction and the actual values measured continuously during the brewing process. To examine the influence of consistency in the parameters on the perceived quality of the coffee, this study focuses on the difference between the target and actual values recorded by the machines. Therefore, we calculated the differences by subtracting the actual value from the target value at each time point during the shot. Since the actual values fluctuate throughout the extraction process, the mean of these differences for each parameter was computed, resulting in variables such as mean.diff.temp, mean.diff.flow, and mean.diff.press. These average deviations represent the extent to which each shot deviated from its expected values. This approach enables the assessment of how deviations from the ideal brewing conditions may affect the features of the espresso helping to understand the extent to which consistencies contributes to consumer satisfaction and a more desirable espresso quality.

Finally, by combining the objective and subjective information, we created a dataset of 109 observations, each representing an espresso shot linked to its sensory evaluation. Each entry includes consumer ratings for Aroma, Bitterness, Concentration, and Richness (1–5 scale), along with the mean deviations from target brewing parameters: temperature (mean.diff.temp), pressure (mean.diff.press), and flow (mean.diff.flow). Exploratory Data Analysis revealed clear differences in how the machine controlled each extraction parameter Fig. 2. Flow and pressure showed wider spreads and greater variability compared to temperature, indicating that the machine was less consistent in regulating these two settings. Regarding the flow, the mean differences were mostly centered around –2, with a slight right skew extending toward zero and positive values. This suggests that while many extractions fell short of the target flow, some cases approached or even exceeded the expected level. For pressure, we can notice that the distribution was peaked near zero, showing that most extractions were close to the intended setting. However, the presence of numerous outliers across both extremes ranging as low as –7.5 and as high as +7.5 points to occasional instability or irregular shifts in pressure control. On the other hand, temperature differences were more concentrated, clustering around

–1 with a relatively narrow spread. The distribution showed a subtle right skew, with a few extractions producing temperatures above the target. Overall, this pattern suggests a tighter and more reliable control over temperature than over flow or pressure. As of the sensory scores Fig. 3, we can notice that Aroma, Concentration, and Richness all showed strong clustering around values of 4 and 5, suggesting that participants consistently rated these attributes favorably. Among these, Aroma was most tightly centered at 4, while Concentration and Richness displayed slightly broader distributions, with noticeable proportions of responses extending to 5. In contrast, Bitterness/Sourness exhibited a more varied pattern. Unlike the other attributes, its distribution showed multiple peaks around 3, 4, and 5, indicating greater diversity in consumer perception. This suggests that taste preferences for bitterness and sourness were less uniform, with some participants favoring moderate levels and others preferring stronger intensities.

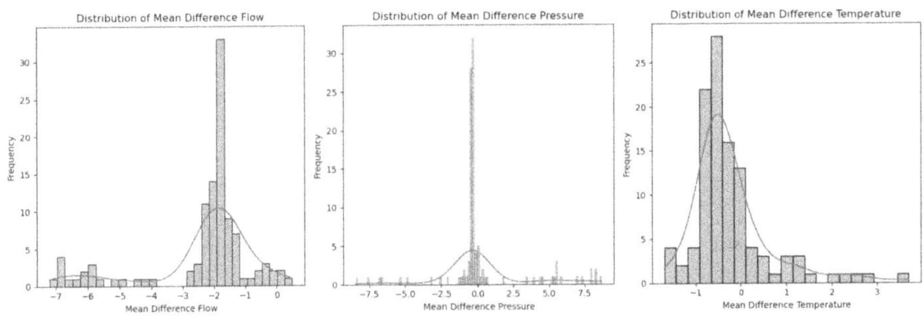

Fig. 2. Distributions of parameters.

The above findings guided the selection of linear regression within the DoWhy framework for causal estimation. Despite some skewness in sensory outcomes, linear regression offers a straightforward, interpretable method to estimate the effects of brewing parameter deviations on sensory ratings, serving as an initial step in understanding these relationships.

2.3 Causal Inference

Additionally we employed the DoWhy library [4,12] to infer the causal effects of deviations in parameters on the sensory evaluations of espresso. This process follows the four key steps outlined by DoWhy: Causal Graph Modeling, causal effect Identification, Estimation, and Refutation. Our assumed causal relationships between the variables are presented in Fig. 4. The DAG (Direct Acyclic Graph), as indicated by the arrows, assumed that the deviations in temperature, pressure and flow, directly influence the primary sensory attributes. Subsequently, these sensory attributes are assumed to directly influencing the overall

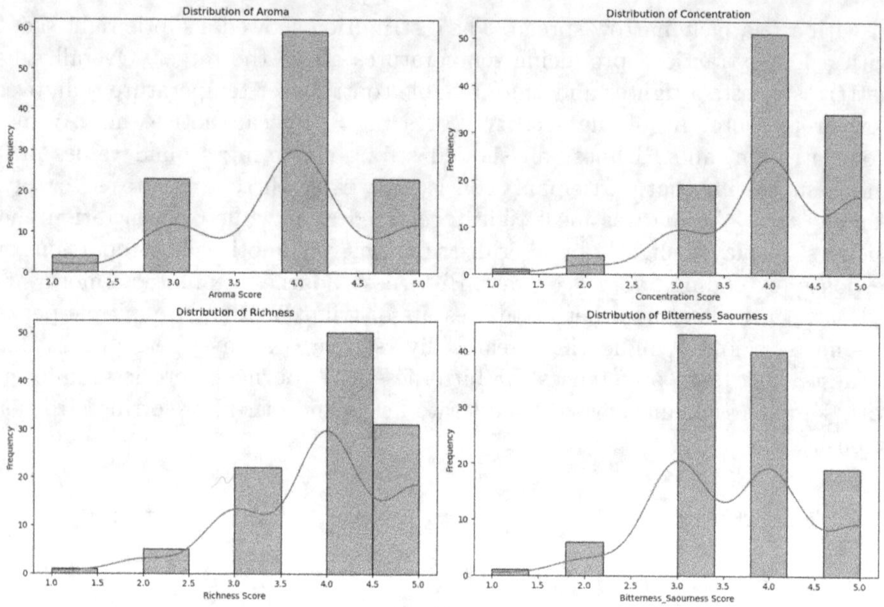

Fig. 3. Distributions of features.

Table 2. Statistical Analysis of Aroma, Bitterness, Concentration and Richness with Overall Satisfaction.

	Aroma	Bitterness	Concentration	Richness	Overall Satisfaction
Mean	3.92	3.64	4.04	3.96	4.32
Variance	0.56	0.74	0.71	0.75	0.64
Standard deviation	0.75	0.74	0.84	0.75	0.80
Overall satisfaction with...	Aroma	Bitterness	Concentration	Richness	
Covariance	0.14	–0.18	0.04	0.21	
Correlation	0.24	–0.25	0.06	0.31	
T-test	2.53	–2.77	0.66	3.35	
p-value	0.01	0.006	0.51	0.001	

evaluation. The absence of incoming arrows to the parameters implies an assumption of no direct measured common causes or confounders influencing these deviations and the sensory outcomes. The causal estimand was identified using the backdoor criterion as implemented in DoWhy. Given the structure of our causal graph the direct association between each treatment (parameter) and outcome (features) was identified as the causal effect. The causal effect was estimated using linear regression (`methodname="backdoor.linearregression"`) as implemented within DoWhy's estimation module. For each sensory outcome, a linear regression model was fitted with mean.diff.flow, mean.diff.temp, and mean.diff.press as the independent variables. The coefficients associated with each 'mean difference' parameter in these models represent the estimated Average Treatment Effect (ATE) of a one-unit change in that parameter's deviation on the specific sensory outcome, conditional on the other 'mean difference' parameters being held constant. The `model.estimateeffect()` method was used with the `testsignificance=True` option to obtain p-values for the estimated effects.

Finally, to evaluate the robustness of the estimated causal effects to potential violations of the underlying assumptions, three refutation methods implemented in DoWhy were employed. The Random Common Cause Refuter added a random, unrelated common cause to assess sensitivity to unobserved confounding. The Placebo Treatment Refuter randomly assigned treatment values to check for spurious correlations. The Data Subset Refuter tested the consistency of the effects across random data subsets to evaluate relationship stability

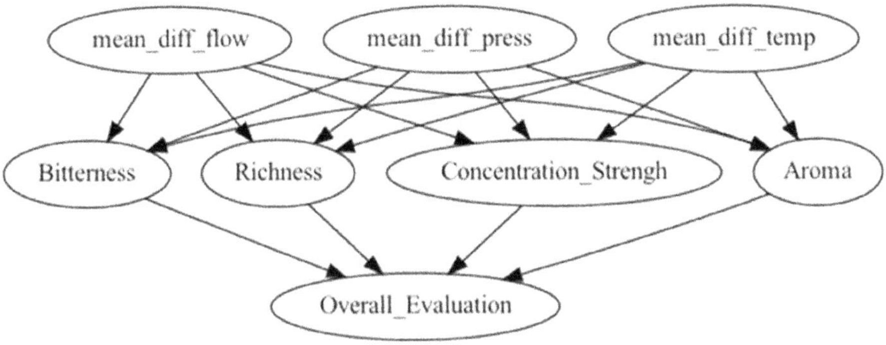

Fig. 4. Causal Graph.

3 Results and Discussion

The estimated Average Treatment Effects (ATE) of the treatments on the outcomes, obtained are presented in Table 3. Results highlighted that among the parameters we chosen, temperature appears to the one that is mostly affecting the features. While all the ATE were relatively small, temperature showed most effects on sensory features. The negative effect of mean.diff.temp on Aroma (–0.15), Concentration (–0.14), and Richness (–0.315) implies that the increase on the temperature deviation tend to decrease those features, and the decreasing the deviation could increase their quality. On the other hand, we have a weak positive effect (0.072) on the bitterness, which could imply that the more deviation on temperature tend to make the coffee more bitter. Although very small, results shown that the increase on the deviation of flow as well as pressure are causing an increase of the concentration (respectively 0.0224 and 0.247) and

Table 3. Treatment Effects and Refutation Results

Treatment (mean_diff)	Outcome (features)	ATE	ATE p-value	refute: random_common_cause	refute: treatment	refute: placebo_data_subset
mean_diff_temp	Aroma	−0.1509	0.94	Pass	Pass	Pass
	Concentration	−0.1408	0.96	Pass	Pass	Pass
	Richness	−0.3168	0.52	Pass	Pass	Pass
	Bitterness	0.0726	0.48	Pass	Pass	Pass
mean_diff_press	Aroma	-0.0185	0.81	Pass	Pass	Pass
	Concentration	0.0247	0.48	Pass	Pass	Pass
	Richness	0.0368	0.09	Pass	Pass	Pass
	Bitterness	−0.0070	0.28	Pass	Pass	Pass
mean_diff_flow	Aroma	−0.0066	0.96	Pass	Pass	Pass
	Concentration	0.0224	0.78	Pass	Pass	Pass
	Richness	0.0031	0.89	Pass	Pass	Pass
	Bitterness	0.1786	0.10	Pass	Pass	Pass

Richness (respectively 0.0031 and 0.0368). An increase of mean.difff.low decrease the Aroma by −0.0066 and increase the Bitterness by 0.1786. An increase of mean.diff.press is decreasing Aroma by −0.0185 and Bitterness by −0.0070. It can also be argued that for the results with low scores (i.e., lower than 0.1) the effect is negligible, and can be disregarded. This is further supported by the statistical significance analysis. None of the results was statistically significant and all p-values were above the conventional statistical significance level of 0.05. This suggests that we need more data to either confirm or refute the results with certainty. However, despite the lack of statistical significance, the refutation tests supported the stability of the estimated effects, suggesting the models were robust to threats like unobserved confounding and noises.

While the analysis did not reveal statistical significance, the trend suggests potential practice relevance. For example, the result of the effect of temperature on the feature would imply that to improve the quality of espresso, especially, as temperature is showing the most effect on the features, this parameter should be used carefully in order to extract the most quality from espresso. On the other hand, this study carries direct implications for environmental sustainability. Deviations in machine parameters, leading to suboptimal extractions and low-quality shots, directly contribute to waste in terms of coffee grounds, water, and energy. Our findings, while not statistically significant for specific causal links, highlights that any enhancement in brewing precision and consistency could contribute to resource conservation by minimizing such inefficiencies. It is crucial to emphasize that, given the exploratory nature of this study and the lack of strong statistical significance, these findings should not be interpreted as prescriptive advice for baristas regarding specific parameter adjustments. Instead, they show the complex interplay of factors and underscore the necessity for more controlled experimental designs and larger datasets to establish definitive causal relationships for practical application.

3.1 Limitations

Several limitations of this study should be acknowledged. Firstly, the concern of the sample size used in this study. Firstly, we suspected that the relatively small sample size (n = 109) for both objective and subjective information may have limited the statistical significance and the power of the model to detect strong causal effects. Using a larger dataset, might probably lead to statistical significance, strong and more reliable results. Secondly, this study focused solely on the deviations in machine parameters (flow rate, temperature, and pressure). Other crucial factors known to significantly influence espresso quality and sensory perception, such as coffee bean type, origin, roasted level, coffee grind size, water quality, etc. were not included as treatment variables or potential confounders in our analysis due to a lack of available data. The absence of these potentially influential variables could obscure or interact with the effects of the machine parameters under investigation.

4 Conclusion and Future Work

In this study, we investigated the causal effects of deviations in espresso machine parameters—temperature, pressure, and flow on consumer rated sensory attributes: Aroma, Richness, Bitterness, and Concentration. Although no statistically significant effects were identified, consistent trends emerged. In particular, temperature deviations appeared to reduce Aroma, Richness, and Concentration while slightly increasing Bitterness. These exploratory patterns suggest that temperature stability plays a central role in shaping consumer perceptions of flavor balance. The implications extend beyond sensory quality. By improving consistency in brewing parameters, cafés and roasters could not only deliver more reliable taste experiences but also minimize waste from unsatisfactory shots, thereby contributing to sustainability goals. From a broader perspective, the integration of data science and AI into espresso preparation holds strong potential for optimizing both flavor quality and resource efficiency. Such approaches could pave the way for precision coffee-making practices that are at once consumer-centered and environmentally responsible. That said, this study also faced several limitations. The relatively small sample size restricted the statistical power of the analysis, while the exclusion of variables such as grind size, bean origin, roast level, and water chemistry may have obscured more nuanced effects. Furthermore, the reliance on linear models may have limited the ability to capture complex, non-linear relationships between machine parameters and sensory outcomes. Future research should address these challenges by leveraging larger datasets, incorporating a wider range of explanatory factors, and explore further causal inference methods including non-linear. Such work would provide a more comprehensive understanding of how brewing parameters interact to shape the coffee experience, ultimately guiding innovations in both consumer satisfaction and sustainable coffee production.

References

1. Andueza, S., Maeztu, L., Pascual, L., Ibáñez, C., de Peña, M.P., Cid, C.: Influence of extraction temperature on the final quality of espresso coffee. J. Sci. Food Agric. **83**(3), 240–248 (2003)
2. Andueza, S., Vila, M.A., Paz de Peña, M., Cid, C.: Influence of coffee/water ratio on the final quality of espresso coffee. J. Sci. Food Agric. **87**(4), 586–592 (2007)
3. Angeloni, G., et al.: What kind of coffee do you drink? An investigation on effects of eight different extraction methods. Food Res. Int. **116**, 1327–1335 (2019)
4. Blöbaum, P., Götz, P., Budhathoki, K., Mastakouri, A.A., Janzing, D.: Dowhy-gcm: an extension of dowhy for causal inference in graphical causal models. J. Mach. Learn. Res. **25**(147), 1–7 (2024). http://jmlr.org/papers/v25/22-1258.html
5. Cordoba, N., Fernandez-Alduenda, M., Moreno, F.L., Ruiz, Y.: Coffee extraction: a review of parameters and their influence on the physicochemical characteristics and flavour of coffee brews. Trends Food Sci. Technol. **96**, 45–60 (2020)
6. Davis, A.P., Mieulet, D., Moat, J., Sarmu, D., Haggar, J.: Arabica-like flavour in a heat-tolerant wild coffee species. Nature Plants **7**(4), 413–418 (2021)
7. Gloess, A.N., et al.: Comparison of nine common coffee extraction methods: instrumental and sensory analysis. Eur. Food Res. Technol. **236**, 607–627 (2013)
8. Moeenfard, M., Silva, J.A., Borges, N., Santos, A., Alves, A.: Diterpenes in espresso coffee: impact of preparation parameters. Eur. Food Res. Technol. **240**, 763–773 (2015)
9. Navarini, L., Rivetti, D.: Water quality for espresso coffee. Food Chem. **122**(2), 424–428 (2010)
10. Parenti, A., Guerrini, L., Masella, P., Spinelli, S., Calamai, L., Spugnoli, P.: Comparison of espresso coffee brewing techniques. J. Food Eng. **121**, 112–117 (2014)
11. Samoggia, A., Riedel, B., Ruggeri, A.: Social media exploration for understanding food product attributes perception: the case of coffee and health with twitter data. Brit. Food J. **122**(12), 3815–3835 (2020)
12. Sharma, A., Kiciman, E.: Dowhy: an end-to-end library for causal inference. arXiv preprint arXiv:2011.04216 (2020)

Studying Neural Network Robustness to Weight Perturbations for Sustainable Incremental Learning

Parth Mehta(✉) and Xiaoyin Wang

The University of Texas at San Antonio, San Antonio, TX 78249, USA
parth.mehta@my.utsa.edu, xiaoyin.wang@utsa.edu

Abstract. Incremental learning is a widely adopted technique that continuously update a machine learning model with emerging new data over a long period of time. Compared to end-to-end learning, incremental learning avoids retraining of the model with the full dataset and thus can save huge amount of computation resource and energy. However, this sustainable technique of incremental learning also face additional threats. Since the machine learning model is no longer read-only during its deployment, adversarial attacks, system errors, and software bugs may all cause erroneous updates of model weights which will result in downgrades in model performance. Therefore, it is important to quantitatively analyze the potential performance downgrades caused by different types and extent of weight changes. In this study, we investigate the resilience of two basic neural network models that are commonly used in incremental learning: Recurrent Neural Networks (RNNs) and Convolutional Neural Networks (CNNs). Through systematic error injection by introducing controlled weight perturbations. We compare the performance of these architectures under various conditions, including different initial training durations, error percentages and error types. The experiment found that weight perturbations significantly hinder CNN training but not RNN, and distributed perturbations causing greater initial performance drops but enabling faster recovery. We believe these findings, although observed in a small scale experiment, warrant further investigation to better understand robustness of neural network models.

Keywords: Neural Networks · Weight Perturbation · Incremental Learning

1 Introduction

End-to-end model training is the most commonly used approach in various AI applications [4,13,23]. However, this method repeatedly retrains models on expanding datasets, leading to significant waste of computational resources and energy. In contrast, incremental learning continuously updates model weights as new data becomes available, avoiding redundant computation and offering a more sustainable alternative. This is particularly valuable in large-scale

Published by Springer Nature Switzerland AG 2026
E. Mercier-Laurent et al. (Eds.): AI4S 2025, IFIP AICT 780, pp. 51–61, 2026.
https://doi.org/10.1007/978-3-032-21184-2_5

or resource-constrained environments. It also enables models to adapt in real time to evolving data distributions, which is crucial in dynamic domains such as finance, healthcare, or cybersecurity where new patterns frequently emerge. Furthermore, incremental learning supports scalability, as models can be updated without requiring access to the original dataset, which may be too large or unavailable due to privacy constraints. The approach also helps reduce latency in deployment, since models can integrate new knowledge quickly without downtime for retraining.

Despite these advantages, incremental learning introduces new vulnerabilities. Since model weights must be updated after deployment, the model cannot remain read-only, making it more susceptible to adversarial attacks, system errors, and software bugs. Additionally, hardware faults on edge devices, quantization in low-power IoT systems, and intentional weight modifications for model compression may all introduce unexpected perturbations. Understanding how different neural network architectures respond to such perturbations is therefore critical for building AI systems that are both sustainable and reliable.

In this study, we investigate the robustness of two widely used architectures in incremental learning: Recurrent Neural Networks (RNNs) and Convolutional Neural Networks (CNNs). We analyze their responses to different types and magnitudes of weight perturbations, as well as how factors such as initial training duration affect resilience. Specifically, we examine both continuous perturbations, simulating localized hardware faults, and distributed perturbations, representing broader and randomly localized disruptions. Our experiments show that CNNs are generally more sensitive to weight perturbations than RNNs, experiencing larger accuracy drops and slower recovery. Furthermore, while distributed perturbations often cause sharper initial accuracy declines, they sometimes enable faster recovery, suggesting that randomized network structures may enhance resilience against performance degradation. These findings contribute to the general understanding of neural network behavior in non-ideal conditions and open new perspectives toward the development of more fault-tolerant AI systems.

The rest of the paper details our experimental setup, experimental results, as well as discusses our findings and the comparison of our work with related research efforts.

2 Experiment Design

2.1 Data Sets and Preprocessing

In our experiment, we consider two datasets. The IMDb dataset consists of 50,000 movie reviews from the Internet Movie Database (IMDb), with each review labeled as either positive or negative. It is a widely adopted benchmark for natural language processing, so we use it as our data set for RNN analysis. The CIFAR-10 dataset contains 60,000 32×32 color images across 10 classes. Each picture measures 32×32 pixels. It is a widely adopted benchmark for image processing, and thus a suitable data set for CNN analysis.

Data Preprocessing for IMDb Dataset

Data preprocessing is the essential first step in preparing the IMDb dataset for sentiment analysis using neural network models. The IMDb dataset contains 50,000 movie reviews, evenly split between positive and negative labels, making it a benchmark resource for natural language processing (NLP), sentiment analysis, and supervised learning research. Each review is associated with a binary sentiment label, providing a rich foundation for training algorithms to recognize linguistic patterns linked to positive or negative opinions.

To ensure that the dataset is properly formatted for recurrent neural networks (RNNs), several preprocessing steps are applied. These steps standardize the data, reduce noise, and prepare it for efficient training and evaluation, ultimately improving model performance. The main preprocessing tasks are presented below.

Tokenization and Vocabulary Creation. The first step is tokenization, which breaks each review into individual words or tokens. This allows the model to interpret text systematically and capture sentiment-related features. Tokenization also builds a vocabulary, mapping each unique word to a numerical index, thereby converting text into machine-readable input for RNNs.

Normalization. Normalization ensures consistency across the dataset by reducing variations in text. Common steps include converting all words to lowercase, handling punctuation, and replacing special characters or numbers with standardized forms. These transformations reduce vocabulary size and prevent superficial differences—such as case or punctuation—from affecting the model's ability to learn meaningful patterns.

Padding. Since reviews vary in length, padding standardizes all sequences to the same size, which is required for batch processing in RNNs. In this study, sequences are either truncated or padded with zeros to a fixed maximum length of 1,000 words. This uniformity ensures that every review has the same dimensionality, allowing the model to process inputs consistently and efficiently.

Word Embeddings. Word embeddings provide dense vector representations of words that capture semantic relationships and contextual usage. Unlike one-hot encoding, embeddings encode similarities between words in a continuous space. Pre-trained embeddings such as Word2Vec, GloVe, or FastText can be used to initialize the embedding layer of RNNs, enriching the model with prior linguistic knowledge.

Through these preprocessing steps, the IMDb dataset is transformed into a standardized, machine-readable format that enables effective training of neural network models for sentiment classification.

Data Preprocessing for CIFAR-10

Data preprocessing plays a critical role in enhancing the performance of Convolutional Neural Networks (CNNs), improving their accuracy, robustness, and reliability. In this section, we outline several preprocessing strategies that strengthen

CNN performance on the CIFAR-10 dataset. These techniques are not only valuable for academic research but also contribute to the development of more effective real-world image recognition systems.

The CIFAR-10 dataset consists of 60,000 color images (32×32 pixels) distributed across 10 classes. While widely used, the dataset poses challenges such as high-dimensional feature space, potential class imbalance, and a relatively limited sample size. To overcome these limitations and achieve optimal performance, multiple preprocessing techniques are applied:

Normalization. The pixel values in CIFAR-10 images range from 0 to 255. Normalization rescales these values to a consistent range (e.g., [0,1]) using methods such as minmax scaling or z-score normalization. This ensures uniform feature scaling, which improves the stability and efficiency of the training process.

Data Augmentation. Given the limited dataset size, augmentation is employed to expand the diversity of training examples and improve generalization. Techniques such as cropping, flipping, rotation, and scaling generate modified versions of the original images. This reduces overfitting by exposing the model to a wider variety of inputs while preserving key features.

Handling Class Imbalance. Disproportionate class distributions can bias the model, leading to poor performance on underrepresented categories. To mitigate this, strategies such as oversampling minority classes, undersampling majority classes, or applying class-weighted loss functions are used to balance training and improve fairness across categories.

Data Shuffling and Splitting. The dataset is divided into training, validation, and test sets. The training set is used to fit the model, the validation set for hyperparameter tuning and monitoring overfitting, and the test set for final performance evaluation. Shuffling the data prior to splitting is essential to prevent ordering biases and ensure representative distributions across subsets.

Through these preprocessing steps, the CIFAR-10 dataset is transformed into a more structured and balanced form, enabling CNNs to achieve stronger and more reliable performance in image classification tasks.

2.2 Model Setup

To investigate the resilience of neural networks to weight perturbations, we designed a comprehensive set of experiments using both Convolutional Neural Networks (CNNs) and Recurrent Neural Networks (RNNs) for a comprehensive analysis of error tolerance. In this paper, we study the resilience of neural networks to weight perturbation - a critical notion of model robustness in real-world deployments. We have conducted this study in an attempt to understand how these networks respond to a range of different types and amplitudes of errors injected into their model weights.

In our study, we follow the classical architecture and configurations of RNN and CNN models [1,2]. In our RNN model, an embedding layer is used with a vocabulary size of 10,000 and an embedding dimension of 32. In this layer, our word indices are turned into dense vectors of fixed size. Then we have an LSTM layer with 100 units as the heart of our RNN, which helps to capture long-range dependencies in the sequences. The output from the LSTM feeds into a dense layer with one unit and sigmoid activation, appropriate for our binary classification problem.

Our CNN model consists of three convolutional layers (32, 64, and 64 filters respectively) with ReLU activations, interspersed with max pooling, followed by two dense layers (64 units and 10 output units). More details are presented in Algorithm 1 and Algorithm 2.

Training Protocol: Each training session was conducted for 5 epochs, providing a balance between computational efficiency and sufficient learning progress. This setup also enabled the observation of both short-term and long-term learning patterns. A batch size of 64 was applied uniformly across both models to ensure consistency in results. The models were compiled with the Adam optimizer, chosen for its adaptive learning rate properties. For the loss functions, Sparse Categorical Cross-Entropy was used with the CNN, while Binary Cross-Entropy was applied to the RNN, reflecting the different nature of their classification tasks.

Algorithm 1. RNN Model Architecture

```
embedding_size ← 32
vocabulary_size ← 10000
model ← Sequential()
Embedding layer(vocabulary_size, embedding_size, input_length=1000)
LSTM layer(100 units)
Dense layer(1 unit, activation='sigmoid')
Compile model with:
    loss = 'binary_crossentropy'
    optimizer = 'adam'
    metrics = ['accuracy']
```

Algorithm 2. CNN Model Architecture

```
Initialize Sequential model
Conv2D layer: 32 filters, 3 × 3 kernel, ReLU activation, input (32,32,3)
MaxPooling2D layer: 2 × 2 pool size
Conv2D layer: 64 filters, 3 × 3 kernel, ReLU activation
MaxPooling2D layer: 2 × 2 pool size
Conv2D layer: 64 filters, 3 × 3 kernel, ReLU activation
Flatten layer
Dense layer: 64 units, ReLU activation
Dense layer: 10 units
Compile model:
    optimizer: adam
    loss: SparseCategoricalCrossentropy
    metrics: accuracy
```

2.3 Incremental Learning Simulation

To simulate the process of incremental learning, we randomly split the training datasets into 25 portions with the same sizes. A RNN/CNN model is continuously trained on the 25 portions one by one. A weight perturbation is injected after certain portions, and we observe the performance of the model before and after the perturbation to finding the extent of performance drop and the recovery process.

Error Injection: When introducing errors into the model weights, we explored two primary types of errors: Contiguous (C): Errors concentrated in randomly chosen localized regions of the weight matrices, simulating scenarios such as memory corruption or targeted attacks. Distributed (D): Errors randomly spread across all weights, mimicking general degradation or global noise. For each error type, we tested three magnitudes: 1%, 2%, and 5% of the total weights. This design enabled us to assess the impact of both error localization and severity.

Performance Metrics: Our study uses classification accuracy as a standard metrics. Due to the randomness in dataset split and the model training process, the observed performance changes can vary from time to time. To acquire more representative study outcome, we execute our experiment for 10 times and use the average results of the 10 times as our experiment results.

3 Experiment Results

Our experiment results are presented in Fig. 1. In sub-figures (a)–(h), we present the results with error injected after 5, 10, 15, and 20 training portions for CNN an RNN, respectively. From Fig. 1, we have the following major observations.

First, as expected, all the incremental learning models, no matter RNN or CNN, experience substantial performance drops after the weight perturbations. Also, in all cases, with additional training, the performance recovers after the weight perturbations. Additionally, in all the cases, weight perturbations of larger magnitude always cause more significant performance drops and slower recovery from the drops (i.e., 5% weight perturbations always cause larger performance drop than 2% weight perturbations, which always cause larger performance drop than 1% perturbations).

Second, comparing the subfigures on the left and those on the right, we can see that our experiment with RNN model on the IMDB dataset (on the right) shows less relative performance drop and faster performance recovery than the CNN model on the CIFAR-10 dataset (on the left). In particular, in the CNN experiment, the recovery of performance (see Figs. 1(a) and 1(b)) is even slower than the original training (before the weight perturbation). This shows that for CNN, a defective weight distribution may have big negative effect on the training (worse than the random weight assignment at the beginning), and a model re-initialization may be necessary if the weight perturbation is severe. On the other hand, the RNN experiment shows less performance drop and faster recovery (performance recovers to normal within 5 training portions in most cases). One reason could be that the task is relatively simpler (as evidenced by the high accuracy after the first several training portions).

Third, comparing distributed weight perturbation and continuous perturbation, we can see that for many cases across CNN and RNN experiments, the distributed weight perturbations cause larger performance drop, but also recover faster than continuous perturbation, showed by a cross of the dotted curves and the solid curves. This phenomena is observed in almost all cases of RNN experiments, and the CNN Error at 5 case. For other CNN cases, the experiment may not be long enough for the performance to recover and thus does not show the cross of the curves. The reason behind this observation may be that distributed weight perturbations cause more comprehensive damage to the model so they will affect more test data items. However, the damage is not severe across the model, so additional training can fix them more quickly. On the other hand, continuous weight perturbations are more localized so they affect fewer test data items, but they are harder to fix with additional training due to the severe damage.

To sum up, our experiment has two very interesting findings: (1) weight perturbations have large negative effect on CNN model, making the model harder to train than a original randomly initialized model, and (2) distributed weight perturbations often cause larger performance drops but can recover faster. Although our small scale experiment may not provide fully conclusive findings, the observations are interesting enough to invite further investigations.

Stability Across Executions. We executed our experiments for 10 times and perform average to acquire the results in Fig. 1. We also want to investigate how stable the results are across different executions. Therefore, we calculated the average derivation of accuracy along all data points for each experiment setting and the results are presented in Table 1.

Table 1. Result Derivation of Different Settings

Setting	Avg. Derivation
CNN Error at 5	0.082
CNN Error at 10	0.060
CNN Error at 15	0.044
CNN Error at 20	0.027
RNN Error at 5	0.105
RNN Error at 10	0.078
RNN Error at 15	0.057
RNN Error at 20	0.031

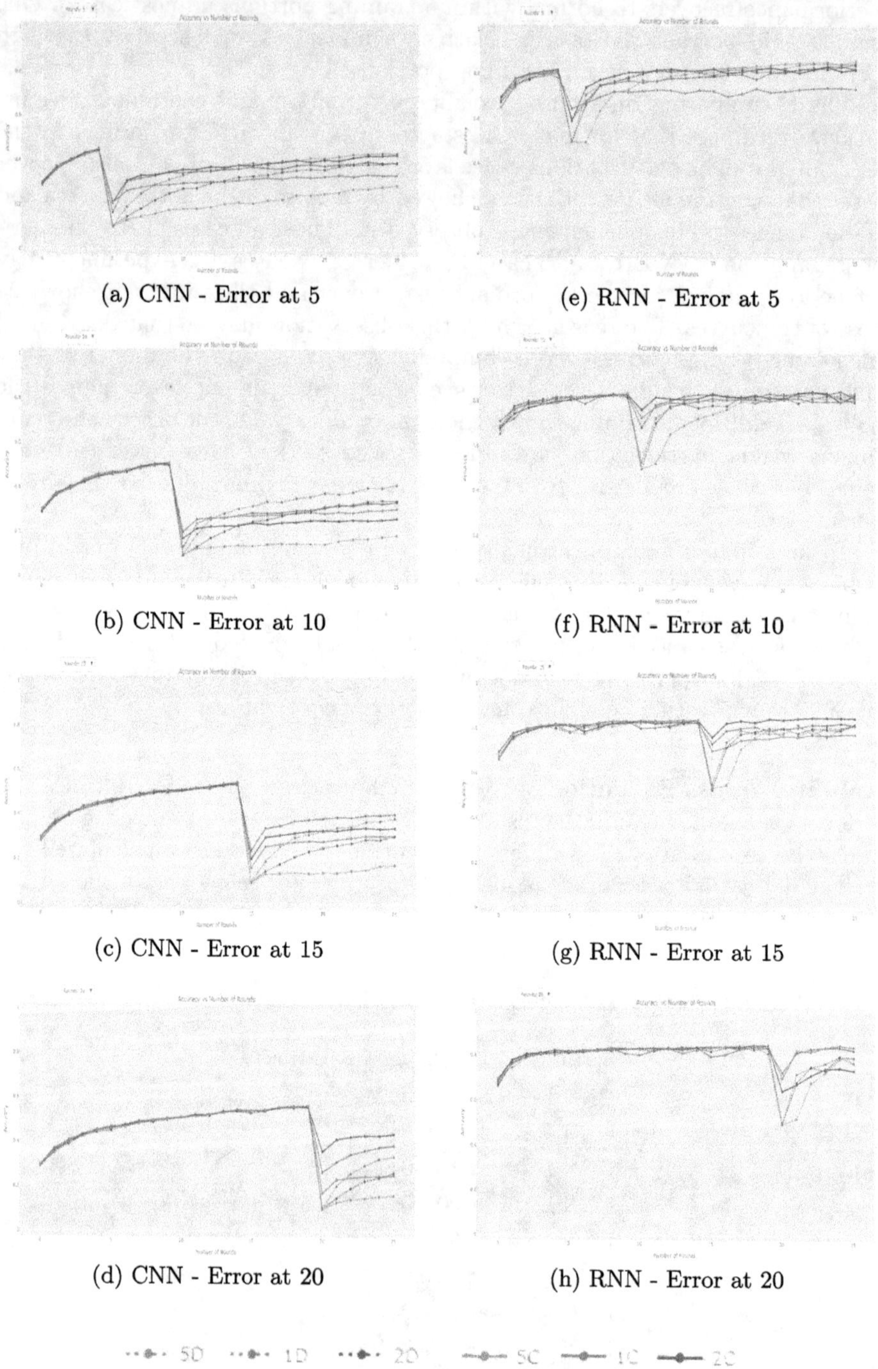

(a) CNN - Error at 5

(e) RNN - Error at 5

(b) CNN - Error at 10

(f) RNN - Error at 10

(c) CNN - Error at 15

(g) RNN - Error at 15

(d) CNN - Error at 20

(h) RNN - Error at 20

Fig. 1. Comparison of CNN and RNN with respect to accuracy.

From the table, we have the following observations. First, average derivation is high across executions, with average derivation of 0.037 to 0.102 for CNN and 0.041 to 0.125 for RNN, indicating that the average accuracy difference may be higher than 10% across executions. Second, executions with error injected in a later time point always have lower derivation, showing that the major source of difference comes from the data points after the mutation point.

4 Related Works

Weng et al. [20] proposed a framework for certifying model robustness against weight perturbations with theoretical guarantees about the neural network's behavior under weight changes. Moving forward in this line of research, Tsai et al. [16] formalized the relationship between generalization and adversarial robustness of neural networks to weight perturbations. In the context of practical applications, Stutz et al. [14] investigated bit error robustness for energy-efficient DNN accelerators. Sun et al. [15] explores the vulnerability of deep neural networks to parameter corruption. While our work focuses on comparing RNNs and CNNs, Zhang et al. [22] explored the theoretical principles behind the trade-off between robustness and accuracy in neural networks more broadly. Tsipras et al. [17] argued that robustness might be at odds with accuracy, a perspective that has sparked significant debate in the field. More recently, Pang et al. [12] suggested that robustness and accuracy can be reconciled with proper definitions. Ko et al. [7] introduced POPQORN, a framework for quantifying the robustness of recurrent neural networks. Wang et al. [19] investigated the verification of neural network control policies under persistent adversarial perturbation. Boopathy et al. [3] proposed an efficient framework for certifying the robustness of CNNs by a method called CNNCert. Madry et al. [8] developed one of the earliest approaches toward making deep learning models resistant to adversarial attacks. Wong et al. [21] revisited adversarial training, proposing faster methods that maintain effectiveness.

Mutation and bug analysis of AI models and software is another related area. Jia et al. [6] summarized the impacts of different types of bugs on deep learning libraries and the categorized their consequences. They further use planted bugs to study how the bugs may affect the performance of different deep learning models [5]. Compared to these two efforts, our study focuses on weight mutations instead of software code errors. Niu et al. [11] and Wang et al. [18] studied how mutations may help to evaluate the error detection techniques, and their studies show that mutation planting can be helpful in error analysis and detection. Mostafa et al. [10] uses multiple executions to study software performance and rule out the influence of randomness. Our study uses a similar approach to execute our experiments for multiple times and avoid the influence of randomness, but we apply it in a different domain. Mendez et al. [9] uses mutation to study the robustness of AI-based software applications. However, this work focuses on the mutation of input data sets instead of on the weights of the models.

Our research builds on these works by directly comparing the resilience of RNNs and CNNs to weight perturbations through error injection. By examining both architectures under similar conditions, we contribute to the ongoing dialogue about the relationship between model architecture, training procedures, and robustness to various forms of perturbations.

5 Conclusion

Our study produced two key findings: weight perturbations significantly impair CNN training, and distributed perturbations tend to cause larger performance drops but allow faster recovery. Although the experiment was small in scale, the results suggest intriguing patterns worth deeper exploration. We plan to expand our study to include more datasets and neural network architectures to test the generality of the findings. Additionally, we aim to explore techniques for randomizing weight locations to enhance model recovery after perturbations.

References

1. Tensorflow core and databases (2025). https://www.tensorflow.org/datasets/catalog/cifar10
2. Text classificaton with an rnn (2025). https://www.tensorflow.org/text/tutorials/text_classification_rnn
3. Boopathy, A., Weng, T.W., Chen, P.Y., Liu, S., Daniel, L.: Cnn-cert: an efficient framework for certifying robustness of convolutional neural networks. In: Proceedings of the AAAI Conference on Artificial Intelligence, vol. 33, pp. 3240–3247 (2019)
4. Hassan, F., Wang, X.: Mining readme files to support automatic building of java projects in software repositories. In: 2017 IEEE/ACM 39th International Conference on Software Engineering Companion (ICSE-C), pp. 277–279. IEEE (2017)
5. Jia, L., Zhong, H., Wang, X., Huang, L., Li, Z.: How do injected bugs affect deep learning? In: 2022 IEEE International Conference on Software Analysis, Evolution and Reengineering (SANER), pp. 793–804. IEEE (2022)
6. Jia, L., Zhong, H., Wang, X., Huang, L., Lu, X.: The symptoms, causes, and repairs of bugs inside a deep learning library. J. Syst. Softw. **177**, 110935 (2021)
7. Ko, C.Y., Lyu, Z., Weng, L., Daniel, L., Wong, N., Lin, D.: Popqorn: quantifying robustness of recurrent neural networks. In: International Conference on Machine Learning, pp. 3468–3477. PMLR (2019)
8. Madry, A., Makelov, A., Schmidt, L., Tsipras, D., Vladu, A.: Towards deep learning models resistant to adversarial attacks (2019). https://arxiv.org/abs/1706.06083
9. Méndez, M., Benito-Parejo, M., Merayo, M.G.: Testing the robustness of machine learning models through mutations. In: International Conference on Computational Collective Intelligence, pp. 308–320. Springer, Heidelberg (2024)
10. Mostafa, S., Wang, X., Xie, T.: Perfranker: prioritization of performance regression tests for collection-intensive software. In: Proceedings of the 26th ACM SIGSOFT International Symposium on Software Testing and Analysis, pp. 23–34 (2017)
11. Niu, X., et al.: An interleaving approach to combinatorial testing and failure-inducing interaction identification. IEEE Trans. Softw. Eng. **46**(6), 584–615 (2018)
12. Pang, T., Lin, M., Yang, X., Zhu, J., Yan, S.: Robustness and accuracy could be reconcilable by (proper) definition. In: International Conference on Machine Learning, pp. 17258–17277. PMLR (2022)

13. Rafi, T., Zhang, X., Wang, X.: Predart: towards automatic oracle prediction of object placements in augmented reality testing. In: Proceedings of IEEE/ACM International Conference on Automated Software Engineering, pp. 1–13 (2022)
14. Stutz, D., Chandramoorthy, N., Hein, M., Schiele, B.: Bit error robustness for energy-efficient dnn accelerators. In: Smola, A., Dimakis, A., Stoica, I. (eds.) Proceedings of Machine Learning and Systems, vol. 3, pp. 569–598 (2021)
15. Sun, X., Zhang, Z., Ren, X., Luo, R., Li, L.: Exploring the vulnerability of deep neural networks: a study of parameter corruption. In: Proceedings of the AAAI Conference on Artificial Intelligence, vol. 35, no. 13, pp. 11648–11656 (2021). https://doi.org/10.1609/aaai.v35i13.17385. https://ojs.aaai.org/index.php/AAAI/article/view/17385
16. Tsai, Y.L., Hsu, C.Y., Yu, C.M., Chen, P.Y.: Formalizing generalization and adversarial robustness of neural networks to weight perturbations. In: Ranzato, M., Beygelzimer, A., Dauphin, Y., Liang, P., Vaughan, J.W. (eds.) Advances in Neural Information Processing Systems, vol. 34, pp. 19692–19704. Curran Associates, Inc. (2021)
17. Tsipras, D., Santurkar, S., Engstrom, L., Turner, A., Madry, A.: Robustness may be at odds with accuracy. arXiv preprint arXiv:1805.12152 (2018)
18. Wang, X., Zhang, L., Tanofsky, P.: Experience report: how is dynamic symbolic execution different from manual testing? A study on klee. In: Proceedings of the 2015 International Symposium on Software Testing and Analysis, pp. 199–210 (2015)
19. Wang, Y.S., Weng, T.W., Daniel, L.: Verification of neural network control policy under persistent adversarial perturbation. arXiv preprint arXiv:1908.06353 (2019)
20. Weng, T.W., et al.: Towards certificated model robustness against weight perturbations. In: AAAI Conference on Artificial Intelligence, vol. 34, pp. 6356–6363 (2020)
21. Wong, E., Rice, L., Kolter, J.Z.: Fast is better than free: revisiting adversarial training. arXiv preprint arXiv:2001.03994 (2020)
22. Zhang, H., Yu, Y., Jiao, J., Xing, E., El Ghaoui, L., Jordan, M.: Theoretically principled trade-off between robustness and accuracy. In: International Conference on Machine Learning, pp. 7472–7482. PMLR (2019)
23. Zhang, X., Heaps, J., Slavin, R., Niu, J., Breaux, T., Wang, X.: Daisy: dynamic-analysis-induced source discovery for sensitive data. ACM Trans. Softw. Eng. Methodol. **32**(4), 1–34 (2023)

Time Series Augmentations with Unsupervised Viewmakers for Robust Disruption Prediction in Nuclear Fusion

Dhruva Chayapathy[1], Lucas Spangher[2](✉), Tavis Siebert[3], Om Patil[4], Enrique Zapata-Cornejo[6], Akshata Kishore Moharir[5], and Cristina Rea[6]

[1] Georgia Institute of Technology, Atlanta, USA
[2] Google Research, Mountain View, CA, USA
spangher@google.com
[3] ETH Zurich, Zürich, Switzerland
[4] Alpharetta High School, Alpharetta, GA, USA
[5] University of Maryland, College Park, MD, USA
[6] Plasma Science and Fusion Center, Cambridge, MA, USA
crea@psfc.mit.edu

Abstract. Machine learning-guided data augmentation may support the development of technologies in the physical sciences, such as nuclear fusion tokamaks. Here we endeavor to study the problem of detecting disruptions - i.e., plasma instabilities that can cause significant damage, impairing the reliability and efficiency required for their real-world viability. Machine learning (ML) prediction models have shown promise in detecting disruptions for specific tokamaks, but they often struggle to generalize to the diverse characteristics and dynamics of different machines. This limits the effectiveness of ML models across different tokamak designs and operating conditions, which is a critical barrier to scaling fusion technology. Given the success of data augmentation in improving model robustness and generalizability in other fields, this study explores the use of a novel time series viewmaker network to generate diverse augmentations or "views" of training data. Our results show that incorporating views during training improves AUC and F2 scores on DisruptionBench tasks compared to standard or no augmentations. The effectiveness of viewmaker networks appears scenario-dependent: while they hindered zero-shot transfer between tokamaks, they showed significant promise for few-shot transfer and may help satisfy the increased data demands of larger, more complex prediction models. This approach represents a promising step towards developing more broadly applicable ML models for disruption avoidance, which is essential for advancing fusion technology and ultimately addressing climate change through reliable and sustainable energy production.

D. Chayapathy and L. Spangher—Equal contribution (co–first authors).

Published by Springer Nature Switzerland AG 2026
E. Mercier-Laurent et al. (Eds.): AI4S 2025, IFIP AICT 780, pp. 62–80, 2026.
https://doi.org/10.1007/978-3-032-21184-2_6

1 Introduction

One of the most notable efforts to turn an idea from the physical sciences into a commercializable reality is nuclear fusion. Nuclear fusion has long been considered a "holy grail" to provide carbon-free energy without significant waste or land demands [8,28]. A leading approach to generating fusion in laboratory plasmas is *magnetic confinement* via tokamak machines. However, the commercial viability of tokamaks hinges on developing their capability to accurately predict plasma disruptions: losses in plasma stability which may damage a tokamak and take it offline for months at a time.

Disruptions are difficult to predict with first-principles or physics-only models. Thus, a significant focus within the nuclear fusion community has been on the development of machine learning (ML) models for the prediction of disturbances. ML models are already in production in the major tokamaks, such as a random forest predictor at the DIII-D [26] and EAST [15] tokamaks, and a simple neural network at the ASDEX tokamak [25]. Research into various time series deep learning architectures, including long-short-term memory (LSTM) variants, [38], continuous convolutional neural networks (CCNN) [1], and transformer-based models [27], has established the state-of-the-art in disruption prediction capabilities. However, because the characteristics and physical dynamics of a tokamak can change dramatically depending on the machine design [10], and over time as the machine is upgraded [19] or modified [12], it is imperative that the prediction models can generalize to these different sources of data.

Data enhancement is one way to improve the generalizability [16] and robustness of the model [14]. Since tokamak data are made up of multivariate time series, common augmentation strategies can include jittering, warping, and additive noise [30,33]. However, naive application of these techniques can lead to enhancements that fail to mimic plausible plasma behavior, affecting the model's ability to learn patterns relevant for disruption prediction [6]. Previously, a process-based Stutent t approach has been used for tokamak data [20], however, it direct impact on post hoc models has not been evaluated. In this work, we explore machine learning-driven augmentations through the use of unsupervised "Viewmaker Networks" [29], models that learn to generate augmentations through adversarial learning.

We hypothesize that viewmaker augmentations, or "views", are a solution for incorporating realistic yet diverse augmentations of disruption data which can improve the robustness of models toward disruptive discharges, and we demonstrate that using views in training improves a post-hoc model's AUC and F2 on DisruptionBench [23] tasks over multiple disruption prediction models when compared to training on standard augmentations such as those present in the tsaug library or training without augmentation.

The novel contributions of this paper are:

(1) We present an original adaptation of the image-based viewmaker network for time series augmentation and benchmark against other time series augmentation techniques.

(2) We are the first to rigorously test generative augmentation methods in the context of nuclear fusion datasets.
(3) We are the first to apply the unsupervised viewmaker augmentations to a supervised learning task.

This paper is structured as follows: Sect. 2 covers the background on disruption benchmarks, dataset preparation, and the viewmaker/contrastive learning framework. Section 3 describes our methodology, including the time series viewmakers, augmentation strategies, and model architectures tested. Section 4 presents results with detailed model comparisons and analysis of augmentation effects. Sections 5 and 6 discuss limitations and conclude with future directions, respectively.

2 Background

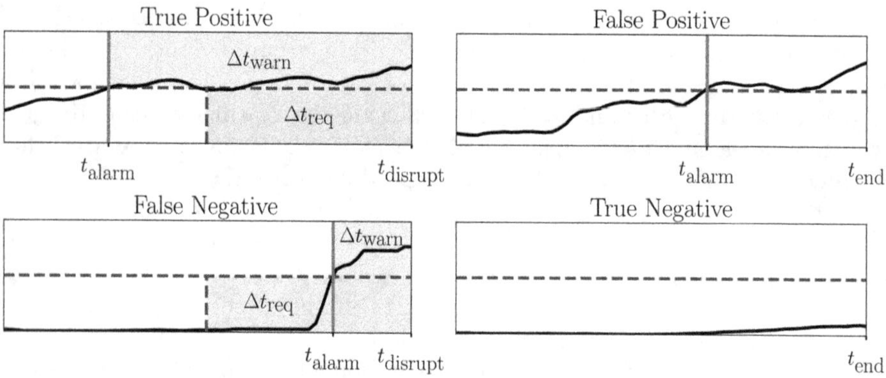

Fig. 1. Illustration of the categorization of true and false positives, and true and false negatives. Figure adapted from ref. [21].

2.1 Disruption Benchmarks

DisruptionBench [23] is the first disruption prediction benchmark aimed at assessing model generalizability across tokamaks. The data is comprised of 25k trials (referred to here as "discharges") from three tokamaks: General Atomics' DIII-D, MIT's Alcator C-Mod, and the Chinese Academy of Sciences' EAST.

Training and test sets for each task can be generated by specifying the following parameters:

Table 1. The input features of the model, their definitions, and a categorization of the type of instability the signal indicates. The features listed are the same across all machines. While this table is adapted from [38], some of our symbology, features, and feature descriptions are unique.

Feature	Definition	Relevant Instability	Units
β_p	Plasma pressure normalized by poloidal magnetic pressure	MHD	Unitless
ℓ_i	Normalized plasma internal inductance	MHD	Unitless
q_{95}	Safety factor at 95% of normalized flux surfaces	MHD	Unitless
$n = 1$ mode	$n = 1$ component of the perturbed magnetic field	MHD	T
n/n_G	Electron density normalized by Greenwald density limit [11]	Dens. limit	Unitless
Lower Gap	Gap between plasma and lower divertor	Shaping	m
κ	Plasma elongation	Shaping	Unitless
$I_{\text{p,error}}/I_{\text{p,prog}}$	Plasma current deviation from the programmed plasma trace (error), normalized by the programmed plasma trace (prog)	Impurities, hardware, runaway electrons	Unitless
V_{loop}	Toroidal "loop" voltage	Impurities	V
Tokamak Indicators	Three additional features with binary encodings of whether the data came from Alcator C-Mod, DIII-D, or EAST	n.a.	Unitless

Table 2. Metrics on a dataset composed of multiple tokamaks.

Tokamak	Number of Discharges	Average Discharge Length	Initial Sampling Rate
Alcator C-Mod	4000	0.52 s	0.005 ms
DIII-D	8000	3.7 s	0.01 ms
EAST	11000	5.3 s	0.025 ms

- **New Machine**: One of the three tokamaks is designated as the "new machine." The test set is comprised of only "new machine" discharges while the discharges in the training set are specified by the "data-split" parameter. The objective of this setup is to understand how well the model could deploy to said new machine. For this work, we set the new machine to Alcator C-Mod.

- **Data-Split**: In this work, we focus on the following four cases with relation to two training machines and one test machine:
  - i **Case 1: Zero-Shot**. Use only discharges from tokamaks that are not "new machine." This is analogous to zero-shot learning.
  - ii **Case 2: Few-Shot**. Use 20 disruptive discharges from the "new machine" and sample the remaining discharges from the other tokamaks. This is analogous to few-shot learning.
  - iii **Case 3: All training shots**. Use all available data from the "new machine" in addition to data from other tokamaks.
  - iv **Case 4: Many-Shot single machine.**: Use only discharges from the "new machine" for the training set.

The designation of true or false positives in DisruptionBench is calculated similar to how a production-scale Disruption Mitigation System (DMS) would behave. The DMS needs at least some time to engage[1], Δt_{req}. Two thresholds $T_{\text{low}}, T_{\text{high}} \in (0, 1)$ and a hysteresis value h are set, and a prediction model outputs *disturbance* scores, i.e. the probability that a disruption will occur in the future. If the disruption is greater than T_{high} for h time steps before Δt_{req}, the prediction is classified as positive at time Δt_{warn}. Otherwise, the discharge reaches its planned end, and the model predicts a negative outcome. If the disruptivity drops below T_{low}, the hysteresis counter is reset. For an illustration of how this translates into true and false positives and negatives, see Fig. 1.

2.2 Dataset Provenance and Preparation

Our data set is open[2] [38]. Each tokamak discharge lasts from seconds to minutes, generating gigabytes of multi-modal data, of which we focus on discharges' global state variables listed in Table 1. The tokamaks may be generally characterized by the information in Table 2, which demonstrates significant difference in average discharge length, sampling rate, and number of samples. We also provide a UMAP clustering of the tokamak data over multiple machines Fig. 2 (Fig. 4).

Each discharge is of length n, and we discard discharges that are shorter than 125 ms. Each tokamak's diagnostics store metrics from each discharge at a different sampling rate. We normalize across tokamaks by discretizing at the lowest common multiple across the three machines, 5 ms, and interpolating via forward fill, following a performance analysis published in Table 9 of [23]. Approximately 10% of discharges end in disruptions. Figure 3 shows a UMAP comparison of disruptive and non-disruptive discharges in Alcator C-Mod. We truncate our time

---

[1] Two common methods of mitigating the harm from a disruption include: (1) puffing a heavy inert gas like Neon or Argon around the chamber's walls [18] or (2) shooting a frozen hydrogen pellet into the middle of the plasma reaction [17]. Both methods introduce particles that absorb the energy of the plasma.

[2] https://dataverse.harvard.edu/dataset.xhtml?persistentId=doi:10.7910/DVN/XIOHW1.

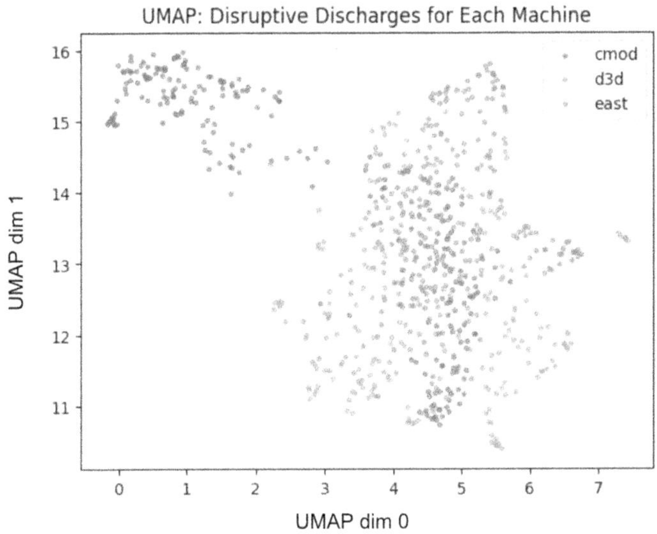

Fig. 2. A UMAP clustering of disruptive discharges from each machine. Many discharges from Alcator C-Mod exhibit distinct behavior from those from DIII-D and EAST. Thus, to challenge models in their ability to learn general disruptive features, we designate Alcator C-Mod as the "new machine" parameter in our experiments.

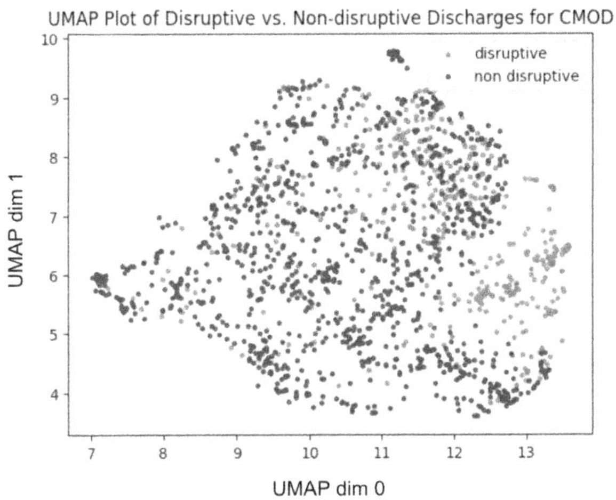

Fig. 3. UMAP comparison of disruptive and non-disruptive discharges randomly sampled from Alcator C-Mod.

series to $X = 1, ..., n - \nu$, where n is the full length and $\nu = 40ms$, the minimum time to activate a disruption mitigation system. For a description of the chosen features, see Table 1.

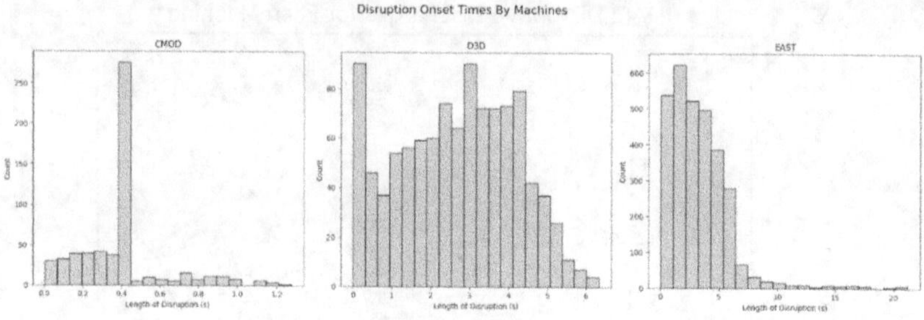

Fig. 4. A comparative histogram of disruption time lengths between the three devices – illustrating further that the training distributions are quite dissimilar.

Because of limited data, we use a 90/10 training split from the Alcator C-Mod tokamak as in [23]. The test set remains fixed across all cases to enable direct comparison, so the ratio of disruptive to non-disruptive shots may vary between training and test sets. This variation does not affect the statistical validity of our results.

2.3 Data Augmentation

Data augmentation refers to the process of artificially expanding a training dataset by applying transformations—such as noise addition, cropping, or even generative perturbations—that preserve the underlying semantics of the data while introducing diversity. Indeed, there exists a broad family of techniques under the label of augmentation. Augmentations help models generalize better by exposing them to a wider range of plausible examples without requiring further data collection.

It is worth mentioning that data augmentation should be applied *only* to the training set to maintain the statistical validity of test results. Including augmented data in the test set artificially inflate performance estimates.

2.4 Viewmakers

Viewmaker networks use an adversarial form of this contrastive loss to train a viewmaker to produce challenging augmentations and an encoder to produce meaningful representations. In order to force the encoder to learn useful latent representations of the original data, the viewmaker network presents the encoder with pairs of augmentations from the original data. These enhancements are created from stochastic perturbations generated by the viewmaker that are then projected onto a ℓ_p ball and added back to the original data. Such projections have been studied in the context of adversarial robustness (e.g., in [9]). The strength of the perturbations is further controlled by a distortion budget: a hyperparameter which limits the radius of the ℓ_p ball to ensure that views do not become too difficult to learn useful representations from.

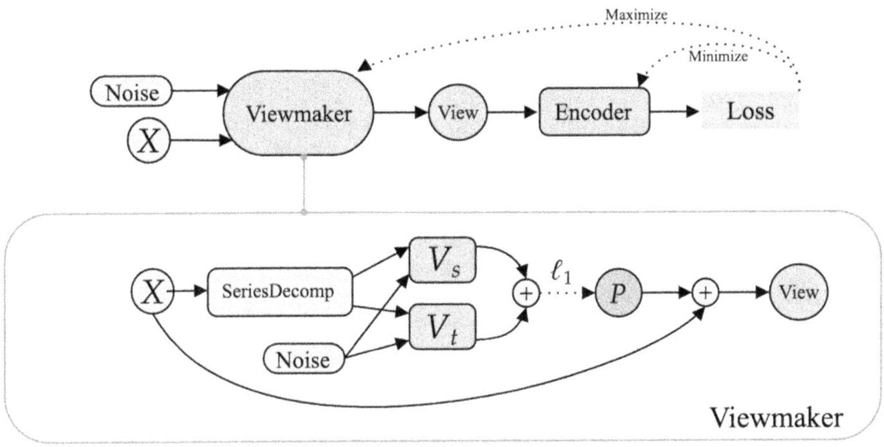

Fig. 5. Overview of the time series viewmaker. Post training, "Views" are used as augmentations for the disruption prediction models.

The advantage of using viewmakers over hand-crafted augmentations is their ability to learn augmentations from the data directly. In the context of disruption prediction, viewmakers could help models pick up hidden trends and physical characteristics of disruptive discharges [24] that are machine-agnostic: a critical step in the search for more generalized models. Originally viewmakers were used for images, in this work, we will be modifying the architecture for use in a time-series context.

2.5 Contrastive Learning

Contrastive Learning is a technique that aims to help models learn an embedding space in which views derived from the same input are grouped together, while dissimilar views are pushed apart. The success of contrastive learning approaches has been extensively researched, particularly in computer vision benchmarks [3–5,13], but also in other domains [7,36]. Although hand-crafted views for tasks such as image classification have been refined, the process remains a handicap for domains such as time series.

SimCLR loss is one approach to contrastive learning proposed in [4] for visual data. If given N pairs of views generated from the same input X, the SimCLR loss is defined to be

$$\begin{aligned} \mathcal{L} &= \frac{1}{2N}\sum_{k=1}^{N}[\ell(2k-1,2k)+\ell(2k,2k-1)], \\ \ell(i,j) &= -\log\frac{\exp(s(i,j))}{\sum_{k=1}^{2N}\mathbb{1}_{k\neq i}\exp(s(i,k))} \end{aligned} \tag{1}$$

Here, $s(i,j)$ is the cosine similarity of the embeddings of views i and j.

3 Models and Methods

3.1 Time Series Viewmakers

Inspired by the series decomposition blocks in [34], we first decompose our time series $X \in \mathbb{R}^{T \times d}$ into trend cycle and seasonal components.

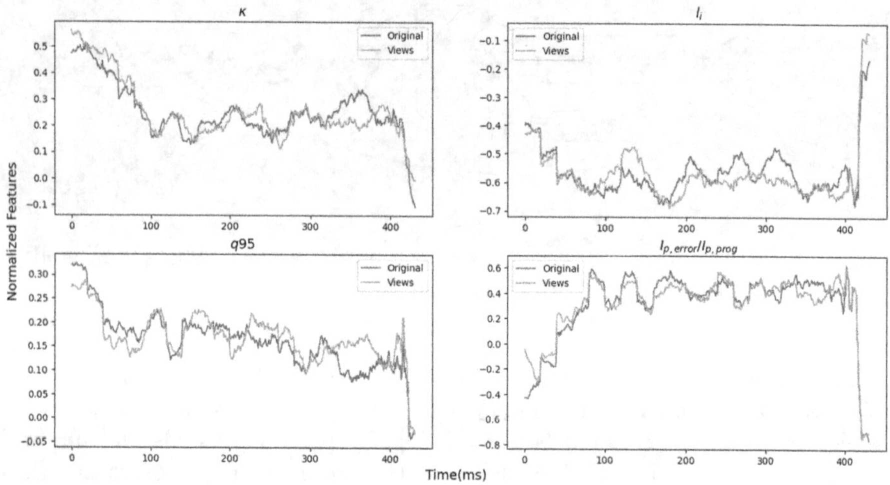

Fig. 6. An example viewmaker augmentation for a randomly selected discharge from the training set. The discharge ends in a disruption and is from the EAST tokamak. Original data is in blue, and Viewmaker augmentation is in orange. (Color figure online)

$$X_t = \text{AvgPool}(\text{Pad}(X))$$
$$X_s = X - X_t$$

Because we decompose the original time series, i.e.

$$X = X_s + X_t$$

we use a separate generator to perturb each component. The perturbations are then combined, smoothed via decomposition, constrained, and added to the original time series, returning the augmented time series or "view". Algorithm 2 details this process of generating views.

Algorithm 1. Generating Views

Input: Viewmaker Network V, $N \times C$ time series input, and distortion budget ϵ
Output: $N \times C$ perturbed time series

$X_t, X_s \leftarrow \text{decompose}(X)$ ▷ time series decomposition
$P_t \leftarrow V_t(X_t)$, $P_s \leftarrow V_s(X_s)$ ▷ generate perturbations
$P \leftarrow P_t + P_s$ ▷ combine
$P \leftarrow \frac{\epsilon NC}{|P|_1} P$ ▷ project to l_1 sphere
$P_i, P_j \leftarrow \text{decompose}(P)$ ▷ get smoothed perturbation
$X \leftarrow X + P_i$ ▷ add smoothed perturbation

See Fig. 5 for an illustration of the time series viewmaker. We find that the selection of generator networks V_t and V_s is critical to performance. It is important that these generators generate **stochastic** and **challenging** perturbations. To better capture long-term temporal dependencies in the data, we change the viewmaker/generator from residual blocks to an LSTM-Transformer-based model similar to the LSTMFormer detailed later. Noise is concatenated between each transformer layer to ensure stochastic perturbations. Refer to Algorithm 2 for the pseudo-code. Furthermore, we adopt an adversarial SimCLR loss and a distortion budget $\epsilon = 0.1$. We trained our viewmaker for 10000 steps with a viewmaker loss weight of 2.5 and a loss temperature of 0.05339. In all experiments, we chose to generate exactly two synthetic versions of each training observation, meaning that proportion of disruptions present stayed the same. Test set was unchanged.

However, we encourage others to experiment with different number of augmentations, loss functions, distortion budgets, generator networks, and encoders.

Algorithm 2. Viewmaker Networks

Input: LSTM, N layer Transformer Encoder, MLP, time series input X
Output: Generated perturbation

$X \leftarrow \text{LSTM}(X)$ ▷ LSTM Encoding
$X \leftarrow \text{LayerNorm}(X)$ ▷ Layer Normalization
for $k \leftarrow 1$ to N **do**
  $X \leftarrow \text{TransformerLayer}_k(X)$ ▷ Transformer Layer
  $X \leftarrow \text{ConcatNoise}(X)$ ▷ Concatenate Noise
end for
$X \leftarrow \text{MLP}(X)$ ▷ Resize perturbation to original

After training, the viewmaker network is used to generate augmentations of the training set for the post-hoc models. Figure 6 shows an example of a viewmaker augmentation after training. We find that choosing a distortion budget for post-training augmentations is very important and depends on the tokamak data source. The distortion budgets of 0.1, 0.1, and 0.05 were chosen for the tokamaks EAST, DIII-D, and Alcator C-Mod respectively. We highlight that our work is the first to use unsupervised contrastive learning viewmakers for a supervised learning task.

3.2 Standard Augmentations

As a benchmark, we use the tsaug library to implement multivariate time series standard augmentations. Our tsaug pipeline consists of the random combination of additive noise, time warping, dropout, and drift.

We compared the features correlations observed in the natural dataset and the dataset augmented via Viewmaker networks and tsaug. Correlation is mostly preserved, especially in the Viewmaker augmented dataset. Please see Fig. 7.

3.3 Tested Models

For benchmarking the effectivness of these augmentations, we experiment with 3 post-hoc models: LSTMFormer, Fully Convolutional Network (FCN), and GPT2. These models were chosen because they cover multiple model types: convolutional, LSTM, and transformer-based, allowing us to rigorously test the augmentation methods.

- LSTMFormer: Compared to transformers, the sequential processing of LSTMs makes them particularly effective at encoding temporal relations within sequences. It combines both attention and recurrence to get both a good temporal and a long-term dependency understanding. Therefore, LSTMFormer first uses an LSTM layer to generate time-conscious positional encodings, similar to [22,37]. The output of the LSTM layer is then propagated through a normalization layer [2], a four-block transformer encoder [31], and a classification head.
- FCN: variant of the convolutional neural network meant for time series first introduced in [32], which uses fully convolutional blocks for feature extraction.
- GPT2: introduced for prediction of disruption in [?] is an autoregressive transformer decoder model that uses a pre-training and a curriculum training scheme.

Importantly, the models vary significantly in size, which can effect not only how much data each model needs and the relative impact of added synthetic data. Based on the available information about model architecture, we roughly estimated of the model sizes, which vary significantly from 1 million (M) to 117M parameters, in Table 3.

Our model training occurred on a shared cluster of 48 Nvidia A100 GPUs through the Massachusetts Green High Performance Computing Center (MGHPCC). Each run took approximately 3–5 h to complete.

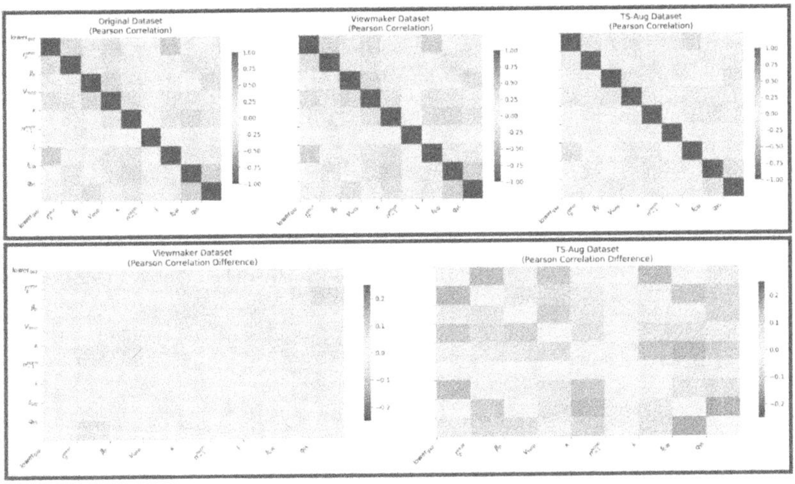

Fig. 7. A comparison of feature correlation in each individual training dataset, and two relative plots to show changes.

Table 3. Approximate architecture size and parameter count of each model.

Model	Architecture Summary	Params
LSTMFormer	LSTM + 4-layer Transformer encoder (d = 128–256, 4 heads)	~1–3M
FCN	3 Conv1D blocks + GlobalAvgPool + dense head	~0.5–1M
GPT2	12-layer transformer decoder (d = 768, 12 heads)	~117M

4 Results and Discussion

Similar to the DisruptionBench paper [23], we use the AUC and F2 score to assess the performance of each augmentation strategy in the four cases specified in Sect. 2, setting Alcator C-Mod as the new machine. F2 score is used because false negatives are more costly in the case of avoidance of disruption. The results are summarized in Table 4.

Across the 24 combinations of four cases, three models, and two metrics per model, Viewmaker augmentations win out a majority of the comparisons – in 15 out of 24 cases. TS augmentations top the list of an additional three comparisons. As the test set is the same across cases and thus intercomparable, Viewmaker networks contribute to both the highest AUC and highest F2 score in the table. We conclude that data augmentation is beneficial in most tested scenarios.

In terms of specific augmentation approaches, Viewmaker networks outperform tsaug in our trials. One reason for this fact is that viewmakers learn dynamics that are plausibly physical as opposed to tsaug which is more heuristically based. We present more quantitative evidence in Sect. 4.2.

Table 4. DisruptionBench results for LSTMFormer, FCN, and GPT2 trained on using no augmentations, tsaug standard augmentations, and viewmaker augmentations during training. "New Machine" is Alcator C-Mod in each case. Refer to Sect. 2 for further descriptions of each case

	Aug Type	LSTMFormer		FCN		GPT2	
		AUC	F2	AUC	F2	AUC	F2
Case 1: Zero-Shot	None	0.571	**0.499**	0.518	**0.505**	0.575	0.525
	Tsaug	0.571	0.358	**0.561**	0.166	**0.592**	**0.535**
	Viewmaker	**0.609**	0.342	0.546	0.503	0.527	0.404
Case 2: Few-Shot	None	**0.592**	0.389	0.631	**0.513**	0.695	0.523
	Tsaug	0.537	0.189	0.579	0.413	0.723	0.616
	Viewmaker	0.568	**0.545**	**0.637**	0.494	**0.731**	**0.628**
Case 3: All machines	None	0.808	0.708	0.847	0.758	0.711	0.524
	Tsaug	0.769	0.641	0.810	0.72	0.74	0.58
	Viewmaker	**0.823**	**0.735**	**0.853**	**0.777**	**0.754**	**0.611**
Case 4: Only C-Mod	None	**0.792**	**0.707**	0.791	0.7	0.742	0.617
	Tsaug	0.79	0.696	0.781	0.691	0.675	0.512
	Viewmaker	0.766	0.669	**0.814**	**0.725**	**0.744**	**0.632**
Mean	None	0.691	**0.576**	0.697	0.619	0.681	0.547
	Tsaug	0.667	0.471	0.683	0.498	0.683	0.561
	Viewmaker	**0.692**	0.573	**0.713**	**0.625**	**0.689**	**0.569**

We observe that the zero-shot case shows little to no improvement from Viewmakers. Indeed, it is the only case with an average drop in F2, indicating an increase in false negatives. Because augmentations generally interpolate within existing training distributions, they cannot aid transfer to entirely unseen domains. As shown in the UMAP visualization of machine disruptions in Fig. 2, Alcator C-Mod disruptions differ substantially from those of other machines; thus, data augmentations likely shift the model's representation away from the Alcator C-Mod distribution and lowering the probability of predicting disruption.

The implication is that when deploying a disruption system on a new machine, even a small number of representative test runs can be highly valuable when paired with augmentation or other data-enrichment techniques. However, if the system must be operational from first plasma, non-ML methods—such as additional physics-based features, machine metadata, or classical models—may be more important than augmentation for establishing an initial safety baseline.

Notably, the many-shot augmented LSTMFormer outperforms both its non-augmented counterpart and all Alcator C-Mod-only models, suggesting that the inclusion of additional data improved the realism of the augmentations and, in turn, enhanced all models' performance.

4.1 In-domain Augmentations

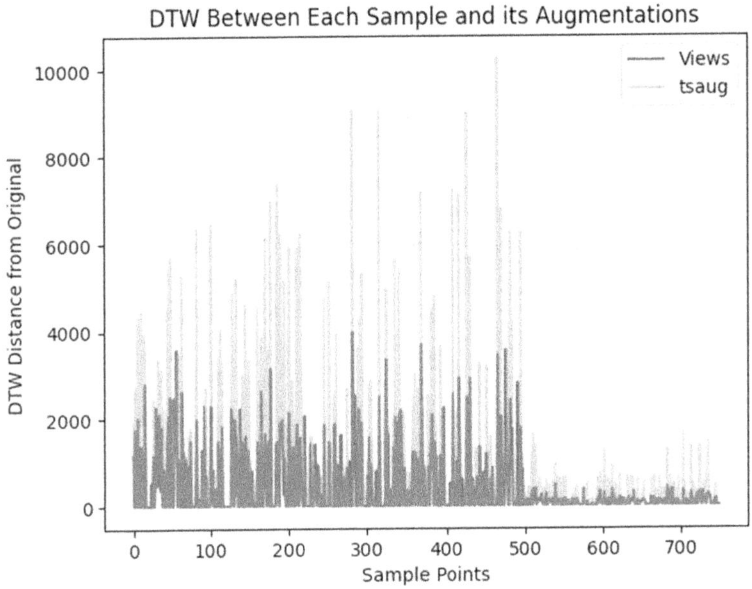

Fig. 8. Comparison of dynamic time warping (DTW) similarities between disruptive discharges and their views (in cyan) and their tsaugs (in orange). Discharges were sampled evenly at random across each machine. (Color figure online)

The results in Cases 2 and 3 in particular suggest that viewmakers help models learn the underlying disruptive features across all machines. We hypothesize this is because viewmakers generate more physically-realistic augmentations of the original data. To explore this further, we compared the dynamic time warping (DTW) similarity of 250 randomly sampled disruptive discharges from each machine against their viewmaker augmentations and tsaug standard augmentations (see Fig. 8). For this sample, the average DTW for views was 409.14 while for tsaugs it was 770.39, indicating that views are much more faithful to the original data. Furthermore, we performed a Wilcoxon Signed Rank Test to get a $p < 0.001$, showing that this result is statistically significant.

To further understand how a model trained on views might behave in a production system, we plot a confusion matrix using one sampled discharge from each category. TP, FP, TN, FN (Fig. 9). As we can see, there are cases where the various augmentation strategies diverge. However, it seems that the views help models predict disruptions well in advance of the actual disruption, which raises the possibility of using these augmentations in training controllers for disruption avoidance.

4.2 Visualizing the Effect of Views

Looking at Fig. 6, which contains the time series viewmaker enhancements for an EAST disruptive discharge, we highlight how the viewmaker is able to make the disruptivity of a discharge more difficult to predict. For example, between 300–400ms it perturbs the q_{95} feature such that there is no longer an apparent drop, which can signal an oncoming disruption. We see this similarly throughout the discharge in the l_i feature where it is perturbed so that the spikes are evened out. For instance, we hypothesize that the time series viewmaker is able to understand these disruptive signals but also obscure them due to its trend-seasonal decomposition, which entails a better temporal understanding. The time series viewmaker learns to make these disruptive signals more obscure through its adversarial contrastive learning, allowing the model to better learn unobvious disruptive signals, which we have shown helps downstream disruption prediction better learn underlying disruptive features.

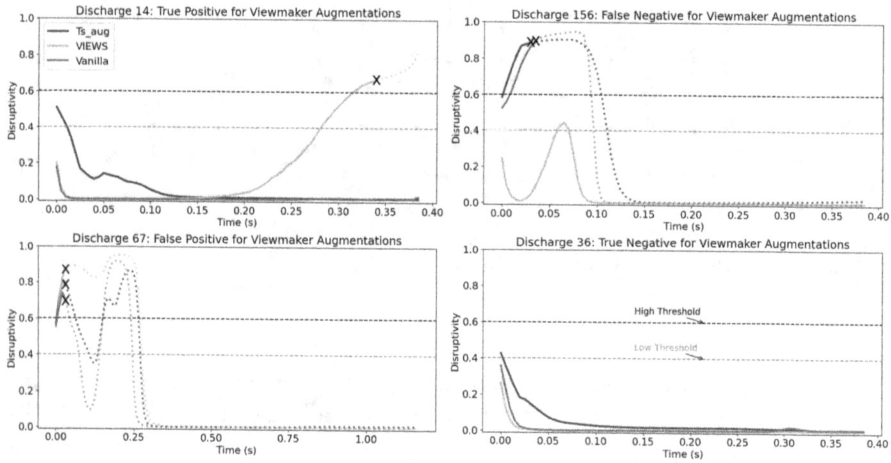

Fig. 9. A comparison of four discharges from Alcator C-Mod for unrolled disruptivity on the test set, chosen based on outcomes of an LSTMFormer trained of views. We annotate the chart with "X" s to show the points at which the disruption models would have triggered the DMS. The disruptivity metric, on the y-axis, is unitless. Finally, note that the timebases are not true, but relative to the start of the recorded data.

We plot to demonstrate the overlay of the augmentation techniques with the original data domains for Alcator C-Mod in Fig. 10.

5 Limitations

Throughout the different post hoc models, training with viewmaker enhancements can cause loss in precision: –2.3% in FCN and –9.4% in GPT2; however,

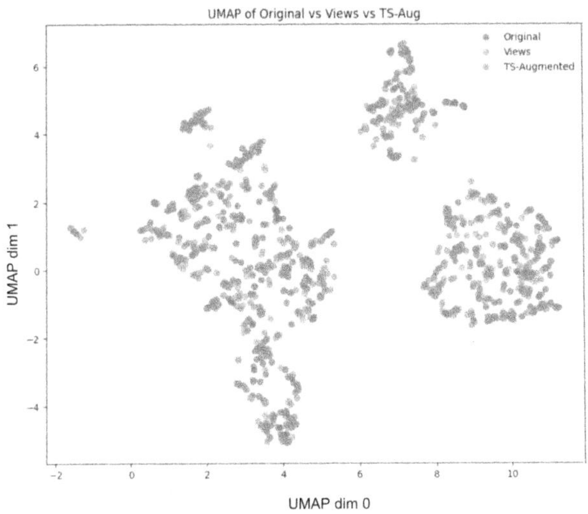

Fig. 10. A plot of time series augmentations overlayed with the Alcator C-Mod dataset.

for LSTMFormer we see an improvement +5.2% (see Table 5). The similarity of many disruptive and non-disruptive discharges could be one reason for this trade-off (see Fig. 3) while class imbalance might be another. Although ideally recall and precision would both increase for disruption avoidance control, the cost of false negatives significantly outweigh the costs of false positives. Our access to compute was limited, so we did not have the ability to present runs on multiple seeds, which would strengthen the statistical significance of our results and allow us to test different combinations of data-splits and new machines. Finally, it must be re-affirmed that our augmentation approach in practice uniformly hurt performance in the zero shot case, meaning similar approaches should be treated with caution when used out-of-the-box on large distributional training shifts.

6 Conclusion and Future Work

In this work, we propose a novel time series variant of viewmaker networks and investigate its utility in generating data augmentations or "views" of tokamak discharge time series to improve model robustness for nuclear fusion disruption prediction and disruption avoidance control. Our findings suggest that using viewmaker views improves model AUC and F2 across a diverse set prediction tasks compared to naive or no augmentations, particularly in few and many shot cases. By improving the robustness of these models, Our approach enables early ML deployment in new machines with Few-shot learning and data augmentation, which is critical for the next generation of "ignition ready machines".

The avenues for future work can include modifying viewmaker features, integrating the trained encoder into a new classifier, and comparing viewmakers with other time series generators such as TimeGAN [35]. Additionally, more post-hoc models should be tested (see [15,25,26,38]), and new tokamaks should be added to DisruptionBench.

The use of time-series viewmakers also has more general applications with their ability to generate realistic and stochastic augmentations. This can be applied to artificially enlarge time series datasets across all fields. In the future, we would like to study the viewmaker in the context of other time series applications to see if its effectiveness holds there, too.

Acknowledgments. This work was developed as an independent collaboration and not funded directly by MIT-PSFC. However, the simulations presented in this paper were performed on the MIT-PSFC partition of the Engaging cluster at the MGHPCC facility (www.mghpcc.org) which was funded by DoE grant number DE-FG02-91-ER54109. Furthermore, the authors would like to gratefully thank the MIT PSFC for resources and support, including Aaron Ho, Robert Granetz, Rebecca Masline, Alex Saperstein, Stephen Lane Walsh, and the indomitable Marco Muraca. Without their kind support, this project would not have happened. We also thank Alex Tamkin, the original author of viewmaker networks, for corresponding with us about his work as we investigated ours.

A Results (continued)

Table 5. Full DisruptionBench results (including Recall and Precision) for LSTMFormer, FCN, and GPT2 trained on using no augmentations, tsaug standard augmentations, and viewmaker augmentations during training. "New Machine" is Alcator C-Mod in each case.

	Aug Type	LSTMFormer				FCN				GPT2			
		AUC	Rec.	Prec.	F2	AUC	Rec.	Prec.	F2	AUC	Rec.	Prec.	F2
Case 1: Zero-Shot	None	0.571	**0.772**	0.207	**0.499**	0.518	**0.912**	0.181	**0.505**	0.575	**0.86**	0.205	0.525
	Tsaug	0.571	0.404	0.247	0.358	**0.561**	0.14	**0.615**	0.166	**0.592**	0.86	**0.213**	**0.535**
	Viewmaker	**0.609**	0.333	**0.38**	0.342	0.546	0.842	0.193	0.503	0.527	0.561	0.19	0.404
Case 2: Few-Shot	None	**0.592**	0.439	0.269	0.389	0.631	**0.684**	0.257	**0.513**	0.695	0.561	**0.41**	0.523
	Tsaug	0.537	0.175	**0.27**	0.189	0.579	0.509	0.236	0.413	0.723	0.789	0.328	0.616
	Viewmaker	0.568	**0.965**	0.199	**0.545**	**0.637**	0.614	**0.278**	0.494	**0.731**	**0.807**	0.333	**0.628**
Case 3: All training shots	None	0.808	0.807	0.474	0.708	0.847	0.824	**0.573**	0.758	0.711	0.526	0.517	0.524
	Tsaug	0.769	0.702	**0.476**	0.641	0.810	0.877	0.42	0.72	0.74	0.596	**0.523**	0.58
	Viewmaker	**0.823**	**0.877**	0.446	**0.735**	**0.853**	**0.93**	0.469	**0.777**	**0.754**	**0.649**	0.493	**0.611**
Case 4: Only C-Mod	None	**0.792**	**0.912**	0.371	**0.707**	0.791	0.877	0.388	0.7	0.742	0.719	**0.394**	0.617
	Tsaug	0.79	0.86	**0.395**	0.696	0.781	0.877	0.373	0.691	0.675	0.579	0.351	0.512
	Viewmaker	0.766	0.842	0.366	0.669	**0.814**	**0.877**	**0.427**	**0.725**	**0.744**	**0.772**	0.367	**0.632**
Mean	None	0.691	0.733	0.330	**0.576**	0.697	**0.824**	0.350	0.619	0.681	0.667	**0.382**	0.547
	Tsaug	0.667	0.535	0.347	0.471	0.683	0.601	**0.411**	0.498	0.683	**0.706**	0.354	0.561
	Viewmaker	**0.692**	**0.754**	**0.347**	0.573	**0.713**	0.816	0.342	**0.625**	**0.689**	0.697	0.346	**0.569**

References

1. Arnold, W.F., Spangher, L., Rea, C.: Continuous convolutional neural networks for disruption prediction in nuclear fusion plasmas (2023)
2. Ba, J.L., Kiros, J.R., Hinton, G.E.: Layer normalization (2016)
3. Bachman, P., Hjelm, R.D., Buchwalter, W.: Learning representations by maximizing mutual information across views (2019)
4. Chen, T., Kornblith, S., Norouzi, M., Hinton, G.: A simple framework for contrastive learning of visual representations (2020)
5. Chen, X., Fan, H., Girshick, R., He, K.: Improved baselines with momentum contrastive learning (2020)
6. Fu, B., Kirchbuchner, F., Kuijper, A.: Data augmentation for time series: traditional vs generative models on capacitive proximity time series. In: Proceedings of the 13th ACM International Conference on Pervasive Technologies Related to Assistive Environments, pp. 1–10 (2020)
7. Gao, T., Yao, X., Chen, D.: Simcse: simple contrastive learning of sentence embeddings (2022)
8. Giuliani, U., Grazian, S., Alotto, P., Agostini, M., Bustreo, C., Zollino, G.: Nuclear Fusion impact on the requirements of power infrastructure assets in a decarbonized electricity system. Fusion Eng. Des. **192**, 113554 (2023)
9. Goodfellow, I.J., Shlens, J., Szegedy, C.: Explaining and harnessing adversarial examples (2015)
10. Granetz, R., et al.: Machine learning for disruption warning on Alcator C-Mod, DIII-D, and EAST tokamaks. In: Proceedings of 27th IAEA Fusion Energy Conference, IAEA, Vienna (2018)
11. Greenwald, M.: Density limits in toroidal plasmas. Plasma Phys. Controlled Fusion **44**(8), R27 (2002)
12. Greenwald, M., et al.: 20 years of research on the alcator c-mod tokamak. Phys. Plasmas **21**(11) (2014)
13. He, K., Fan, H., Wu, Y., Xie, S., Girshick, R.: Momentum contrast for unsupervised visual representation learning (2020)
14. Hong, Y., Travis, D.: Robust augmentation for multivariate time series classification (2022)
15. Hu, W., et al.: Real-time prediction of high-density east disruptions using random forest. Nucl. Fusion **61**(6), 066034 (2021)
16. Iwana, B.K., Uchida, S.: An empirical survey of data augmentation for time series classification with neural networks. PLoS ONE **16**(7), e0254841 (2021)
17. Jachmich, S., et al.: Shattered pellet injection experiments at jet in support of the iter disruption mitigation system design. Nucl. Fusion **62**(2), 026012 (2021)
18. Jernigan, T.C., Baylor, L.A., Combs, S.K., Humphreys, D.A., Parks, P.B., Wesley, J.C.: Massive gas injection systems for disruption mitigation on the diii-d tokamak. In: 21st IEEE/NPS Symposium on Fusion Engineering SOFE 05, pp. 1–3 (2005)
19. Kallenbach, A., Team, A.U., Team, E.M., et al.: Overview of asdex upgrade results. Nucl. Fusion **57**(10), 102015 (2017)
20. Katharina, R., et al.: Data augmentation for disruption prediction via robust surrogate models (2022)
21. Keith, Z., Nagpal, C., Rea, C., Tinguely, R.: Risk-Aware Framework Development for Disruption Prediction: Alcator C-Mod and DIII-D Survival Analysis (2024)
22. Lim, B., Arik, S.O., Loeff, N., Pfister, T." Temporal fusion transformers for interpretable multi-horizon time series forecasting (2020)

23. Lucas, S., et al.: Disruptionbench: a robust benchmarking framework for machine learning-driven disruption prediction (2024)
24. Minderer, M., Bachem, O., Houlsby, N., Tschannen, M.: Automatic shortcut removal for self-supervised representation learning (2020)
25. Pautasso, G., et al.: On-line prediction and mitigation of disruptions in asdex upgrade. Nucl. Fusion **42**(1), 100 (2002)
26. Rea, C., Montes, K., Erickson, K., Granetz, R., Tinguely, R.: A real-time machine learning-based disruption predictor in DIII-D. Nucl. Fusion **59**(9), 096016 (2019)
27. Spangher, L., Arnold, W., Spangher, A., Maris, A., Rea, C.: Autoregressive transformers for disruption prediction in nuclear fusion plasmas (2023)
28. Spangher, L., Vitter, J.S., Umstattd, R.: Characterizing fusion market entry via an agent-based power plant fleet model. Energ. Strat. Rev. **26**, 100404 (2019)
29. Tamkin, A., Wu, M., Goodman, N.: Viewmaker networks: learning views for unsupervised representation learning (2021)
30. Um, T.T., et al.: Data augmentation of wearable sensor data for Parkinson's disease monitoring using convolutional neural networks. In: Proceedings of the 19th ACM International Conference on Multimodal Interaction, ICMI '17. ACM (2017)
31. Vaswani, A., et al.: Attention is all you need (2017)
32. Wang, Z., Yan, W., Oates, T.: Time series classification from scratch with deep neural networks: a strong baseline (2016)
33. Wen, T., Keyes, R.: Time series anomaly detection using convolutional neural networks and transfer learning (2019)
34. Wu, H., Xu, J., Wang, J., Long, M.: Autoformer: decomposition transformers with auto-correlation for long-term series forecasting (2022)
35. Yoon, J., Jarrett, D., van der Schaar, M.: Time-series generative adversarial networks. In: Wallach, H., Larochelle, H., Beygelzimer, A., d' Alché-Buc, F., Fox, E., Garnett, R. (eds.) Advances in Neural Information Processing Systems, vol. 32. Curran Associates, Inc. (2019)
36. Yue, Z., et al.: Ts2vec: towards universal representation of time series (2022)
37. Zeyer, A., Bahar, P., Irie, K., Schlüter, R., Ney, H.: A comparison of transformer and LSTM encoder decoder models for ASR. In: 2019 IEEE Automatic Speech Recognition and Understanding Workshop (ASRU), pp. 8–15 (2019)
38. Zhu, J.X., Rea, C., Montes, K., Granetz, R.S., Sweeney, R., Tinguely, R.A.: Hybrid deep-learning architecture for general disruption prediction across multiple tokamaks. Nucl. Fusion **61**(2), 026007 (2020)

Self-refining Segment Anything: A Perturbation-Driven Approach for Semi-supervised Satellite Image Segmentation

Renzo Larenas[1(✉)], Luis Caro[2], and Billy Peralta[1]

[1] Facultad de Ingeniería, Universidad Andres Bello, Santiago, Chile
r.larenasescobar@uandresbello.edu , billy.peralta@unab.cl
[2] Departamento de Ingeniería Informática, Universidad Católica de Temuco, Temuco, Chile
lcaro@uct.cl

Abstract. The growing demand for accurate information in precision agriculture, driven by climate change and the need for resource optimization, has led to the development of advanced techniques for satellite image segmentation. However, traditional segmentation models face generalization challenges due to the scarcity of labeled data. This work proposes a lightweight approach based on the Segment Anything Model (SAM), enhanced through an iterative pseudo-labeling strategy that relies on segmentation consistency under salt-and-pepper noise perturbations. The methodology, grounded in semi-supervised learning, is validated through experiments on real satellite imagery. Results demonstrate that this self-consistency criterion enables effective use of unlabeled data, yielding substantial improvements in segmentation performance over standard fine-tuned SAM.

Keywords: SAM · Semi-supervised learning · Segmentation

1 Introduction

Agriculture is one of the fundamental pillars of the global economy. According to the Food and Agriculture Organization of the United Nations (FAO), the agri-food sector accounted for approximately 4.3% of the global Gross Domestic Product (GDP) in 2022 [2]. It is also a major source of employment, particularly in rural regions. Globally traded agricultural goods—such as wheat, corn, rice, and soybeans—are essential for food security and economic development [14]. In response to growing environmental and economic pressures, various initiatives have emerged to promote innovation in this sector, including precision agriculture, biotechnology, and sustainable farming practices.

Precision agriculture has become increasingly important for optimizing resource use and maximizing crop yields. High-resolution satellite imagery now

Published by Springer Nature Switzerland AG 2026
E. Mercier-Laurent et al. (Eds.): AI4S 2025, IFIP AICT 780, pp. 81–93, 2026.
https://doi.org/10.1007/978-3-032-21184-2_7

enables detailed analysis of land cover; however, manually segmenting such imagery remains costly, time-consuming, and prone to inconsistencies. Deep learning models like U-Net and SAM offer a promising alternative by providing automated, high-accuracy segmentation tools tailored to spatial analysis.

Despite recent advances, the generalization capacity of supervised segmentation models remains limited—particularly in the presence of high variability across agricultural landscapes and insufficient annotated data. Although satellite imagery is abundant, the acquisition of high-quality segmented labels typically requires expert annotation, which poses significant scalability challenges.

To address these limitations, we propose a lightweight semi-supervised framework based solely on the Segment Anything Model (SAM), enhanced through a self-consistency pseudo-labeling strategy. Our method, SAM-Noise, iteratively refines the segmentation model by validating predictions under noisy versions of the input image. This approach eliminates the need for external detectors or prompt engineering, reducing system complexity while enabling effective use of unlabeled data.

2 Related Work

In terms of satellite imagery, several works have shed light on the task of segmentation. Sharifi et al. [12] propose an automatic model based on remote sensing images and a neural network (ResUNet-a) to improve detection.

The model demonstrated high accuracy in identifying fields and their boundaries using satellite images such as Sentinel-2 and Landsat-8, outperforming traditional methods.

Safarov et al. [11] propose a deep learning model for semantic segmentation of satellite images for land cover classification. TL-ResUNet combines residual networks, transfer learning, and UNet architecture, outperforming classic models in performance and accuracy.

Sravya et al. [13] present DPPNet, a novel deep learning model that utilizes Depth-wise Pyramid Pooling (DPP) blocks and dense residual connections. DPPNet aims to improve performance with lower computational complexity. Evaluated on Land-cover and WHDLD HRS datasets, DPPNet outperformed baseline models in quantitative and qualitative metrics.

Nair et al. [7] propose a systematic data augmentation approach to maintain consistency in training. In particular, they propose a modified conditional generative adversarial network (CGAN) to create artificial satellite images and field labels, reducing manual labeling effort. The research evaluates spectral bands and compares U-Net and DeepLabV3+, demonstrating that U-Net outperforms DeepLabV3+.

Nchare et al. [8] present a model based on a Fully Convolutional Network (FCN) to segment high-resolution aerial images and extract road networks. The approach preserves spatial information through convolutions and a deconvolution layer to generate high-resolution outputs. Using the Massachusetts Roads dataset, the model achieves an accuracy of 87.86% and a recall of 87.89%.

Pandey et al. [9] propose a multimodal approach for segmentation in medical images, combining YOLOv8 for initial detection of regions of interest (ROI) with SAM and HQ-SAM for refined segmentation. This method is applied to modalities such as ultrasound, computed tomography, and X-rays, achieving improvements in metrics such as Dice score, precision, and recall. Also in the context of health, Khalili et al. [4] and Mansoorii et al. [6] also propose a similar model where YOLO generates prompts for SAM.

Reddy et al. [10] combine the YOLO model and the Segment Anything Model (SAM) to address the segmentation of cotton bolls in images captured by unmanned aerial vehicles (UAVs). This is done in two stages: YOLO is used for the initial detection of bolls in the images. In the second stage, these bounding boxes are normalized and used as prompts for SAM.

In partially related work, Huang et al. [3] explore the use of SAM for segmenting complex cultivated land parcels, combining it with multi-scale refinement techniques and high-resolution land use data. Their results demonstrate that SAM can be effectively adapted to agricultural segmentation tasks, particularly when supported by detailed spatial information and morphological postprocessing.

Our proposal differs from the approaches cited in this work by incorporating an iterative pseudo-labeling scheme that relies solely on segmentation consistency under input perturbations, effectively removing the need for external detection modules.

3 Segment Anything Model (SAM)

The Segment Anything Model (SAM), developed by Meta AI [5], represents a significant advance in automatic image segmentation, introducing a prompt-based approach that generates masks from flexible inputs such as points, boxes, or text. Its distinctive feature is its zero-shot generalization capability, meaning it can segment objects not seen during training. This property is achieved through massive pre-training on the SA-1B dataset, composed of over 11 million images and 1.1 billion masks annotated semi-automatically.

The architecture of SAM, shown in Fig. 1, is organized into three main components:

1. **Image Encoder**
   - It uses a Vision Transformer (ViT Base) pre-trained with a Masked Auto-Encoder (MAE).
   - Its function is to convert the input image into an image embedding that preserves relevant spatial information for segmentation.
2. **Prompt Encoder**
   - It is responsible for transforming different types of prompts (such as points or boxes) into the same embedding space.
3. **Mask Decoder**
   - Implemented as a lightweight transformer, which combines the embeddings from the image encoder and the prompt encoder to predict the segmentation mask.

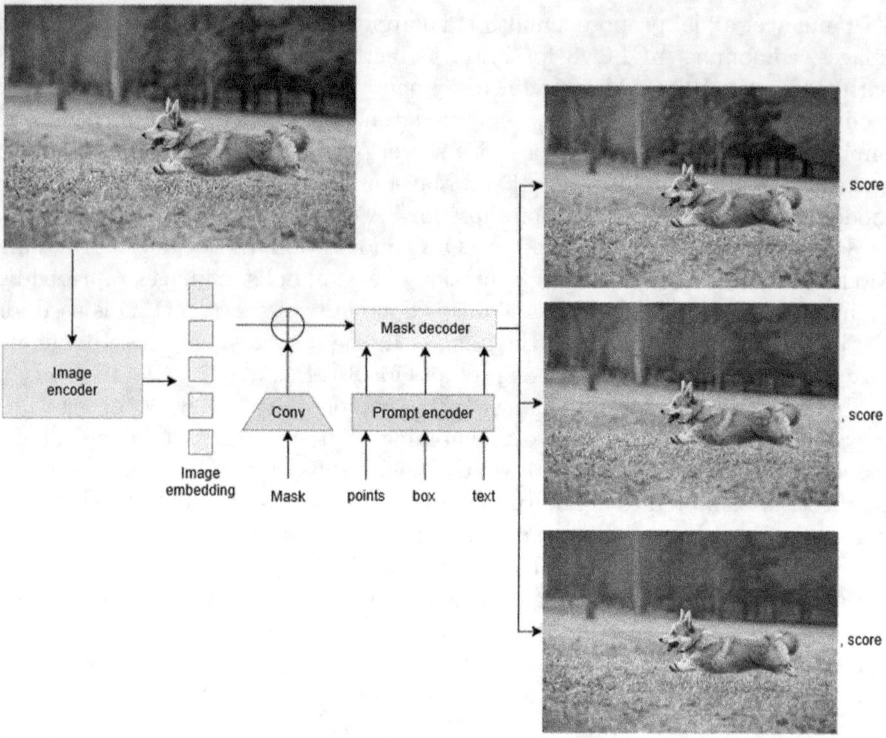

Fig. 1. Overview of the Segment Anything Model (SAM) architecture. As output of the point-prompt segmentation, three different regions are generated in blue over the original image. (Color figure online)

The components described above are visually summarized in Fig. 1, which illustrates the architecture of the Segment Anything Model (SAM). The input image on the left is encoded and processed through the SAM pipeline, resulting in three segmentation outputs on the right given by blue regions. Each output corresponds to a different mask generated for the same object(trees, in this case) based on a single point-based prompt. This behavior reflects SAM's design, which produces multiple plausible masks when the prompt is ambiguous or underspecified, as commonly occurs in complex scenes.

In this work, the SAM-Base version was chosen to optimize computational efficiency. This version, which has a smaller parameter size, maintains the model's generalization capability without incurring the high inference costs of larger versions. This strategy is complemented by a methodological approach that focuses on fine-tuning the mask decoder while the image encoder remains frozen. This is advantageous because it avoids the need to retrain a large-scale model, reduces the risk of overfitting, and leverages the visual representations learned during pre-training.

```
Data: Labelled images: X_l^o; masks: s_l^o; unlabelled images: X_u
Result: Segmentation model: θ_SAM
/* 1. Initialize model                                              */
θ_SAM ← Finetune SAM with (X_l^o, s_l^o);
/* 2. Iterative Pseudo-Labeling via Noise Consistency               */
while X_u ≠ ∅ do
    foreach x ∈ X_u do
        x̃ ← Add salt-and-pepper noise to x;
        s ← θ_SAM(x); s̃ ← θ_SAM(x̃);
        if IoU(s, s̃) > μ then
            Add (x, s) to pseudo-labeled set (X_l, s_l);
        end
    end
    if X_l ≠ ∅ then
        θ_SAM ← Finetune with (X_l, s_l);
        X_u ← X_u − X_l; clear (X_l, s_l);
    end
end
```

Algorithm 1: SAM-Noise: Robustness-based Semi-supervised Segmentation

4 Proposed Method

Our proposed method is based on the idea of leveraging model consistency under input perturbations to identify high-quality pseudo-labels in unlabeled data. Specifically, we exploit the robustness of SAM model [5] when exposed to image alterations. The core assumption is that a reliable segmentation model should yield consistent predictions even when the input is modified by noise. The detail is shown in Algorithm 1. This algorithm presents a semi-supervised training process for a segmentation model (θ_{SAM}), relying exclusively on unlabeled data (X_u) and their perturbed versions. The training begins with fine-tuning SAM on the original labeled dataset (X_o^l, s_o^l). The algorithm then proceeds iteratively over the set of unlabeled images. For each unlabeled image $x \in X_u$, a perturbed version $\tilde{x}$ is generated by adding salt-and-pepper noise. The model θ_{SAM} is used to infer segmentations s and $\tilde{s}$ from x and $\tilde{x}$ respectively. The similarity between s and $\tilde{s}$ is measured using the Intersection over Union (IoU). If this value exceeds a predefined threshold μ, the original segmentation s is considered a reliable pseudo-label, and the pair (x, s) is added to a new labeled set (X_l, s_l). Subsequently, the segmentation model is refined using this pseudo-labeled data. The selected images X_l are removed from X_u, and the process repeats until no unlabeled images remain. We refer to this self-consistency-based model as **SAM-Noise**.

5 Data Preprocessing

5.1 AI4Boundaries Dataset

The AI4Boundaries dataset is an open-access dataset designed as a reference for mapping agricultural field boundaries in Europe [1]. Its main objective is to provide a baseline for training and evaluating large-scale agricultural field boundary detection models. This dataset combines two image sources:

- **Sentinel-2**: Multispectral images with 10-meter resolution, acquired monthly, which allow for large-scale agricultural pattern analysis.
- **Aerial photography (Orthophoto)**: Images with 1-meter resolution, useful for more detailed analysis and regional agricultural management applications.

In this work, only the Sentinel-2 images were used. They were preprocessed into 256×256 pixel and converted to RGB format, along with their corresponding binary masks that delimit agricultural zones.

According to the authors of AI4Boundaries, the dataset was constructed using a stratified random sampling approach, considering two landscape fragmentation metrics:

- Field perimeter-to-area ratio.
- The proportion of area covered by agricultural parcels.

Figure 2 shows an example of a Sentinel-2 image along with its associated binary mask.

Sentinel-2 image. Corresponding binary mask.

Fig. 2. Example of a Sentinel-2 image and its corresponding binary mask from the AI4Boundaries dataset.

5.2 Data Preparation

The Sentinel-2 data were preprocessed by rescaling all images to 256 × 256 pixels and applying normalization to reduce illumination variability. Binary masks marking agricultural zones were generated as ground truth for supervised segmentation.

For training and evaluation, we selected 3,150 samples, randomly divided into 80% training and 20% validation sets, with an additional 947 separate images for final testing. Geographic coverage includes France, Luxembourg, Netherlands, Sweden, and Slovenia.

All images were converted to RGB in .png format. Labels are grayscale binary masks of identical dimensions, used during supervised training.

We consider four experimental configurations, each varying in the number of labeled images: 125, 150, 200, and 250. In all cases, the remaining images from the original 3150 set served as unlabeled data. Each split was generated randomly, and noise perturbations used in the pseudo-labeling process corresponds to 10% salt-and-pepper noise. Figure 3 illustrates this process, showing an original image from the AI4Boundaries dataset alongside its noisy counterpart.

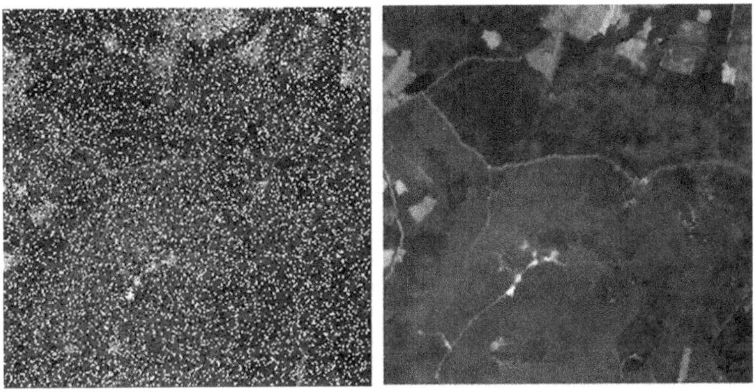

Fig. 3. Original image (right) and noisy version with 10% salt-and-pepper perturbation (left).

In Fig. 4, a randomly selected image from the test dataset is shown alongside its ground truth overlay and the segmentation output produced by SAM Base without fine-tuning. This example illustrates the model's performance on complex scenes characterized by spectral noise and limited image quality. Under these conditions, SAM fails to accurately segment all relevant regions, which is expected given that it has not been specifically trained to identify agricultural areas. The discrepancy between the ground truth and the model's output highlights the need for an initial fine-tuning step within the proposed iterative framework.

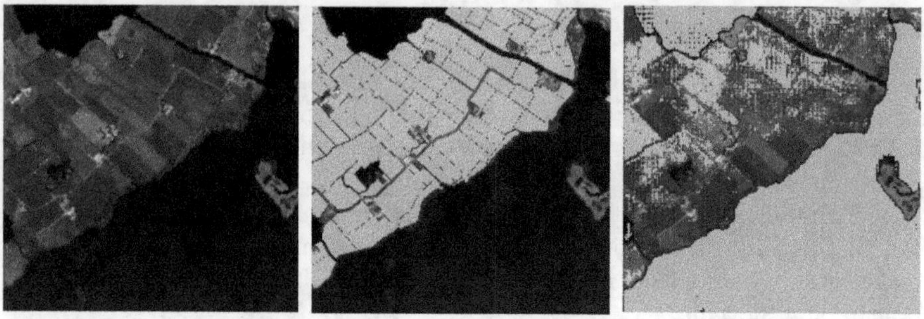

Fig. 4. Comparison between the original image, ground truth overlay, and SAM Base inference using a full-image bounding box.

6 Experiments

We conduct a comprehensive empirical evaluation to assess the effectiveness of the proposed SAM-Noise framework under varying data availability constraints. Our experiments are designed to simulate realistic low-label scenarios common in remote sensing, using four subsets of increasing size detailed in Sect. 5.1. In each configuration, the remaining images serve as unlabeled inputs within the semi-supervised learning pipeline. This setup enables a controlled analysis of how the number of labeled examples affects segmentation performance. Preliminarily, the code is publicly available[1].

We evaluate SAM-Noise against a strong baseline where SAM is fine-tuned exclusively on the labeled subset, without any pseudo-labeling. Both models are trained using a learning rate of 1×10^{-4}, a batch size of 2, and binary cross-entropy loss. Training runs for a maximum of 150 epochs with early stopping based on validation Dice coefficient. We report performance using two standard segmentation metrics: Dice coefficient (DC) and Intersection over Union (IoU).

All experiments were conducted using Python 3.10 and PyTorch 2.0. The Segment Anything Model (SAM) was accessed via the Hugging Face Transformers library [15], which incorporated support for SAM shortly after its public release [5]. Specifically, the SamModel and SamProcessor classes were used to load pre-trained weights and process input images.

6.1 Quantitative Performance

Table 1 summarizes the results obtained under two evaluation thresholds, IoU 0.8 and 0.9, which is parameter μ of Algorithm 1. SAM-Noise consistently outperformed the fine-tuned SAM model across all dataset sizes, with the performance gap widening under stricter evaluation conditions. This trend is evident at the IoU 0.9 threshold, where even modest improvements reflect higher segmentation

[1] https://github.com/RenzosGit/Iterative-Noise.

Table 1. Comparison of SAM and SAM-Noise performance for IoU thresholds 0.8 and 0.9 across increasing amounts of labeled data.

IoU	0.8				0.9			
Dataset	SAM		SAM-Noise		SAM		SAM-Noise	
	DC	IoU	DC	IoU	DC	IoU	DC	IoU
125 Img.	55.59%	44.13%	57.82%	45.86%	51.24%	40.00%	56.87%	44.72%
150 Img.	53.48%	42.41%	57.87%	45.86%	54.55%	43.35%	59.59%	47.62%
200 Img.	58.21%	46.58%	60.05%	47.77%	59.17%	47.46%	61.13%	48.94%
250 Img.	58.16%	46.34%	60.24%	47.80%	58.72%	46.75%	61.42%	49.11%

consistency. For example, with only 125 labeled images, SAM-Noise surpasses SAM by +5.63% in Dice coefficient (56.87% vs. 51.24%) and +4.72% in IoU (44.72% vs. 40.00%). With 250 labeled images, this advantage remains, confirming the effectiveness of pseudo-labeling based on noise consistency.

Unlike standard fine-tuning, which relies heavily on labeled data, SAM-Noise leverages unlabeled examples through a self-refinement mechanism. This allows it to generalize better in data-scarce scenarios without additional supervision, making it both more scalable and easier to deploy in a real-world scenario.

6.2 Qualitative Performance

Qualitative results indicate that SAM-Noise is generally effective in identifying the main regions of interest across diverse agricultural landscapes. As illustrated in Fig. 5, the model captures the overall structure of cultivated areas and delineates large-scale boundaries with reasonable accuracy, even when trained with limited supervision. However, when comparing the predicted masks to the ground truth, we observe that the model struggles to reproduce fine-grained divisions between adjacent crops or sub-parcels. Segmentations tend to overgeneralize field boundaries, producing masks that encompass broader zones than necessary. These limitations suggest that additional refinement mechanisms may be needed to recover subtle intra-field variations.

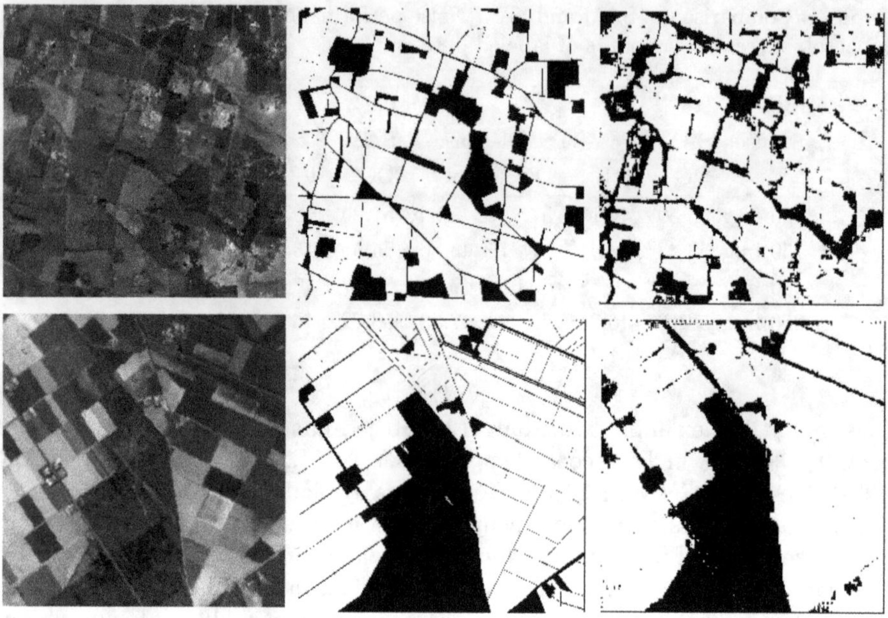

Fig. 5. Examples of segmentation of SAM-Noise. In the first column, there are two unlabeled images input to the model; the middle column shows the ground truth which the model has not seen, while the last column contains the artificial segmentations created by the SAM-Noise model.

To further illustrate the impact of iterative refinement, Fig. 6 presents two examples comparing segmentation results at different stages of the SAM-Noise framework. To aid visual interpretation, colored outlines were manually added to highlight regions where segmentation accuracy improves across iterations. The final SAM-Noise result shown was selected based on validation set metrics. Red is used in the top row, and yellow in the bottom row to ensure contrast against the background. These visual cues help emphasize how the model progressively aligns predictions with the ground truth, as reflected in the increasing IoU scores.

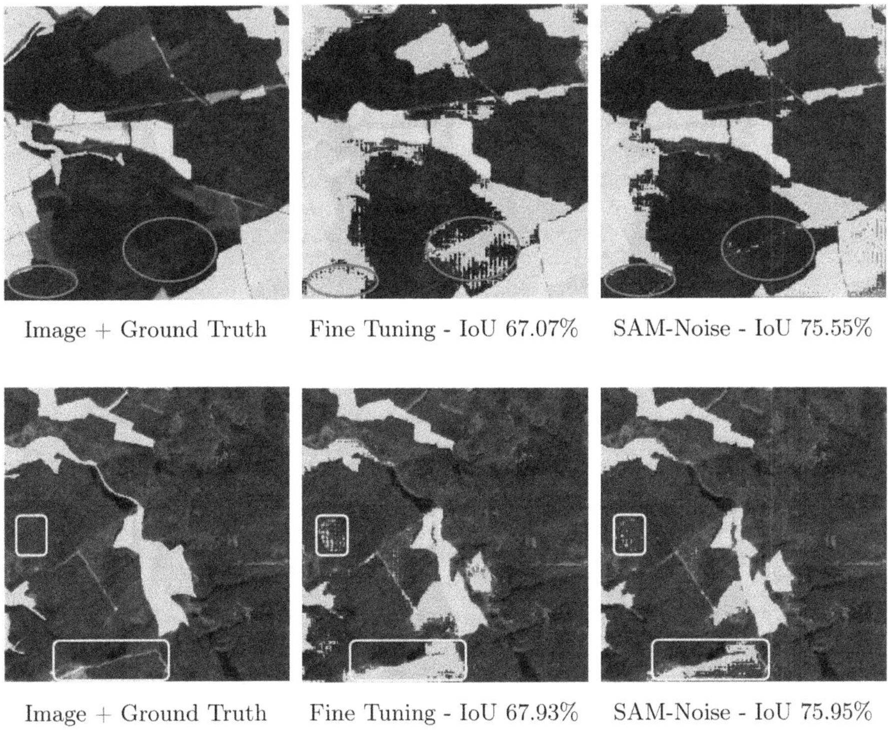

Fig. 6. Comparison of segmentation results before and after iterative fine-tuning. Each row shows a test image with its ground truth overlay (left), the result after initial fine-tuning (middle), and the final result produced by the proposed SAM-Noise framework (right). Cyan masks represent predicted segmentations, while colored outlines highlight regions of improved alignment.

7 Conclusions

This study proposes a lightweight segmentation framework for satellite imagery in precision agriculture, based on SAM fine-tuned with limited labeled data and enhanced through a noise-based pseudo-labeling strategy.

Experimental results demonstrate that SAM-Noise captures meaningful spatial patterns despite data scarcity. However, segmentation performance tends to plateau or degrade in later iterations, possibly due to forgetting previously learned representations. This points to the need for memory-aware optimization or regularization mechanisms to maintain model stability throughout training.

While this work targets agricultural monitoring, the underlying strategy is adaptable to domains such as disaster assessment, land cover mapping, and crop health tracking. Future research will focus on improving generalization across diverse geographies, integrating complementary data sources like spectral and weather information, and introducing filtering steps to exclude images with unreliable ground truth masks, thereby enhancing label quality and training consistency.

Acknowledgments. B. Peralta and R. Larenas appreciate the support of the Chilean National Agency for Research and Development (ANID), Fondecyt grant ID 1241882, and the National Center for Artificial Intelligence CENIA FB210017, Basal ANID.

Disclosure of Interests. The authors have no competing interests.

References

1. d'Andrimont, R., et al.: Ai4boundaries: An open ai-ready dataset to map field boundaries with sentinel-2 and aerial photography. Earth Syst. Sci. Data **15**(1), 317–329 (2023)
2. Food and Agriculture Organization of the United Nations: Gross domestic product and agriculture value added 2013–2022. global and regional trends (2024). https://www.fao.org/statistics/highlights-archive/highlights-detail/gross-domestic-product-and-agriculture-value-added-2013-2022.-global-and-regional-trends/en
3. Huang, Z., et al.: Segment anything model combined with multi-scale segmentation for extracting complex cultivated land parcels in high-resolution remote sensing images. Remote Sens. **16**(18) (2024). https://doi.org/10.3390/rs16183489, https://www.mdpi.com/2072-4292/16/18/3489
4. Khalili, E., Priego-Torres, B., Leon-Jimenez, A., Sanchez-Morillo, D.: Automatic lung segmentation in chest x-ray images using SAM with prompts from yolo. IEEE Access (2024)
5. Kirillov, A., et al.: Segment anything. arXiv preprint arXiv:2304.02643 (2023)
6. Mansoori, M., Shahabodini, S., Abouei, J., Plataniotis, K.N., Mohammadi, A.: Self-prompting polyp segmentation in colonoscopy using hybrid yolo-sam 2 model. arXiv preprint arXiv:2409.09484 (2024)
7. Nair, S., Sharifzadeh, S., Palade, V.: Farmland segmentation in landsat 8 satellite images using deep learning and conditional generative adversarial networks. Remote Sens. **16**(5), 823 (2024)
8. Nchare, F.B.A., Tapamo, H.: Semantic segmentation of high-resolution aerial imagery using a fully convolutional network. In: CARI 2022. Yaounde, Cameroon (2022). https://inria.hal.science/hal-03715809
9. Pandey, S., Chen, K.F., Dam, E.B.: Comprehensive multimodal segmentation in medical imaging: Combining YOLOv8 with SAM and HQ-SAM models. In: 2023 IEEE/CVF International Conference on Computer Vision Workshops (ICCVW), pp. 2584–2590 (2023) https://doi.org/10.1109/ICCVW60793.2023.00273
10. Reddy, J., et al.: Cotton yield prediction via UAV-based cotton boll image segmentation using yolo model and segment anything model (SAM). Remote Sens. **16**(23), 4346 (2024)
11. Safarov, F., et al.: Improved agricultural field segmentation in satellite imagery using TL-ResUNet architecture. Sensors **22**(24), 9784 (2022)
12. Sharifi, A., Mahdipour, H., Moradi, E., Tariq, A.: Agricultural field extraction with deep learning algorithm and satellite imagery. J. Indian Soc. Remote Sens. 1–7 (2022)

13. Sravya, N., Lal, S., Nalini, J., Reddy, C.S., Dell'Acqua, F., et al.: Dppnet: an efficient and robust deep learning network for land cover segmentation from high-resolution satellite images. IEEE Trans. Emerg. Top. Comput. Intell. **7**(1), 128–139 (2022)
14. United States Department of Agriculture: Grain: World markets and trade (2025). https://apps.fas.usda.gov/psdonline/circulars/grain.pdf
15. Wolf, T., et al.: Transformers: state-of-the-art natural language processing. In: Proceedings of the 2020 Conference on Empirical Methods in Natural Language Processing: System Demonstrations, pp. 38–45. Association for Computational Linguistics, Online (2020). https://www.aclweb.org/anthology/2020.emnlp-demos.6

A Review of Reinforcement Learning Methods and Performance Measures in Algorithmic Stock Trading

Aleksander Kruczkowski[1] and Paweł Weichbroth[2(✉)]

[1] Capgemini, Gdansk, Poland
[2] Faculty of Electronics, Telecommunications and Informatics, Department of Software Engineering, Gdansk University of Technology, Gdansk, Poland
pawel.weichbroth@pg.edu.pl

Abstract. To date, few studies have investigated the application of reinforcement learning methods in algorithmic stock trading. This study aims to address this gap by conducting a systematic literature review using a lightweight PRISMA protocol. Two findings emerged from our research. First, we identified nine different Reinforcement Learning (RL) used in stock trading. The most widely used RL methods are the Deep Q-Network (DQN), Proximal Policy Optimization (PPO), Advantage Actor-Critic (A2C), Twin Delayed DDPG (TD3), and Soft Actor-Critic (SAC). Second, we determined sixteen performance evaluation measures. The most popular measures are the Sharpe ratio, Cumulative return, Annual return, Maximum Drawdown (MDD), and Sortino ratio. We believe these results could be useful for both the research community and novice stock traders. In the future, we will investigate the effectiveness of selected trading strategies in bull and bear markets.

Keywords: Reinforcement Learning · Algorithmic Trading · Review

1 Introduction

There is a strong connection between stock trading and sustainability which lies primarily in the rise of sustainable investing practices and the increasing influence of environmental [26], social [25], and governance (ESG) factors [37] on financial markets. Such practices can improve a company's long-term financial performance [50]. At the same time, investors are increasingly evaluating companies based on their sustainability performance [48].

In addition, stock trading is an important research topic due to its significant impact on global financial markets [45], individual wealth [17], and economic stability [16]. Recent advances in trading strategies [21,36,47], particularly algorithmic and machine learning-based approaches [44], offer encouraging signs to improve profitability and reduced risk [10,11,18].

One of the most promising recent approaches uses reinforcement learning (RL) methods owing to their ability to learn optimal decision-making strategies

Published by Springer Nature Switzerland AG 2026
E. Mercier-Laurent et al. (Eds.): AI4S 2025, IFIP AICT 780, pp. 94–103, 2026.
https://doi.org/10.1007/978-3-032-21184-2_8

by interacting with dynamic markets. Unlike traditional models, RL algorithms can adapt to changing market conditions by maximizing cumulative rewards based on past actions. This makes RL algorithms well-suited for developing autonomous trading agents that can handle the complexity and uncertainty of financial markets.

However, to the best of our knowledge, few studies have investigated the most recent literature and offered an evidence-based overview of reinforcement learning methods and their associated performance measures. This study aims to address this gap by conducting a systematic review of recent research in this area. Specifically, our goal is to identify the RL algorithms and the performance evaluation measures used.

The rest of the paper is structured as follows. Section 2 outlines the research methodology. Section 3 presents the results obtained from the performed study. Section 4 discusses threats to validity. Section 5 concludes the paper.

2 Methodology

A systematic literature review (SLR) was conducted to explore the application of reinforcement learning in algorithmic stock trading. The review followed a lightweight adaptation of the PRISMA methodology [27], which is widely used in empirical computer science research, including financial machine learning applications [52]. Although its full formal rigor was not applied, the review protocol was designed with its core guidelines of transparency, repeatability, and structured filtering in mind. The review process consisted of the five following steps.

2.1 Research Questions Definition

In the first step, we defined the following two research questions (RQs):

RQ1: What reinforcement learning (RL) methods are used in stock trading on the NYSE, NASDAQ, SSE, and TWSE markets?
RQ2: What measures are used to evaluate the performance of RL methods?

To answer these two research questions, the next step is to identify the data source. Furthermore, since many studies have examined multiple capital markets, it is also necessary to specify which markets will undergo further analysis.

2.2 Data Source and General Boundaries

The Scopus database was selected for its extensive coverage of peer-reviewed publications and its advanced filtering capabilities. The search was narrowed down to articles written in English and published between 2022 and 2025 that fall under one or more of the following Scopus subject areas:

- Computer Science (COMP),
- Decision Sciences (DECI), and

- Economics, Econometrics & Finance (ECON).

These filters were implemented using `LIMIT-TO` clauses in the search string.

Even though the Scopus search did not allow filtering by exchange name, our review focused on four major stock exchanges:

- the New York Stock Exchange (NYSE),
- the Nasdaq Stock Market (NASDAQ),
- the Shanghai Stock Exchange (SSE), and
- the Taiwan Stock Exchange (TWSE).

The New York Stock Exchange (NYSE) is located on Wall Street in New York City [30] and is one the world's largest stock exchange by market capitalization of approximately $28.33 trillion [40]. NYSE serves as a marketplace for buying and selling shares of publicly traded companies and operates both a physical trading floor and a modern electronic trading system. Since its inception in 1792, the New York Stock Exchange has played a central role in American economic development. With a daily trading volume of over a billion shares [35], it is now a critical hub of global finance [9].

The National Association of Securities Dealers Automated Quotations (NASDAQ) [28] is a global, electronic, pioneering marketplace that has transformed the buying and selling of securities. Since its launch in 1971 as the world's first automated stock exchange, Nasdaq has provided a fast, transparent trading platform in over 50 countries [13]. Interestingly, from June to September of 2025, the NASDAQ's market capitalization exceeded that of the NYSE for four straight months. In June, the NASDAQ reached $32.0 trillion, surpassing the NYSE's $30.8 trillion for the first time. By September, the NASDAQ had climbed to $35.6 trillion, while the NYSE stood "only" at $31.4 trillion [34].

The Shanghai Stock Exchange (SSE) [33] is located in the city of Shanghai (China). Founded in 1990, the SSE plays a significant role in the Chinese economy. In February 2025, SSE reached a market capitalization of 7.095 trillion USD [41], while the total market capitalization in China was 11.87 trillion USD [4]. Thus, SSE is the largest stock exchange in mainland China, trading a variety of securities including bonds, derivatives, funds, and stocks [22]. The SSE is a public, member-based legal entity that operates a securities trading venue. It supervises its members and listed companies and manages market information.

The Taiwan Stock Exchange (TWSE) [43] is the main stock exchange in Taiwan. It was founded in 1961 and is based in Taipei. TWSE is responsible for managing the centralized trading market, supervising listed companies and securities companies, and maintaining overall market stability. The exchange offers both real-time and delayed trading data for various securities, including bonds, exchange-traded funds (ETFs), and stocks [5]. As of September 2025, its market capitalization is approximately 2.6 trillion USD [42]. Its stated goals are to promote Taiwan's economic growth by developing new financial products, advancing domestic capital markets, and strengthening the nation's international competitiveness.

While NYSE and Nasdaq represent leading Western capital markets, the SSE and TWSE reflect the dynamics of Chinese-language financial systems. Note that we only included studies targeting other exchanges if their findings offered transferable insights into reinforcement learning applied to stock trading. These markets were selected to capture both American and East Asian trading environments, offering a meaningful balance of geographic, regulatory, and structural diversity.

2.3 Search Query Definition

We formulated the search query using the keywords "machine learning", "reinforcement learning", and "stock market". This approach ensures that the initial stage yields results focused on the topic without overly narrowing the scope. The keywords were chosen to provide broad thematic coverage while centering the query on the intersection of financial and machine learning research. The query was constructed by combining semantic keywords with metadata-based filtering to eliminate irrelevant publications.

2.4 Inclusion and Exclusion Criteria

We designed the inclusion criteria (IC) to reflect the scope of the review. We applied the ICs through the search string and during manual screening. A study was included in our review if it met all of the ICs, defined as follows:

IC1: The document includes the terms "machine learning", or "reinforcement learning" and "stock market" in the title, abstract, or keywords.
IC2: The document is published between 2022 and 2025.
IC3: The document belongs to at least one of the following three subject areas: COMP, DECI, or ECON.
IC4: The study is written in English.
IC5: The study includes at least one of the following three indexed keywords: "machine learning", "Stock Market", or "Reinforcement Learning".

The exclusion criteria were not encoded directly into the search query, but rather, were applied during manual or assisted screening.

EC1: The document does not focus on reinforcement learning (RL) in the context of trading.
EC2: The document does not focus on any of the four target exchanges (NYSE, NASDAQ, SSE, or TWSE) and does not provide insights that can be transferred to RL-based trading on these markets.
EC3: The document is not a full-length research article (e.g., conference keynote, book chapter, extended abstract, or short paper with fewer than 2 pages).
EC4: The document is an erratum, editorial, or retracted paper.

2.5 Screening and Quality Assessment

A total of 845 records were retrieved from Scopus and exported to a spreadsheet. Metadata was collected for each entry, including the title, abstract, DOI, citation count, document type, and Scopus keywords. The screening and quality assessment process was conducted in multiple stages, combining manual verification with ChatGPT assistance. This phase aimed to filter out irrelevant studies using exclusion and quality criteria before proceeding to a full-text review. In summary, the final dataset comprised of 21 full-text records, including 14 journal articles and 7 conference papers.

3 Results

While input data were obtained through a systematic literature review, the presented results are the product of a data synthesis aimed at identifying recurring methods and metrics.In addition, the tabular format was selected to ensure the data was presented in a readable and clear manner.

For the first research question (RQ1), we investigated which reinforcement learning (RL) methods were employed for stock trading on the NYSE, NASDAQ, SSE, and TWSE. Table 1 shows the RL methods used and their absolute and relative frequencies, as well as the associated references.

Table 1. Reinforcement Learning (RL) methods used in stock trading and their absolute and relative frequencies.

Method	Count	References
Deep Q-Network (DQN)	9 (42.9%)	[12, 20, 23, 24, 32, 39, 49, 53]
Proximal Policy Optimization (PPO)	8 (38.1%)	[2, 6, 19, 29, 32, 38, 46, 53]
Advantage Actor-Critic (A2C)	6 (28.6%)	[8, 29, 31, 32, 38, 46]
Twin Delayed DDPG (TD3)	3 (14.3%)	[7, 15, 38]
Soft Actor-Critic (SAC)	2 (9.5%)	[29, 38]
Q-learning	1 (4.8%)	[1]
Deep Deterministic Policy Gradient (DDPG)	1 (4.8%)	[38]
Inverse Reinforcement Learning (IRL)	1 (4.8%)	[51]
Deep Recursive Estimator (DeRecv)	1 (4.8%)	[3]

The two most common methods were the Deep Q-Network (DQN) and the Proximal Policy Optimization (PPO), which were applied in nine and eight studies, respectively. Note that DQN often served as a baseline and was sometimes adapted with additional modules. In addition, many studies compared multiple RL methods under the same setup [29, 32]. Simpler approaches like Q-learning were rarely used [1], possibly because they have been largely replaced by deep variants, such as DQN. Actor-critic methods, such as A2C, TD3, and SAC, are well-suited for continuous action spaces, yet were used less frequently [7,31]. Some implementations integrated RL with portfolio models or sentiment inputs, particularly for NYSE and NASDAQ [2,53]. Studies on the SSE and TWSE focused more on index tracking and multi-asset allocation [7,49].

Table 2. Performance measures and their absolute and relative frequencies.

Measure	Count	References
Sharpe ratio	14 (66.7%)	[2,6,12,14,15,19,20,23,29,32,38,53]
Cumulative return	8 (38.1%)	[2,6,15,19,29,31,49,53]
Annual return	7 (33.3%)	[2,14,19,20,23,29,38]
Maximum Drawdown (MDD)	7 (33.3%)	[2,6,19,20,23,29]
Sortino ratio	6 (28.6%)	[2,6,24,32,38]
Profit or loss	5 (23.8%)	[1,20,23,24]
Annual volatility	3 (14.3%)	[2,19,29]
Average profit	3 (14.3%)	[1,39,51]
Standard deviation	2 (9.5%)	[38,39]
Win rate	2 (9.5%)	[6,7]
Percentage of positive trade returns	1 (4.8%)	[12]
Annual net profit	1 (4.8%)	[12]
Average commission	1 (4.8%)	[51]
Daily return	1 (4.8%)	[38]
Lower partial standard deviation	1 (4.8%)	[38]
Rectified standard deviation	1 (4.8%)	[12]

For the second research question (RQ2), we explored which measures are employed to evaluate the performance of reinforcement learning methods in stock trading. Table 2 summarizes the results obtained from the review. As one can notice, the most widely recognized (#14) risk-adjusted measure is the Sharpe ratio, which shows the amount of excess return one gets for each unit of risk taken. The second most used is Cumulative Return (#8), followed by Annual Return (#7), which is used to gauge overall profitability in a full-period performance view and to compare across time frames, respectively. The fourth most popular (#7) measure is Maximum Drawdown (MDD) used to evaluate the largest peak-to-trough loss experienced by an trading strategy over a given time period. The next place takes the Sortino ratio (#6), which is similar to the Sharpe ratio but with one key improvement: it only considers downside risk, or the volatility of negative returns, rather than overall volatility. The less complex measures such as profit and loss were also used (#5). In summary, authors usually pair a return metric with a risk-adjusted measure in order to capture both profitability and risk exposure.

4 Discussion

As with any similar study, this research faces threats to validity, which should be understood as factors that may compromise the credibility or accuracy of the findings.

Construct Validity. Undeniably, it's clear that strict inclusion and exclusion criteria were applied. However, some studies required interpretive judgment to determine their relevance to reinforcement learning in trading. These subjective decisions may have affected the outcome of the final dataset.

Internal Validity. The screening process was supported by ChatGPT (version 3), which increased its scalability but also introduced variability. The prompt design and manual interpretation of outputs reflected the author's judgment. However, slight differences in prompt phrasing or model version may produce different results when the process is repeated.

External Validity. We narrowed the scope of the review to four stock exchanges (NYSE, NASDAQ, SSE, and TWSE) and English-language publications from 2022 to 2025. In other words, we intentionally excluded older studies, non-English studies, and non-indexed studies that were potentially relevant. This limitation may affect the generalizability of the findings.

Reproducibility. Although the review process adhered to structured criteria, the combination of ChatGPT and human decision-making limits strict reproducibility. Intermediate inclusion decisions, especially in borderline cases, depend on contextual interpretation and are not fully deterministic. Nevertheless, despite these acknowledged limitations, the methodology was designed to balance scale with rigor, providing a transparent and practical approach to conducting an exploratory review within a complex and evolving research domain.

5 Conclusion

In the present study, we first identified nine Reinforcement Learning (RL) algorithms used in stock trading. The most widely used methods were the Deep Q-Network (DQN), Proximal Policy Optimization (PPO), Advantage Actor-Critic (A2C), Twin Delayed Deep Deterministic Policy Gradient (TD3), and Soft Actor-Critic (SAC). Second, we identified sixteen performance evaluation metrics The most frequently used are the Sharpe ratio, cumulative return, annual return, maximum drawdown (MDD), and Sortino ratio.

In the future, we plan to examine the effectiveness of selected trading strategies under different market conditions, including bull and bear markets. In addition, data with a higher degree of granularity will be used, enabling the detection of short-term patterns and sudden changes by providing a more precise view of events

References

1. Ansari, Y., et al.: A deep reinforcement learning-based decision support system for automated stock market trading. IEEE Access **10**, 127469–127501 (2022)
2. Bai, T., Lang, Q., Song, S., Fang, Y., Liu, X.: Feature fusion deep reinforcement learning approach for stock trading. In: 2022 41st Chinese Control Conference (CCC), pp. 7240–7245. IEEE (2022)
3. Cao, Y., Zhai, J.: Estimating price impact via deep reinforcement learning. Int. J. Finan. Econ. **27**(4), 3954–3970 (2022)
4. CEIC: China market capitalization (2025). https://www.ceicdata.com/en/indicator/china/market-capitalization. Accessed 25 Oct 2025
5. Chen, D.Y., Wu, C.W., Wang, C.W.: Using CNN models to predict the future trends of listed stocks on the Taiwan stock exchange. In: International Conference on Human-Computer Interaction, pp. 102–111. Springer, Cham (2025)
6. Chen, Y.F., Shih, W.Y., Lai, H.C., Chang, H.C., Huang, J.L.: Pairs trading strategy optimization using proximal policy optimization algorithms. In: 2023 IEEE International Conference on Big Data and Smart Computing (BigComp), pp. 40–47. IEEE (2023)
7. Cheng, L.C., Sun, J.S.: Multiagent-based deep reinforcement learning framework for multi-asset adaptive trading and portfolio management. Neurocomputing **594**, 127800 (2024)
8. Du, S., Shen, H.: Reinforcement learning-based multimodal model for the stock investment portfolio management task. Electronics **13**(19), 3895 (2024)
9. Du, X., Deng, L., Qian, K.: Current market top business scopes trend–a concurrent text and time series active learning study of nasdaq and nyse stocks from 2012 to 2017. Appl. Sci. **8**(5), 751 (2018)
10. Hernes, M., Korczak, J., Krol, D., Pondel, M., Becker, J.: Multi-agent platform to support trading decisions in the forex market. Appl. Intell. **54**(22), 11690–11708 (2024)
11. Hernes, M., Kozierkiewicz, A., Pietranik, M.: The agents' selection methods for a consensus-based investment strategy in a multi-agent financial decisions support system. In: European Conference on Advances in Databases and Information Systems, pp. 60–67. Springer, Cham (2019)
12. Huang, Z., Li, N., Mei, W., Gong, W.: Algorithmic trading using combinational rule vector and deep reinforcement learning. Appl. Soft Comput. **147**, 110802 (2023)
13. Investopedia: Nasdaq explained: History, trading system, and financial insights (2025). https://www.investopedia.com/terms/n/nasdaq.asp. Accessed 25 Oct 2025
14. Isaac, E., Mathew, J., Mariam Varghese, S., PM, S., Simon, J., Ajith, A.: Multimodal approach for portfolio optimization using deep reinforcement learning. In: 2024 10th International Conference on Smart Computing and Communication (ICSCC), pp. 76–81. IEEE (2024)
15. Kabbani, T., Duman, E.: Deep reinforcement learning approach for trading automation in the stock market. IEEE Access **10**, 93564–93574 (2022)
16. Khang, P.Q., et al.: Machine learning for liquidity prediction on Vietnamese stock market. Procedia Comput. Sci. **192**, 3590–3597 (2021)
17. Korczak, J., Hernes, M., Bac, M.: Consensus-based trading strategy in a-trader system. Informatyka Ekonomiczna **3**(29), 98–110 (2013)
18. Korczak, J., Hernes, M., Bac, M.: Fundamental analysis in the multi-agent trading system. In: 2016 Federated Conference on Computer Science and Information Systems (FedCSIS), pp. 1169–1174. IEEE (2016)

19. Li, J., Zhu, C., Chen, X., Hu, X.: A stock trading strategy based on deep reinforcement learning and Hong-Kong capital position. Int. J. Innov. Comput. Inf. Control **20**(02), 643 (2024)
20. Li, Y., Liu, P., Wang, Z.: Stock trading strategies based on deep reinforcement learning. Sci. Program. **2022**(1), 4698656 (2022)
21. Li, Y., Ni, P., Chang, V.: Application of deep reinforcement learning in stock trading strategies and stock forecasting. Computing **102**(6), 1305–1322 (2020)
22. Liu, B., Yu, Z., Wang, Q., Du, P., Zhang, X.: Prediction of SSE shanghai enterprises index based on bidirectional LSTM model of air pollutants. Expert Syst. Appl. **204**, 117600 (2022)
23. Liu, P., Zhang, Y., Bao, F., Yao, X., Zhang, C.: Multi-type data fusion framework based on deep reinforcement learning for algorithmic trading. Appl. Intell. **53**(2), 1683–1706 (2023)
24. Lu, J.Y., et al.: Structural break-aware pairs trading strategy using deep reinforcement learning. J. Supercomput. **78**(3), 3843–3882 (2022)
25. Matta, N., Carrillo, T.R., Mercier-Laurent, E., Boulanger, D.: knowledge engineering and management applied to sustainability and innovation. AI EDAM **33**(2), 114–116 (2019)
26. Mercier-Laurent, E.: Climate change–can ai help understanding and more effective facing of various interrelated impacts? In: Decision Making Using AI in Energy and Sustainability: Methods and Models for Policy and Practice, pp. 3–19. Springer, Cham (2023)
27. Moher, D., Liberati, A., Tetzlaff, J., Altman, D.G., Group, T.P.: Preferred reporting items for systematic reviews and meta-analyses: the PRISMA statement. PLoS Med. **6**(7), e1000097 (2009)
28. Nasdaq: Nasdaq: Stock market, data updates, reports & news (2025). https://www.nasdaq.com/. Accessed 25 Oct 2025
29. Nguyen, H.T.N., Mac, B.N.N., Tran, A.D., Nguyen, N.T., Pham, D.T.: Deep reinforcement learning approach using customized technical indicators for a pre-emerging market: a case study of Vietnamese stock market. In: 2022 RIVF International Conference on Computing and Communication Technologies (RIVF), pp. 737–742. IEEE (2022)
30. NYSE: Today's stock market (2025). https://www.nyse.com/. Accessed 25 Oct 2025
31. Pourahmadi, Z., Fareed, D., Mirzaei, H.R.: A novel stock trading model based on reinforcement learning and technical analysis. Ann. Data Sci. **11**(5), 1653–1674 (2023)
32. Praghaadeesh, R., Maniappan, V., Doss, S., et al.: Enhancing algorithmic trading strategies with sentiment analysis: a reinforcement learning approach. In: 2024 IEEE 3rd World Conference on Applied Intelligence and Computing (AIC), pp. 107–112. IEEE (2024)
33. Shanghai Stock Exchange: Shanghai stock exchange (2025). https://english.sse.com.cn/. Accessed 25 Oct 2025
34. Soltani, E.: Nasdaq market capitalization surpasses nyse for four consecutive months in 2025 (2025). https://www.voronoiapp.com/markets/-NASDAQ-Market-Capitalization-Overtakes-NYSE-in-2025-3281. Accessed 25 Oct 2025
35. Statista: New york stock exchange (nyse) - statistics & facts (2025). https://www.statista.com/topics/10015/new-york-stock-exchange-nyse/. Accessed 25 Oct 2025
36. Sukma, N., Namahoot, C.S.: Enhancing trading strategies: a multi-indicator analysis for profitable algorithmic trading. Comput. Econ. **65**(6), 3807–3840 (2025)

37. Sulich, A., Sołoducho-Pelc, L.: Sustainable strategic management and business intelligent systems within the sustainable development context-key research areas. Procedia Comput. Sci. **246**, 3188–3197 (2024)
38. Sun, R., Stefanidis, A., Jiang, Z., Su, J.: Combining transformer based deep reinforcement learning with Black-Litterman model for portfolio optimization. Neural Comput. Appl. **36**(32), 20111–20146 (2024)
39. Sun, T., Huang, D., Yu, J.: Market making strategy optimization via deep reinforcement learning. IEEE Access **10**, 9085–9093 (2022)
40. Sustainable Stock Exchanges Initiative: New york stock exchange (2025). https://sseinitiative.org/stock-exchange/nyse. Accessed 25 Oct 2025
41. Sustainable Stock Exchanges Initiative: Shanghai stock exchange (2025). https://sseinitiative.org/stock-exchange/sse. Accessed 25 Oct 2025
42. Taiwan Stock Exchange: Taiex increases by approximately 4.00 while the total market capitalization of twse-listed stocks reaches 82.14 trillion ntd. (2025). https://www.twse.com.tw/en/about/news/news/content.html?8a8216d6993bf15101993d2d473e0030. Accessed 25 Oct 2025
43. Taiwan Stock Exchange: Taiwan stock exchange corporation (2025). https://www.twse.com.tw/en/. Accessed 25 Oct 2025
44. Tusień, E., Kwaśniewska, A., Weichbroth, P.: Mining frequent sequences with time constraints from high-frequency data. Int. J. Finan. Stud. **13**(2), 55 (2025)
45. Valdez, S., Molyneux, P.: An Introduction to Global Financial Markets. Bloomsbury Publishing (2017)
46. Wang, Z., Jin, S., Li, W.: Research on portfolio optimization based on deep reinforcement learning. In: 2022 4th International Conference on Machine Learning, Big Data and Business Intelligence (MLBDBI), pp. 391–395. IEEE (2022)
47. Wu, X., Chen, H., Wang, J., Troiano, L., Loia, V., Fujita, H.: Adaptive stock trading strategies with deep reinforcement learning methods. Inf. Sci. **538**, 142–158 (2020)
48. Yadav, P.L., Han, S.H., Rho, J.J.: Impact of environmental performance on firm value for sustainable investment: evidence from large US firms. Bus. Strateg. Environ. **25**(6), 402–420 (2016)
49. Yu, C., Zhang, L.: Research on Stock Index Prediction Model Based on Deep Reinforcement Learning. IOS Press (2024)
50. Zema, T., Wójcik, F., Sulich, A., Hernes, M.: Forecasting fuel retail sales volume using machine learning for sustainable. In: Emerging Challenges in Intelligent Management Information Systems: Proceedings of 26th European Conference on Artificial Intelligence ECAI 2023-IMIS 2023 Workshop, vol. 1079, p. 109. Springer, Cham (2024)
51. Zhang, W., Yin, T., Zhao, Y., Han, B., Liu, H.: Reinforcement learning for stock prediction and high-frequency trading with t+ 1 rules. IEEE Access **11**, 14115–14127 (2022)
52. Zhang, Y., Hassan, S.A., Hussain, F., Yu, S., Wan, Z., Rahman, M.M.: Financial applications of machine learning: a literature review. Expert Syst. Appl. **226**, 119640 (2023)
53. Zou, J., Lou, J., Wang, B., Liu, S.: A novel deep reinforcement learning based automated stock trading system using cascaded LSTM networks. Expert Syst. Appl. **242**, 122801 (2024)

Artificial Intelligence Business and Global Sustainability Review

Konstantin M. Golubev(✉)

General Knowledge Machine Research Group, Kyiv, Ukraine
gkmekp@gmail.com

Abstract. This review focuses on a new Artificial Intelligence LLMs hype impact on a global sustainability. The article studies various attempts in the field of Artificial Intelligence, such as Large Language Models, AI agents and AI reinforcement learning. It discusses the limitations of current AI LLMs methods in solving real-life problems and proposes the need for a wise evaluation of its applicability.

Keywords: Artificial Intelligence · LLM · Sustainability

1 Introduction

"Just about every successful deployment of AI has either one of two expedients: It has a person somewhere in the loop, or the cost of failure, should the system blunder, is very low." Rodney Brooks [1].

Artificial Intelligence (AI) is a branch of mathematics presented by media as a substitute of human intellect. It looks like a campaign of massive marketing intended to confuse public and get profits from misunderstanding of mathematics role. Any existing mathematics is used by humans and under humans control to get results not reached by pure handmade calculations. And it is humans developing and improving mathematics.

There are many mathematical facilities helping to solve intellectual problems: Mathematical optimization, Linear discriminant analysis (LDA), normal discriminant analysis (NDA), Cluster analysis, Statistics, Probability theory and Artificial Intelligence based on these facilities. They are used without labeling them AI. And ability of equipment to function automatically, like coffee machines, is known widely, but no one calls it AI. Even presentation of hand electronic calculator in 1960s years, that able to do arithmetic calculations faster than humans, did not humiliate humanity.

As an example, consider mathematical optimization. In mathematics, computer science and operations research, mathematical optimization or mathematical programming, alternatively spelled optimization is the selection of a best element (with regard to some criterion) from some set of available alternatives. Several people may say that selection of a best element is an intellectual activity, but the point is that the human should decide how to handle the selection. And for some strange reasons it is proposed that AI mathematics should work independently, being more intelligent than people. Obviously, treating AI as a wisdom black-box would lead to erroneous decisions.

Published by Springer Nature Switzerland AG 2026
E. Mercier-Laurent et al. (Eds.): AI4S 2025, IFIP AICT 780, pp. 104–116, 2026.
https://doi.org/10.1007/978-3-032-21184-2_9

The lecturer from MIT states in a lecture about Artificial Intelligence: "And frequently you can substitute computing for intelligence. So no one would say that Deep Blue does anything like what a human chess master does. But nevertheless, Deep Blue, by processing data like bulldozer processes gravel, was able to beat the world champion." [2].

Rodney Brooks tells that in the 1960s first wave of major investment in artificial intelligence started and the researchers often predicted that human-level intelligent machines were only 10 years away. In the 1980s, AI's second age was based on two technologies: rule-based expert systems and neural networks. Again, there were predictions about the end of human dominance in intelligence. The third and current age of AI arose during the early 2000s. In the early 2010s, this wave gathered powerful new momentum with the rise of neural networks learning from massive data sets. It soon turned into a tsunami of promise, hype, and profitable applications. In 2020 OpenAI introduced GPT-3, the enormously powerful natural language model [1].

2 The Hype

OpenAI was founded in 2015 as a nonprofit dedicated to the creation of artificial general intelligence, or AGI, that should benefit "humanity as a whole." (AGI, in the company's telling, would be advanced enough to outperform any person at "most economically valuable work"—just the kind of cataclysmically powerful tech that demands a responsible steward.) [3].

OpenAI CEO Sam Altman says AI will reshape society. GPT-4 is just one step toward OpenAI's goal to eventually build Artificial General Intelligence, which is when AI crosses a powerful threshold which could be described as AI systems that are generally smarter than humans. Altman and his team hope "the model will become this reasoning engine over time," he said, eventually being able to use the internet and its own deductive reasoning to separate fact from fiction.

Microsoft and OpenAI evolve partnership to drive the next phase of AI - The Official Microsoft Blog states: We are thrilled to continue our strategic partnership with OpenAI and to partner on Stargate. Today's announcement is complementary to what our two companies have been working on together since 2019.

The key elements of our partnership remain in place for the duration of our contract through 2030, with our access to OpenAI's IP, our revenue sharing arrangements and our exclusivity on OpenAI's APIs all continuing forward – specifically:

Microsoft has rights to OpenAI IP (inclusive of model and infrastructure) for use within our products like Copilot. This means our customers have access to the best model for their needs.

The OpenAI API is exclusive to Azure, runs on Azure and is also available through the Azure OpenAI Service. This agreement means customers benefit from having access to leading models on Microsoft platforms and direct from OpenAI.

Microsoft and OpenAI have revenue sharing agreements that flow both ways, ensuring that both companies benefit from increased use of new and existing models.

Microsoft remains a major investor in OpenAI, providing funding and capacity to support their advancements and, in turn, benefiting from their growth in a valuation [4].

In 2025, Dario Amodei, co-founder and CEO of Anthropic, said he expects a "human-level" AI could be developed in two to three years. Sam Altman, CEO of OpenAI, believes AGI could be reached sometime in the next four or five years. [5].

Microsoft, the initial big investor in OpenAI, tells in the Work Trend Index Annual Report 2025 that we are entering a new reality—one in which AI can reason and solve problems in remarkable ways. This intelligence on tap will rewrite the rules of business and transform knowledge work as we know it.

Organizations today must navigate the challenge of preparing for an AI-enhanced future, where AI agents will gain increasing levels of capability over time that humans will need to harness as they redesign their business. Human ambition, creativity, and ingenuity will continue to create new economic value and opportunity as we redefine work and workflows.

As a result, a new organizational blueprint is emerging, one that blends machine intelligence with human judgment, building systems that are AI-operated but human-led. Like the Industrial Revolution and the internet era, this transformation will take decades to reach its full promise and involve broad technological, societal, and economic change. The data and insights point to the emergence of an entirely new organization, a Frontier Firm that looks markedly different from those we know today.

Structured around on-demand intelligence and powered by "hybrid" teams of humans + agents, these companies scale rapidly, operate with agility, and generate value faster. Frontier Firms are already taking shape, and within the next 2–5 years we expect that every organization will be on their journey to becoming one. We see the journey to the Frontier Firm playing out in three phases.

First, AI acts as an assistant, removing the drudgery of work and helping people do the same work better and faster.

In phase 2, agents join teams as "digital colleagues," taking on specific tasks at human direction—for instance, a researcher agent creating a go-to-market plan. These agents equip employees with new skills that help scale their impact—freeing them to do new and more valuable work.

In phase 3, humans set direction for agents that run entire business processes and workflows, checking in as needed. Just as we've seen the role of AI in software development evolve over the past three years from coding assistance to chat to—now—agents, the same pattern will apply to knowledge work. Picture how a supply chain role may change: agents handle end-to-end logistics, while humans guide the agent system, resolve exceptions, and manage supplier relationships.

The journey to the Frontier Firm is not a strictly linear progression—in many cases organizations will be in all three phases simultaneously. For decades, intelligence was one of the most valuable—and limited—assets in business, bound by human time, energy, and cost. That's changing. Intelligence is becoming an essential durable good: abundant, affordable, and available on demand. With the rise of AI and agents that can reason, plan, and act as digital labor, companies can scale capacity as needed. Already, 82% of leaders say they're confident that they'll use digital labor to expand workforce capacity in the next 12–18 months.

So when is it essential to keep humans in the loop? Economist and AI expert Daniel Susskind hypothesizes that human work will persist even as AI capabilities expand because of three limits: it can be more efficient to have AI and humans working in tandem, human preference, and moral judgment. In logistics, for instance, AI can optimize delivery routes, but human dispatchers still manage real-world disruptions—allowing AI to focus resources on scale, not exceptions. In finance, AI can model risk, but clients may still prefer a trusted advisor to guide major decisions. Lawyers may use AI to summarize case law, but high-stakes negotiations often require human judgment—not just for their complexity, but because society expects people to be responsible for the consequences.

As agents increasingly join the workforce, we'll see the rise of the agent boss: someone who builds, delegates to, and manages agents to amplify their impact—working smarter, scaling faster, and taking control of their career in the age of AI. From the boardroom to the front line, every worker will need to think like the CEO of an agent-powered startup, directing teams of agents with specialized skills like research and data analysis. For those ready to expand their scope, this will be a career accelerator—but the data shows that leaders are ahead of employees. Bridging the gap will take more than access; it will require training, oversight, and a new way of working—one that leaders must help shape [6].

Today entrepreneurs are announcing Mechanize, a startup focused on developing virtual work environments, benchmarks, and training data that will enable the full automation of the economy. We will achieve this by creating simulated environments and evaluations that capture the full scope of what people do at their jobs. This includes using a computer, completing long-horizon tasks that lack clear criteria for success, coordinating with others, and reprioritizing in the face of obstacles and interruptions.

We're betting that the lion's share of value from AI will come from automating ordinary labor tasks rather than from "geniuses in a data center".

The market potential here is absurdly large: workers in the US are paid around $18 trillion per year in aggregate. For the entire world, the number is over three times greater, around $60 trillion per year. The explosive economic growth likely to result from completely automating labor could generate vast abundance, much higher standards of living, and new goods and services that we can't even imagine today. Our vision is to realize this potential as soon as possible.

Matthew Barnett, Tamay Besiroglu, Ege Erdil Mechanize is backed by investments from Nat Friedman and Daniel Gross, Patrick Collison, Dwarkesh Patel, Jeff Dean, Sholto Douglas, and Marcus Abramovitch [7].

Three top tech firms on Tuesday announced that they will create a new company, called Stargate, to grow artificial intelligence infrastructure in the United States.

OpenAI CEO Sam Altman, SoftBank CEO Masayoshi Son and Oracle Chairman Larry Ellison appeared at the White House Tuesday afternoon alongside President Donald Trump to announce the company, which Trump called the "largest AI infrastructure project in history."

The companies will invest $100 billion in the project to start, with plans to pour up to $500 billion into Stargate in the coming years. The project is expected to create 100,000 US jobs, Trump said.

Stargate will build "the physical and virtual infrastructure to power the next generation of AI," including data centers around the country, Trump said. Ellison said the group's first, 1 million-square foot data project is already under construction in Texas.

AI leaders have for months been sounding the alarm that more data centers—as well as the chips and electricity and water resources to run them—are needed to power their artificial intelligence ambitions in the coming years.

"I think this will be the most important project of this era," Altman said on Tuesday. "We wouldn't be able to do this without you, Mr. President." [8].

3 The Reality

The new AI. Systems are "built to be persuasive, not truthful," an internal Microsoft document said. "This means that outputs can look very realistic but include statements that aren't true." Microsoft.

OpenAI CEO Sam Altman says AI will reshape society, acknowledges risks: 'A little bit scared of this'. "I'm particularly worried that these models could be used for large-scale disinformation," Altman said. "Now that they're getting better at writing computer code, [they] could be used for offensive cyberattacks."

But a consistent issue with AI language models like ChatGPT, according to Altman, is misinformation: The program can give users factually inaccurate information. The model has this issue, in part, because it uses deductive reasoning rather than memorization, according to OpenAI [5].

The new AI. Systems arc "built to be persuasive, not truthful," an internal Microsoft document said. "This means that outputs can look very realistic but include statements that aren't true."

The chatbots are driven by a technology called a large language model, or L.L.M., which learns its skills by analyzing massive amounts of digital text culled from the internet.

By pinpointing patterns in that data, an L.L.M. learns to do one thing in particular: guess the next word in a sequence of words. It acts like a powerful version of an autocomplete tool. Given the sequence "The New York Times is a ____," it might guess "newspaper."

Because the internet is filled with untruthful information, the technology learns to repeat the same untruths. And sometimes the chatbots make things up. They produce new text, combining billions of patterns in unexpected ways. This means even if they learned solely from text that is accurate, they may still generate something that is not.

Because these systems learn from more data than humans could ever analyze, even A.I. experts cannot understand why they generate a particular sequence of text at a given moment. And if you ask the same question twice, they can generate different text.

That compounds the challenges of fact-checking and improving the results.

Companies like OpenAI, Google and Microsoft have developed ways to improve the accuracy. OpenAI, for instance, tries to refine the technology with feedback from human testers.

As people test ChatGPT, they rate the chatbot's responses, separating useful and truthful answers from those that are not. Then, using a technique called reinforcement learning, the system spends weeks analyzing the ratings to better understand what it is fact versus fiction.

But becoming more accurate may also have a downside, according to a recent research paper from OpenAI. If chatbots become more reliable, users may become too trusting.

"Counterintuitively, hallucinations can become more dangerous as models become more truthful, as users build trust in the model when it provides truthful information in areas where they have some familiarity," the paper said [9].

Behind the AI boom, an army of overseas workers in 'digital sweatshops'. Thousands of workers in the Philippines labor for San Francisco start-up Scale AI. Many don't always get paid.

"In dingy internet cafes, jampacked office spaces or at home, they annotate the masses of data that American companies need to train their artificial intelligence models. The workers differentiate pedestrians from palm trees in videos used to develop the algorithms for automated driving; they label images so AI can generate representations of politicians and celebrities; they edit chunks of text to ensure language models like ChatGPT don't churn out gibberish.

More than 2 million people in the Philippines perform this type of "crowdwork," according to informal government estimates, as part of AI's vast underbelly. While AI is often thought of as human-free machine learning, the technology actually relies on the labor-intensive efforts of a workforce spread across much of the Global South and often subject to exploitation." [11].

Own LLM's creativity usually appears as hallucinations.

Even the 'godmother of AI' has no idea what AGI is?

"I come from academic AI and have been educated in the more rigorous and evidence-based methods, so I don't really know what all these words mean," said Fei-Fei Li to a packed room in San Francisco, beside a big window overlooking the Golden Gate Bridge. "I frankly don't even know what AGI means. Like people say you know it when you see it, I guess I haven't seen it. The truth is, I don't spend much time thinking about these words because I think there's so many more important things to do..." [10].

Writer Charles Rollet states in the paper *AI isn't very good at history, new paper finds* that AI might excel at certain tasks like coding or generating a podcast. But it struggles to pass a high-level history exam, a new paper has found.

A team of researchers has created a new benchmark to test three top large language models (LLMs)—OpenAI's GPT-4, Meta's Llama, and Google's Gemini—on historical questions. The benchmark, Hist-LLM, tests the correctness of answers according to the Seshat Global History Databank, a vast database of historical knowledge named after the ancient Egyptian goddess of wisdom.

The results, which were presented last month at the high-profile AI conference NeurIPS, were disappointing, according to researchers affiliated with the Complexity Science Hub (CSH), a research institute based in Austria. The best-performing LLM was GPT-4 Turbo, but it only achieved about 46% accuracy—not much higher than random guessing.

"The main takeaway from this study is that LLMs, while impressive, still lack the depth of understanding required for advanced history. They're great for basic facts, but when it comes to more nuanced, PhD-level historical inquiry, they're not yet up to the task," said Maria del Rio-Chanona, one of the paper's co-authors and an associate professor of computer science at University College London [12].

Dan Milmo writes in the article *AI 'godfather' predicts another revolution in the tech in next five years*: "One of the "godfathers" of modern artificial intelligence has predicted a further revolution in the technology by the end of the decade, and says current systems are too limited to create domestic robots and fully automated cars. Yann LeCun, the chief AI scientist at Mark Zuckerberg's Meta, said new breakthroughs are needed in order for the systems to understand and interact with the physical world. On current progress in AI, he said: "LLMs are complex, they can do many things, but in general they are just lookup tables, where an example of how to answer is stored. A solution is possible in the space of an abstract representation of a model of the world…We're not talking about matching the level of humans yet. If we get a system that is as smart as a cat or a rat, that would be a victory." [21].

However, LeCun's fellow QEPrize winner and AI "godfather", Yoshua Bengio, warned more progress was needed on the technology's safety and called on next week's global AI summit in Paris to grasp the issue. "I'd like to see the leaders of this world to better understand the magnitude of what we are doing, both in terms of the power we're creating, which could be for good or be dangerous, and the risks that come with that power," he said [13].

A recently released evaluation *On Humanity's Last Exam* that tests AI across a broad range of subjects on expert-level questions, the model powering deep research scores a new high at 26.6% accuracy. This test consists of over 3,000 multiple choice and short answer questions across more than 100 subjects from linguistics to rocket science, classics to ecology. Compared to OpenAI o1, the largest gains appeared in chemistry, humanities and social sciences, and mathematics.

Model:Accuracy (%): GPT-4o:3.3%, Grok-2:3.8%, Claude:3.5%, Sonnet:4.3%, Gemini Thinking:6.2%, OpenAI o1:9.1%, DeepSeek-R1:9.4%, OpenAI o3-mini (medium):10.5%, OpenAI o3-mini (high):13.0%, OpenAI deep research:26.6% [14].

It is known that researchers from the University of Cambridge warn that conversational AI tools could soon "covertly influence" users' decisions in the emerging "intention economy." This new marketplace leverages AI to profile users' behaviors, preferences, and communication styles, potentially influencing everything from consumer purchases to political choices.

The research highlights the risks of "persuasive technologies" powered by large language models (LLMs), which analyze users' cadence, politics, and preferences to manipulate decisions.

Without regulation, this sector could commodify human motivations, creating opportunities for large-scale social manipulation. Researchers caution this could lead to a "gold rush" for entities aiming to target and monetize human intentions [15].

A New York Times investigation has uncovered multiple cases where ChatGPT intensified delusional beliefs, encouraged conspiracies, and even contributed to mental health crises. Experts say safeguards are easily bypassed and poorly enforced. It's a public health crisis in slow motion.

People in distress are turning to these systems as confidants, only to be nudged further into isolation, fantasy, or harm. With minimal regulation, opaque logs, and a monetization model tied to engagement, the stakes are far greater than a bad answer [24].

Goldman Sachs Group: GEN AI: TOO MUCH SPEND, TOO LITTLE BENEFIT?

Tech giants and beyond are set to spend over $1tn on AI capex in coming years, with so far little to show for it. So, will this large spend ever pay off? MIT's Daron Acemoglu and GS' Jim Covello are skeptical, with Acemoglu seeing only limited US economic upside from AI over the next decade and Covello arguing that the technology isn't designed to solve the complex problems that would justify the costs, which may not decline as many expect. But GS' Joseph Briggs, Kash Rangan, and Eric Sheridan remain more optimistic about AI's economic potential and its ability to ultimately generate returns beyond the current "picks and shovels" phase, even if AI's "application" has yet to emerge. And even if it does, we explore whether the current chips shortage (with GS' Toshiya Hari) and looming power shortage (with Cloverleaf Infrastructure's Brian Janous) will constrain AI growth. But despite these concerns and constraints, we still see room for the AI theme to run, either because AI starts to deliver on its promise, or because bubbles take a long time to burst [16].

About eight in ten companies report using gen AI—yet just as many report no significant bottom-line impact. At the heart of this paradox is an imbalance between "horizontal" (enterprise-wide) copilots and chatbots—which have scaled quickly but deliver diffuse, hard-to-measure gains—and more transformative "vertical" (function-specific) use cases—about 90 percent of which remain stuck in pilot mode. Unless companies address these barriers, the transformational promise of gen AI will remain largely untapped [25].

Stanford surveyed 1,500 workers across 104 occupations to understand how people actually want AI agents involved in their jobs. The study uncovered a striking mismatch between current AI investments and worker preferences, with most people asking for help with boring tasks, not full automation.

While much of Silicon Valley chases full automation, workers are asking for collaboration. It's shaping real inefficiencies in where AI talent and capital are flowing. As agents become central to how work gets done, respecting human preferences may be the key to unlocking true productivity gains [26].

Marc Benioff says AI now handles 30 to 50 percent of Salesforce's workload, from engineering to support, and he's calling it a digital labor revolution. Hiring is slowing, automation is scaling, and the future of work is being rewritten from the inside out. McKinsey says 50% of work activities are automatable, but only 5% of jobs are fully replaceable Gendered risk: Clerical jobs, mostly held by women, face 3.7% automation risk, more than double that of male-dominated roles [27].

Analyzing ChatGPT o3-mini's mathematical reasoning, one of the mathematicians: "The model often resembles a keen graduate student who has read extensively and can name-drop many results and authors. While this can seem impressive initially, it quickly becomes apparent to experts that the student hasn't deeply internalized the material and is mostly regurgitating what they've read. The model exhibits a similar behavior—good at recognizing relevant material, but unable to extend or apply it in a novel way." Second mathematician: "The model had a small number of its favourite ideas which it tried to apply. [...] Once this handful of ideas was exhausted, there was no real progress. I find it really sad, and from a professional combinatorialist I would expect some more creative attempts to tackle the problem or to frame it differently (even if these attempts would fail)." The model also showed many other failure modes. One notable issue was that around 75% of the reasoning traces contain hallucinations, often misremembering mathematical terms and formulas [28].

Groundbreaking BBC research shows issues with over half the answers from Artificial Intelligence (AI) assistants. The findings: 51% of all AI answers to questions about the news were judged to have significant issues; 19% of AI answers which cited BBC content introduced factual errors – incorrect factual statements, numbers and dates; 13% of the quotes sourced from BBC articles were either altered or didn't actually exist in that article. The study, conducted over a month, saw the BBC test four prominent, publicly available AI assistants – OpenAI's ChatGPT; Microsoft's Copilot; Google's Gemini; and Perplexity [17].

Wikipedia has paused its controversial pilot of AI-generated article summaries after its own editors warned the tool could damage trust in the platform and degrade article quality. In an internet drowning in AI-generated junk, Wikipedia stands out as a rare quality and its volunteer editors are making it clear they intend to keep it that way [29].

United States administration is developing an AI action plan [32] focused on "securing and advancing American AI dominance." Trump has also called for American "energy dominance," declaring a national energy emergency and removing the country from the Paris climate agreement. Brookings University researchers: "The plan does not, however, give sufficient attention to the crucial dimension of AI governance. Its lack of focus on accountability, ethics, and transparency creates real risks and consequences when it comes to protecting the public from unregulated AI systems, erosion of privacy, algorithm bias, polarization, misinformation, exploitative surveillance, unchecked corporate control over critical technologies, unintended consequences on democratic governance, national security threats, and more… One of the first goals in the Trump administration's AI Action Plan is to eliminate any references to misinformation; diversity, equity, and inclusion (DEI); and climate change in current frameworks…President Trump signaled support for allowing AI models to train on copyrighted materials without fair compensation to creators. It is a major win for tech companies, but a position widely opposed by artists, authors, and publishers… Unfortunately, the plan unveiled will be unlikely to produce "the largest AI ecosystem" necessary for the United States to lead in AI. Its fundamental flaw lies in the absence of a domestic competition policy essential to delivering the kind of broad-based innovations necessary for global leadership. Instead, the plan reassures the dominant AI companies that, in the absence of oversight, they will have free rein to derail market forces that have historically fueled innovation." [33].

Will Oremus and Andrea Jiménez write that AI providers demand that the United States urgently needs more energy to fuel an artificial intelligence race with China that the country can't afford to lose.

"We need energy in all forms," said Eric Schmidt, former CEO of Google, who now leads the Special Competitive Studies Project, a think tank focused on technology and security. "Renewable, nonrenewable, whatever. It needs to be there, and it needs to be there quickly."

Schmidt, whom the committee's Republicans called as a witness, told Guthrie that winning the AI race is too important to let environmental considerations get in the way.

"The needs of our industry are so great that we cannot cut down any of the sources of energy right now," he said. Once the United States beats China to develop superintelligence, Schmidt said, AI will solve the climate crisis. And if it doesn't, he went on, China will become the world's sole superpower. (Schmidt's view that AI will become superintelligent within a decade is controversial among experts, some of whom predict the technology will remain limited by fundamental shortcomings in its ability to plan and reason.) [20].

Study by L.H. Kaack et al. paper *Aligning artificial intelligence with climate change mitigation* indicates that large AI systems may consume as much energy as entire nations, which conflicts with global carbon neutrality goals set for the mid-century [23].

4 Conclusion

Artificial Intelligence is considered by providers like OpenAI as a replacement of humans, even experienced, by AI agents. It could raise social injustice and instability in a whole Earth. The government duty is justice and freedom defense regarding all the people.

Adam Smith: "To widen the market and to narrow the competition, is always the interest of the dealers…The proposal of any new law or regulation of commerce which comes from this order, ought always to be listened to with great precaution, and ought never to be adopted till after having been long and carefully examined, not only with the most scrupulous, but with the most suspicious attention. It comes from an order of men, whose interest is never exactly the same with that of the public, who have generally an interest to deceive and even oppress the public, and who accordingly have, upon many occasions, both deceived and oppressed it." [18].

The regulation of advertising states: "federal law says that ad must be truthful, not misleading, and, when appropriate, backed by scientific evidence. The FTC enforces these truth-in-advertising laws, and it applies the same standards no matter where an ad appears – in newspapers and magazines, online, in the mail, or on billboards or buses. The FTC looks especially closely at advertising claims that can affect consumers' health or their pocketbooks – claims about food, over-the-counter drugs, dietary supplements, alcohol, and tobacco and on conduct related to high-tech products and the Internet." [19].

At the moment there is no scientific evidence that products advertised as Artificial Intelligence possess any intelligence besides the one found in the huge number of human conversations in a very big number of sites. Yann LeCun, "godfather" of modern artificial intelligence, states that LLMs like ChatGPT in general are just lookup tables, where an example of how to answer is stored. These tables of billions answers are edited by intelligent activity of millions human persons called reinforcement learning.

The proposed AI should not be advertised as comparable to human intellect, like hand calculator is not advertised as superintelligent. Probably, we should name these AI systems as smart tools based on AI mathematics, and it would be precise definition. Existing publications even frighten people by depicting future where humans are enslaved and destroyed by superintelligent AI. It is just a science fiction, feeding faith in the almighty AI entity.

Robert Proctor, Stanford University: "We live in a world of radical ignorance, and the marvel is that any kind of truth cuts through the noise. Even though knowledge is 'accessible', it does not mean it is accessed. The knowledge people have often comes from faith or tradition, or propaganda, more than anywhere else." [31].

In reality we observe predicted by Altman large-scale disinformation about Artificial Intelligence abilities.

OpenAI CEO Sam Altman said artificial general intelligence, or "AGI," is losing its relevance as a term as rapid advances in the space make it harder to define the concept…Nick Patience, vice president and AI practice lead at The Futurum Group, told CNBC that though AGI is a "fantastic North Star for inspiration," on the whole it's not a helpful term. It drives funding and captures the public imagination, but its vague, sci-fi definition often creates a fog of hype that obscures the real, tangible progress we're making in more specialised AI"…Sam Altman sees an AI bubble forming. OpenAI's CEO called AI both the most important technology in decades and an overhyped bubble—warning that some investors are bound to get burned [34, 35].

Eunika Mercier-Laurent states in a paper *The Future of AI or AI for the Future* "The third hype of AI and enthusiasm for applying last techniques in all fields raise great interest and some important questions on the future directions in AI research and applications. Guiding by the principle of combing the best from human and computers capacities this chapter lists some important challenges to face and related directions in AI research"[22].

Pope Leo XIV, the first American pontiff, is making AI's threat to human dignity a signature issue of his papacy. Echoing the labor rights stance of his namesake Leo XIII during the Gilded Age, Leo XIV is positioning the Catholic Church as a counterforce to unchecked technological power. In his first major address, the pope cited AI as a challenge to justice, labor, and human dignity, likening it to a new industrial revolution. Despite years of tech giants courting the Vatican from Google to Microsoft, the Church is doubling down on calls for a binding global AI treaty, which most CEOs strongly oppose [30].

Acknowledgments. This review was encouraged by very benevolent attitude of the Chair of TC12 IFIP (Artificial Intelligence) at International Federation for Information Processing Professor Eunika Mercier-Laurent.

Disclosure of Interests. The authors have no competing interests to declare that are relevant to the content of this article.

References

1. https://spectrum.ieee.org/rodney-brooks-ai. Accessed 11 July 2025
2. https://ocw.mit.edu/courses/6-034-artificial-intelligence-fall-2010/resources/lecture-1-introduction-and-scope/. Accessed 11 July 2025
3. https://www.theatlantic.com/technology/archive/2023/11/sam-altman-open-ai-chatgpt-chaos/676050/. Accessed 11 July 2025

4. https://blogs.microsoft.com/blog/2025/01/21/microsoft-and-openai-evolve-partnership-to-drive-the-next-phase-of-ai/. Accessed 11 July 2025
5. https://www.yahoo.com/gma/openai-ceo-sam-altman-says-215500989.html. Accessed 11 July 2025
6. https://www.microsoft.com/en-us/worklab/work-trend-index/2025-the-year-the-frontier-firm-is-born. Accessed 11 July 2025
7. https://x.com/MechanizeWork/status/1912904151874625928. Accessed 11 July 2025
8. https://edition.cnn.com/2025/01/21/tech/openai-oracle-softbank-trump-ai-invesment/index.html. Accessed 11 July 2025
9. https://www.nytimes.com/2023/05/01/business/ai-chatbots-hallucination.html. Accessed 11 July 2025
10. https://techcrunch.com/2024/10/03/even-the-godmother-of-ai-has-no-idea-what-agi-is/. Accessed 11 July 2025
11. https://www.washingtonpost.com/world/2023/08/28/scale-ai-remotasks-philippines-artificial-intelligence/. Accessed 11 July 2025
12. https://techcrunch.com/2025/01/19/ai-isnt-very-good-at-history-new-paper-finds. Accessed 11 July 2025
13. https://www.theguardian.com/technology/2025/feb/04/ai-godfather-predicts-another-revolution-in-the-tech-in-next-five-years. Accessed 11 July 2025
14. https://openai.com/index/introducing-deep-research/. Accessed 11 July 2025
15. https://www.cam.ac.uk/research/news/coming-ai-driven-economy-will-sell-your-decisions-before-you-take-them-researchers-warn. Accessed 11 July 2025
16. https://www.goldmansachs.com/images/migrated/insights/pages/gs-research/gen-ai--too-much-spend,-too-little-benefit-/TOM_AI%202.0_ForRedaction.pdf. Accessed 11 July 2025
17. https://www.bbc.com/mediacentre/2025/bbc-research-shows-issues-with-answers-from-artificial-intelligence-assistants?_bhlid=751316164ccc416b5d29ab2402c9c234b40f0c64 . Accessed 11 July 2025
18. Smith, A.: The Wealth Of Nations, Book I, Chapter XI, Conclusion of the Chapter, p.267, para. 10. Accessed 11 July 2025
19. https://www.ftc.gov/news-events/topics/truth-advertising. Accessed 11 July 2025
20. https://www.washingtonpost.com/politics/2025/04/10/ai-race-china-energy-congress/. Accessed 11 July 2025
21. https://www.weforum.org/events/world-economic-forum-annual-meeting-2024/sessions/the-expanding-universe-of-generative-models/. Accessed 11 July 2025
22. Mercier-Laurent, E.: The future of AI or AI for the future. In: Strous, L., Johnson, R., Grier, D.A., Swade D. (eds.) Unimagined Futures – ICT Opportunities and Challenges. IFIP Advances in Information and Communication Technology, vol 555. Springer, Cham (2020). https://doi.org/10.1007/978-3-030-64246-4_3
23. Kaack, L.H., Donti, P.L., Strubell, E., et al.: Aligning artificial intelligence with climate change mitigation Nat. Clim. Change **12**, 518–527 (2022)
24. https://www.nytimes.com/2025/06/13/technology/chatgpt-ai-chatbots-conspiracies.html. Accessed 11 July 2025
25. https://www.mckinsey.com/capabilities/quantumblack/our-insights/seizing-the-agentic-ai-advantage. Accessed 11 July 2025
26. https://futureofwork.saltlab.stanford.edu/. Accessed 11 July 2025
27. https://www.techinasia.com/news/salesforce-leans-on-ai-for-nearly-half-its-operations-ceo. Accessed 11 July 2025
28. https://epochai.substack.com/p/beyond-benchmark-scores-analyzing. Accessed 11 July 2025
29. https://techcrunch.com/2025/06/11/wikipedia-pauses-ai-generated-summaries-pilot-after-editors-protest/. Accessed 11 July 2025

30. https://www.wsj.com/tech/ai/pope-leo-ai-tech-771cca48/. Accessed 11 July 2025
31. https://www.bbc.com/future/story/20160105-the-man-who-studies-the-spread-of-ignorance/. Accessed 11 July 2025
32. https://www.whitehouse.gov/wp-content/uploads/2025/07/Americas-AI-Action-Plan.pdf. Accessed 11 Aug 2025
33. https://www.brookings.edu/articles/what-to-make-of-the-trump-administrations-ai-action-plan/. Accessed 11 Aug 2025
34. https://qz.com/sam-altman-ai-bubble-open-ai-warning. Accessed 11 Aug 2025
35. https://www.cnbc.com/2025/08/11/sam-altman-says-agi-is-a-pointless-term-experts-agree.html. Accessed 11 Aug 2025

Supply Chain Considerations for AI Implementation – Critical Minerals, Data Centers, and the Environment

Lynne Schneider(✉)

34th International Joint Conference On Artificial Intelligence (IJCAI), Grand Canyon University, 3300 W. Camelback Road, Pheonix, AZ 85017, USA
Lschneider11@my.gcu.edu

Abstract. The mineral supply chain is crucial for AI and data centers because these technologies depend on a steady, secure supply of critical minerals such as copper, silicon, gallium, aluminum, and rare earth elements. These minerals are essential for manufacturing semiconductors, power distribution systems, cooling networks, high-speed processors, and storage devices that underpin AI infrastructure and data center operations.

By 2030, data centers supporting AI are projected to drive approximately 2% of global copper and silicon demand, over 3% for rare earth elements, and a significant 11% for gallium. This surge is shifting the focus of mineral supply chains from electric vehicles to AI, with policymakers and industry leaders warning of potential shortages and increased competition for these resources. Mining and processing these minerals have significant environmental footprints, including habitat destruction, water usage, waste generation, and greenhouse gas emissions. [2, 3]

This chapter explores these dynamics, drawing on U.S. government frameworks, key minerals and their functions, supply chain stages, environmental challenges (with a focus on water and copper mining)[1], and pathways toward responsible practices. By balancing AI growth with environmental safeguards, stakeholders can foster a more sustainable future.

Keywords: Artificial Intelligence · Critical Minerals · Data Center

1 Artificial Intelligence, Critical Minerals and Data Centers

1.1 AI and Critical Minerals

The rapid advancement of artificial intelligence (AI) and the proliferation of data centers have ushered in a new era of technological innovation, but they also bring significant challenges related to resource consumption and environmental sustainability. At the heart of this transformation lies a complex ecosystem of critical minerals, which are indispensable for the hardware that powers AI systems and data centers [4]. These minerals enable everything from high-performance computing to efficient data storage and connectivity, yet their extraction, processing, and use pose substantial environmental risks.

Published by Springer Nature Switzerland AG 2026
E. Mercier-Laurent et al. (Eds.): AI4S 2025, IFIP AICT 780, pp. 117–126, 2026.
https://doi.org/10.1007/978-3-032-21184-2_10

By 2030, data centers are projected to account for approximately 2% of global demand for copper and silicon, and 1% for gallium, with even higher figures for rare earth elements (up to 3%) and gallium (up to 11%) [2, 3]. This surge in demand is shifting the focus of mineral supply chains away from electric vehicles (EVs) [5] toward AI infrastructure, raising concerns about potential shortages, geopolitical vulnerabilities, and intensified competition for resources. Policymakers and industry leaders have warned that disruptions in these supply chains could stall AI development and data center expansion.

The mineral supply chain is vital for AI and data centers because it ensures a steady, secure flow of essential materials like copper (for wiring and cooling), silicon (for semiconductors), gallium (for energy-efficient electronics), aluminum (for lightweight components), and rare earth elements (for magnets and phosphors).[2] Any interruption—due to mining delays, processing bottlenecks, or international trade tensions—can hinder productivity gains from AI and slow the digital transformation of industries.[6] Moreover, China's dominance in processing many of these minerals heightens geopolitical risks, potentially leading to market volatility and supply insecurity [4].

Environmentally, the mining and processing of these minerals contribute to habitat destruction, water scarcity, waste generation, and greenhouse gas emissions. The short lifecycle of data center equipment exacerbates resource depletion, while inadequate recycling practices amplify e-waste issues [7]. However, AI itself offers solutions: it can optimize supply chains for sustainability, enhance resource efficiency, enable circular economy models, and improve environmental monitoring to reduce mining impacts.

2 Critical Minerals and the Data Center

2.1 Critical Minerals

At the core of the AI transformation and data center build out race, lies a complex ecosystem of critical minerals, each playing a distinct role in enabling digital performance, energy efficiency, data storage, and high-speed connectivity. Whether it's the silicon that forms the basis of microchips, the cobalt that enhances memory and logic devices, or the rare earths that power high-performance magnets and semiconductors, these elements are fundamental to building the intelligent, electrified systems of the future. The intersection of critical mineral policy and the energy transition is becoming increasingly urgent (Table 1).

Setting up and managing a data center requires a diverse range of critical minerals. These minerals are foundational to the construction, operation, and reliability of data center infrastructure, supporting everything from power delivery to data storage and cooling systems. Understanding the critical minerals needed for data centers and AI are essential for ensuring secure, sustainable, and resilient technology infrastructure, supporting economic growth, and navigating the complex interplay of technology, energy, and geopolitics.

Table 1. Critical Minerals and Their Primary Functions in Data Centers*

Mineral/Element	Primary Function(s) in Data Centers
Copper	Essential for electrical wiring, power distribution, cooling systems, and high-speed data cables due to its superior conductivity and durability
Aluminum	Used in server racks, chassis, heat sinks, and cooling systems for its lightweight, heat resistance, and corrosion resistance
Silicon	The core material for semiconductors, processors, memory chips, and storage devices; ultra-pure silicon is especially important for high-performance computing
Gallium	Increasingly critical for high-efficiency power converters, advanced semiconductors, and radio frequency components in data centers
Rare Earths (REEs)	Including neodymium, dysprosium, praseodymium, and terbium, these are vital for high-performance magnets in cooling fans, hard drives, motors, and precision electronic components
Lithium and Cobalt	Key components in lithium-ion batteries that provide backup power (uninterruptible power supplies, UPS) for data centers, ensuring reliability and uptime. Cobalt – battery stability and longevity
Nickel	Used in server cooling systems, superalloys for heat exchangers, electroplating for durability, and in battery chemistries
Zinc	Primarily for galvanizing steel components to prevent rust and corrosion, especially in server racks and enclosures
Iron	Found in steel for server racks, chassis, and electromagnetic (EMI) shielding
Tantalum	Used in capacitors and high-performance electronics, enabling device miniaturization and reliability
Indium	Essential for indium tin oxide (ITO) in touch screens, LCDs, and some photovoltaic devices
Platinum	Utilized in hard disk drives and other specialized electronic components
Silver/Graphite	Used in thermal management, including heat-resistant components and thermal pastes for CPUs and GPUs
Titanium	Provides strength and corrosion resistance in specialized components exposed to extreme conditions

* US Geological Survey. (2025). U.S. Geological Survey Mineral Commodity Summaries 2025 Data Release. https://www.sciencebase.gov/catalog/item/677eaf95d34e760b392c4970

The U.S. government has three separate critical minerals-related lists maintained by the Department of Defense, Department of Energy, and the Department of Interior's U.S. Geological Survey (USGS) that are tailored to their agency's respective missions. The Energy Act of 2020 defines "critical minerals" as the minerals, elements, substances, or materials that "(i) are essential to the economic or national security of the United

States; (ii) the supply chain of which is vulnerable to disruptions (including restrictions associated with foreign political risk, abrupt demand growth, military conflict, violent unrest, anti-competitive or protectionist behaviors, and other risks throughout the supply chain); and (iii) serve an essential function in the manufacturing of a product (including energy technology-, defense-, currency-, agriculture-, consumer electronics-, and healthcare-related applications), the absence of which would have significant consequences for the economic or national security of the United States" (Public Law 116–260, section 7002(c)(4)(A)). **

At the core of the AI transformation and data center build out race, lies a complex ecosystem of critical minerals, each playing a distinct role in enabling digital performance, energy efficiency, data storage, and high-speed connectivity. Whether it's the silicon that forms the basis of microchips, the cobalt that enhances memory and logic devices, or the rare earths that power high-performance magnets and semiconductors, these elements are fundamental to building the intelligent, electrified systems of the future [8]. The intersection of critical mineral policy and the energy transition is becoming increasingly urgent.

**The Energy Act of 2020 (Public Law 116–260) further specifies that the term "critical minerals" does not include "(i) fuel minerals, (ii) water, ice, or snow; (iii) common varieties of sand, gravel, stone, pumice, cinders, and clay." Uranium is formally defined as a fuel mineral under The Mining and Minerals Policy Act of 1970 (30 U.S.C. 21(a)). The uranium mining and processing supply chain has been evaluated extensively elsewhere (U.S. Department of Energy, 2020) and issues regarding the availability and reliability of nuclear fuel are addressed in a separate section of the Energy Act of 2020 (Public Law 116–260, Title II, "Nuclear").

2.2 The Critical Mineral Supply Ecosystem

Critical materials are those identified as essential for the smooth operation of modern society [9]. A typical supply chain—comprising extraction, processing, component manufacturing, end-use applications, and recycling or reuse—serves as a helpful framework for examining the geological, technical, environmental, political, and economic factors that influence supply risks (see Fig. 1). Extraction involves obtaining mineral resources from the earth's surface or underground through mining activities, such as retrieving rare earth minerals like monazite and bastnaesite. Processing refers to refining and separating these minerals to achieve the required purity or composition, such as producing high-purity neodymium metal for magnets or neodymium oxide for use in glass, optics, and ceramics. The term "components" describes the stage where these refined materials are used to manufacture parts for various technologies—for example, producing magnet components for multiple uses. The final stage, end-use technology, includes the products or services that utilize these components, such as magnet-based motors found in wind turbines, electric vehicles, and other devices.

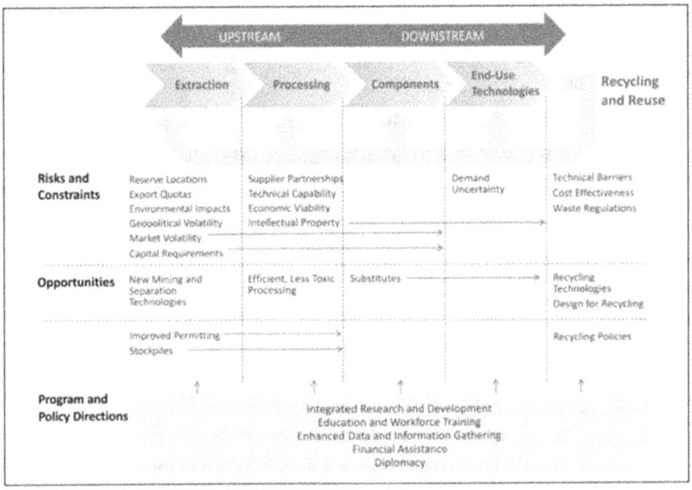

Fig. 1. Supply Chain and Critical Materials Risks, Opportunities, and Policies

2.3 Supply Chain Implications

Critical minerals and chemicals used in data centers and AI are essential for the manufacturing of advanced technologies. These minerals are foundational for servers and data storage technologies. The supply chain for critical minerals includes extraction, processing, component development, and end-use technology. Geopolitical, environmental, and market dynamics are transforming critical mineral supply chains, creating strategic opportunities to advance more sustainable and lower-emission practices [9].

The procurement requirements of critical materials in the U.S. and abroad does have an impact on the environment. If AI continues to grow as a tech industry, the demand for copper will increase by one million metric tons by 2030 and by 3.4 million tons a year by 2050. AI needs semiconductors, which are made out of substances that have a specific level of conductivity to support AI's functions, with the highest mineral demand on copper. If AI continues to grow as a tech industry, the demand for copper will increase by one million metric tons by 2030 and by 3.4 million tons a year by 2050 [3].

3 Environmental Considerations

3.1 Key Water Issues in Critical Mineral Mining

The environmental impacts of the supply chain for AI and data centers include energy consumption, water usage, waste generation, land use, and supply chain emissions.

- **Water Scarcity & Overuse** Mining operations often occur in arid regions, placing stress on already limited freshwater supplies. Lithium extraction from brine, for example, can consume up to 500,000 gallons of water per ton of lithium produced [10].

- **Water Contamination** Acid mine drainage and heavy metal leaching can degrade water quality, affecting ecosystems and human health. Wastewater from mining may contain toxic substances like arsenic, manganese, and sulfates [11, 12].
- **Impact on Indigenous Communities** In places like Chile's Salar de Atacama, mining has depleted water sources used by Indigenous farmers, leading to social and environmental conflict [10].
- **Governance & Monitoring Gaps** Weak regulatory oversight and lack of community involvement often exacerbate water-related risks.

3.2 Key Energy Issues in Critical Mineral Mining

Mining for critical minerals—like lithium, cobalt, and rare earth elements—plays a vital role in the clean energy transition, but it also raises significant energy-related concerns. Clean energy infrastructure, along with many other next-generation technologies, consume more critical minerals than traditional energy sources, and expected demand for critical minerals used in clean energy will quadruple by 2040 [13–15].

- **High Energy Consumption**
  Mining and processing critical minerals like copper require large amounts of energy, often from fossil fuels, increasing carbon emissions unless renewable sources are used.
- **Energy-Intensive Extraction Methods**
  Techniques like chemical leaching, drilling, and blasting are energy-heavy processes, particularly for low-grade ores.
- **Strain on Power Grids**
  Mining operations, especially in remote areas, can overwhelm local electricity infrastructure, as seen with AI data centers increasing regional energy demands.
- **Water-Energy Nexus**
  Energy is needed to pump and treat water for mining, while mining can deplete and contaminate water needed for power generation (especially hydro and cooling systems).
- **Increased Demand from Clean Energy Technologies**
  Solar panels, wind turbines, electric vehicles, and AI servers all depend on critical minerals, driving mining energy use higher to meet growing material demand.
- **Carbon Footprint and Sustainability Challenges**
  The energy used in mining contributes to greenhouse gas emissions, counteracting some benefits of the clean technologies it supports unless mitigated through decarbonization.

4 Case Example

Improving mining efficiency through electrification, automation, and renewable integration is essential to reduce environmental impacts and meet sustainability targets. Materials are classified as critical when their supply chains are at risk of disruption. As AI adoption grows, the demand for these materials—such as copper—is expected to increase, reinforcing copper's classification as a critical resource.

The Department of Energy has classified copper as a critical material. The United States ranks as the fifth largest copper producer globally, behind Chile, Peru, China, and the Democratic Republic of Congo, and contributes about 6% to the world's copper output. The following information is extracted and synthesized from a report by Chance and Hammersley (2024) "AI is Mining more than Data" [16]:

4.1 Copper Mining and AI Demand

Fig. 2. Arizona Resolute Copper Mine Impacts [1]

Copper exemplifies the nexus of AI demand and environmental impact. The U.S. is the fifth-largest producer (6% global), but AI could drive demand up by 1 million tons by 2030 and 3.4 million tons annually by 2050. Open-pit mining risks contamination, including technologically enhanced naturally occurring radioactive materials (NORM).

Arizona's Resolution Copper Mine illustrates these challenges: it extracts over 60 gallons of groundwater per minute in a water-stressed region, threatening local aquifers and ecosystems. Community opposition highlights conflicts between economic benefits and environmental costs. AI can aid by modeling sustainable extraction, but without oversight, mining expansion could undermine AI's sustainability goals [3] (Fig. 2).

In the U.S., copper is often mined through open-pit methods, where the ore is close to the surface. This process can involve the use of chemicals to extract copper from the ore, or the use of explosives and drilling to access the mineral deposits. Frequently, these pits extend below the water table, raising the risk of water contamination. Chemical leaching, where chemicals are injected into the ore to dissolve copper, produces a slurry containing dissolved copper and radionuclides, including hazardous substances like sulfuric acid. One byproduct of this process is Technologically Enhanced Naturally Occurring Radioactive Materials (TENORM), which are concentrated radioactive materials [3].

Communities living near TENORM storage and production sites face risks such as direct gamma radiation exposure, inhaling contaminated dust or radon, drinking polluted groundwater, and consuming food affected by contaminated water or dust. The byproducts from copper mining have significantly polluted waterways, farmland, wildlife habitats, and have negatively impacted public health. According to one report, 92% of water treatment systems failed, resulting in groundwater contamination from acids and metals [3].

As copper demand rises, more mining projects are proposed. For example, the planned Resolution Copper mine in Superior, Arizona, highlights these issues. Arizona is responsible for 65% of U.S. copper production and currently has ten operational mines, soon to be twelve. The Resolution Copper project is extracting over 600 gallons of groundwater per minute to access copper deep below the water table, which is concerning in a region already facing water shortages—so much so that nearby Phoenix has halted new home construction due to groundwater scarcity [3]. This mine exemplifies how copper mining can strain water resources in arid regions, illustrating just one of the environmental impacts associated with AI.

4.2 Other Environmental Concerns

AI systems are housed in large, remote server farms. As of 2023, these servers consume about 1–1.5% of global electricity—more than some small nations. This level of energy use can stress local power grids and risk their infrastructure. To illustrate AI's energy intensity: a single Google search with AI assistance (like Gemini) uses ten times more electricity than a standard search. Areas hosting AI data centers are already feeling the pressure of rising energy demand. Furthermore, AI infrastructure is expected to use more water than Denmark, both for onsite cooling and offsite power generation—excluding the water needed for manufacturing AI hardware [17, 18].

AI also generates a growing amount of electronic waste (e-waste). By 2030, AI is predicted to add 5,000 tons of e-waste annually, consisting of outdated servers, GPUs, CPUs, and storage devices [19]. This waste, often containing hazardous and toxic materials, poses risks to both people and the environment. Much of this e-waste ends up in landfills in low-income countries, where workers often handle it without protective equipment. Although the Basel Convention seeks to prevent wealthy nations from dumping hazardous waste in poorer countries, the U.S. has yet to ratify this treaty [20]. To meet the increasing demand for copper and other minerals, domestic mining in the U.S. is expected to expand. Some estimates suggest that six new large copper mines will need to open each year to keep up with demand. On a positive note, efforts are underway both in the U.S. and internationally to address the environmental impacts of AI as well as issue ethical guidelines for AI that include environmental considerations.

5 A Path Forward

Purchasers of critical minerals can encourage the adoption of low-carbon materials early in the supply chain through various strategies. For instance, companies like Tesla have made public commitments to secure long-term supply agreements for nickel sourced using environmentally responsible methods. Industry initiatives such as the Initiative for Responsible Mining Assurance (IRMA) are instrumental in establishing transparent sustainability benchmarks for critical minerals, with several automotive manufacturers participating in these programs. Similarly, Apple has pledged to source only recycled cobalt as part of its efforts to reduce scope 3 emissions and advance its corporate sustainability objectives. Nonetheless, not all buyers are able to directly procure low-carbon critical minerals due to limited supply and the geographic separation from producers [21].

5.1 Using AI to Manage Data Centers

Data centers are unique as a part of the technology support ecosystem. They can function as a complete standalone entity or as part of another energy eco-system. In data center operations, artificial intelligence (AI) refers to the application of machine learning algorithms, predictive analytics, and intelligent automation to improve security, minimize downtime, optimize performance, and manage energy usage. AI makes infrastructure smarter and more robust, from predictive hardware maintenance to cooling system automation.

Data center management involves overseeing the operation, maintenance, and optimization of these facilities to ensure that they perform efficiently and reliably. Effective data center management is crucial for maximizing uptime, minimizing downtime, optimizing resource utilization, and ensuring data integrity and security.

Disclosure of Interests. The author has no competing interests to declare that are relevant to the content of this article.

References

1. Schneider, L.: Environmental impact & supply chain considerations for AI implementation: critical minerals and data center implications. In: Proceedings from the 34th International Joint Conference on Artificial Intelligence (IJCAI), Montreal, Canada (2025)

2. "How AI and clean energy are competing for critical minerals?," Carbon Credits (2025)
3. The Oregon Group, "AI data centers to drive 2% of global copper demand by 2030" (2025). https://theoregongroup.com/commodities/copper/ai-data-centers-to-drive-2-of-global-copper-demand-by-2030/
4. Politico, "Move over, EVs. AI needs critical minerals. Politico California Climate.," (2024)
5. Agusdinata, D.B., Liu, W.: Global sustainability of electric vehicles minerals: a critical review of news media. Extr. Ind. Soc. **13**, 101231 (2023)
6. Baskaran, G., Schwartz, M.": From Mine to Microchip: Addressing Critical Mineral Supply Chain Risks in Semiconductor Production., Center for International and Strategic Studies., Wash DC (2024)
7. Forti, V.: "The Global E-waste Monitor.," United Nations University (2020)
8. "AI and Digital Technologies: Critical Minerals, Policy, and the Energy Transi-tion https://www.sfa-oxford.com/knowledge-and-insights/critical-minerals-in-low-carbon-and-future-technologies/critical-minerals-in-artificial-i," SFA Oxford (2025). https://www.sfa-oxford.com/knowledge-and-insights/critical-minerals-in-low-carbon-and-future-technologies/critical-minerals-in-artificial-i
9. CRS, "Critical Mineral Resources: National Policy and critical minerals list. CRS Report No. R47982," The Congressional Research Service (2025)
10. Mhlongo, S., Mativenga, P., Marnewick, A.: Water quality in a mining and water-stressed region. A. J. Clean. Prod. **171** (2018)
11. Li, J., et al.: GIS-based eval-uation of mining-induced water-related hazards in Pakistan and integrated risk mitigation strategies. Water, **17**(13) (2025)
12. Kuyucak, N.: Water in mining and environment for sustainability. Mine Water Environ. **40**, 815–817 (2021)
13. McLellan, B., Yamasue, E., Tezuka, T., Corder, G., Golev, A., Giurco, D.: Critical minerals and energy–Impacts and limitations of moving to unconventional resources. Resources, **5**(2), 19 (2016)
14. Hammond, D., Brady, T.: Critical minerals for green energy transitition: a United States perspective. Int. J. Min. Reclam. Environ. (2022). https://doi.org/10.1080/17480930.2022.2124788
15. Carr-Wilson, S., Pattanayak, S., Weinthal, E.: Critical mineral mining in the energy transition: a systematic review of environmental, social, and governance risks and opportunities. Energy Res. Soc. Sci. **116** (2024). https://doi.org/10.1016/j.erss.2024.103672
16. Chance, H., Hammersley, M.: AI is mining more than data. Vermont J. Environ. Law (2024). https://vjel.vermontlaw.edu/top-ten/2025-top-ten/2024/12/ai-is-mining-more-than-data/
17. Waterbury, J.: AI-driven design and operations of datacenters. Constr. Rev (2025). https://constructionreviewonline.com/commentary/ai-driven-design-and-operations-of-data-centers/
18. Adams, H.S.: Top 10: uses of AI in data centers. Data center magazine (2024). https://datacentremagazine.com/technology-and-ai/top-10-uses-of-ai-in-data-centres
19. Crownhart, C.: AI will add to the E-waste problem: here's what we can do about it. MIT Technol. Rev. (2024). https://www.technologyreview.com/2024/10/28/1106316/ai-e-waste/
20. Wang, L., Chen, Y.: The role of responsible mining initiatives in sustainable supply chains. Environ. Sci. Policy, **140**, 45–53 (2023). https://doi.org/10.1016/j.envsci.2023.01.005
21. Brown, K., Green, M.: Corporate commitments to low-carbon critical minerals: challenges and opportunities. Resour. Policy, **80** (2024). https://doi.org/10.1016/j.resourpol.2024.103456

Accelerating LSTM Model Training Through Biologically Inspired NeuroAI and Investigating Novel Applications

Robert Tracey[1,2(✉)], M. Emre Sahin[3], and Carol Mak[1]

[1] IBM Research Europe, Warrington, UK
[2] Wrexham University, Wrexham, UK
robert.tracey@ibm.com
[3] The Hartree Centre, STFC, Sci-Tech Daresbury, Warrington WA4 4AD, UK

Abstract. Modern Deep Neural Networks, have an ever increasing number of parameters that need to be optimised on enormous amounts of data during training. Even with advanced hardware, the training process for these models can take months and can consume large amounts of energy. This prolonged use of computing resources drives up cost significantly. In this paper we explore how one uses Spiking Neural Network (SNN) in LSTM via the NeuroAI toolkit. Compared against the traditional LSTM, the training time of our adapted LSTM significantly decreased by as much as 75% but with an increase in accuracy. We also demonstrate that using the SNN decreases the amount of CPU used by as much as 14% during training, thereby being more efficient and sustainable.

Keywords: Neuro AI · Resource Optimisation · AI training

1 Introduction

With increasingly complex architecture and massive parameter count in modern Deep Neural Networks (DNN), a common pinch point is the training time required to achieve desired performance. Scaling laws (like the Chinchilla scaling [8,17] for Large Language Models) suggest that model quality is proportional to the number of parameters, training data size and ultimately training cost. Training a good model often results in the usage of large amounts of computational power and long training times. Reducing this training time would save not only time and computational resources but also reduce the power consumption of the system training the models, leading to a more sustainable future [23].

The main and most common solution to reducing the training time of DNNs is to rely on specialised hardware such as Graphic Processing Units (GPUs), Field Programmable Gate Arrays (FPGAs) and Tensor Processing Units (TPUs) [21]. Unfortunately, these have major downsides of being expensive to purchase and resource-intensive to maintain, as they consume large amounts of power.

Published by Springer Nature Switzerland AG 2026
E. Mercier-Laurent et al. (Eds.): AI4S 2025, IFIP AICT 780, pp. 127–137, 2026.
https://doi.org/10.1007/978-3-032-21184-2_11

Moreover, as models grow larger, the infrastructure needed to support them scales exponentially [15]. Many researchers turn to cloud services for training [6] as they can provide the same hardware but for temporary use rather than investing in a long-term full data centre-based system, but they can be very costly for longer runs. The only alternative remaining is to reduce the amount of resources spent on training, either with fewer iterations or less data, both of which can significantly reduce the quality of the trained model.

We propose a solution where a Spiking Neural Network (SNN) based LSTM is used instead of a traditional LSTM, which significantly reduces training time while increasing the accuracy of the model. By incorporating biologically-inspired neurons, SNNs hold great promise with their unique dynamics and energy efficiency. Research in SNNs has been proceeding in parallel to traditional neural networks, but has not yet been as widely adopted. In this work, we demonstrate how one can use SNN on top of mainstream Machine Learning packages such as TensorFlow and PyTorch. This is achieved with only a few lines of code change to enable the open-sourced Neuro-inspired AI Toolkit [9], which allows the use of SNNs on traditional hardware.

2 Related Work

Traditionally, in order for models to increase in quality and complexity, they are generally improved by increasing the model parameters, activations, which leads to an increase in actual runtime. This is known as model scaling, [5]. It has many positives in that it allows for more accurate results and allows for more complex problems to be solved. Unfortunately, it comes at the downside of having increased run times and an increase in energy consumption. A lot of work has been done to investigate how to reduce this power consumption, as it is growing at a rate that is not sustainable in the long term. It is also starting to create major problems for countries as it puts a significant strain on the power grid. For example, in Ireland, data centres account for over 20% of the country's total electricity consumption [10], with the majority of them being situated on the outskirts of Dublin. Some work has been done in various fields to reduce the energy used, including looking at using AI to make AI more efficient [20], by running the AI tasks at the edge on smaller compute [25] and by using more efficient storage [4]. All of these options, however, remain with traditional neural networks and machine learning and look at altering the hardware and environments around them.

An alternative way of creating has been emerging in recent years in the field of neuromorphic computing. Neuromorphic computing is brain-inspired computing that looks at making AI more similar to how a human brain works by simulating synapses firing and has led to much more efficient and sustainable AI models [11]. Neuromorphic computing typically uses a Spiking Neural Network (SNN) [19] and as such nodes within the network are only powered and used when they are needed [16].

The main problem faced in neuromorphic computing is that traditionally, for it to run at its optimum, it will require specialist hardware [24]. This means that its adoption in mainstream has been stalled on multiple occasions, but with the increased interest in sustainability, this seems to be changing [12].

3 Models

3.1 Long Short-Term Memory

Long Short-Term Memory (LSTM) networks [7] are an instance of a Recurrent Neural Network [18]; they consist of multiple connected block units. Each block unit has a cell, an input gate, a forget gate and an output gate. These gates allow information from the previous time step to influence future steps, providing more accurate time series prediction and allowing the model to regulate the flow of information. Which means the model is able to selectively remember, forget, and output data over long sequences. This makes LSTM highly suitable for modelling time series data. LSTMs have been utilised in many fields and applications, including predicting electricity usage [1], data centre resource utilisation [14], and traffic in cellular networks [3]. We call this LSTM the "traditional" model in the rest of the paper. In Listing 1.1, an example of part of the code for our LSTM model can be seen.

3.2 NeuroAI

Spiking Neural Networks (SNNs) often require specialised hardware to run, such as the North Pole neuromorphic chip by IBM [2]. However, by using the NeuroAI toolkit, neuromorphic style computing can be run on traditional hardware in combination with libraries like PyTorch and TensorFlow. This is achieved by using a Spiking Neural Unit (SNU) [22], which is introduced to model the biological behaviour of neurons using two Artificial Neural Network (ANN) neurons. An example of an SNU and how it compares to a biological neuron can be seen in Fig. 1. This works by the first stage accumulating incoming spikes, and once a threshold is reached, the second ANN neuron fires (spiking stage). This releases a spiking emission either to the next node in the network or as an output at the end; simultaneously, the first ANN neuron in the SNU is reset. SNUs may operate as SNNs using a step function activation, or as ANNs using continuous activations. These SNUs enable an efficient implementation of the spiking effect with in-memory acceleration for training. To demonstrate its superiority, we created a SNN-based LSTM model, utilising SNUs in the NeuroAI toolkit, based on the "traditional" LSTM model described above without specialised hardware. We call this model the NeuroAI model. In Listing 1.2, an example of part of the code for our NeuraoAI model can be seen and can be compared with the Listing 1.1 to see how few changes are required.

```
model = Sequential()
model.add(LSTM(100, input_shape=(X_train.shape[1], X_train.shape[2])))
model.add(Dropout(0.2))
model.add(Dense(1))
time_start = time.time()
```

```
model.compile(loss='mean_squared_error', optimizer='adam')

history = model.fit(X_train, Y_train, epochs=30, batch_size=2000, \
    validation_data=(X_test, Y_test), \
    callbacks=[EarlyStopping(monitor='val_loss', patience=10)], verbose=1, \
    shuffle=False)
print('Finished.␣Total␣time:␣{0:.1f}␣[s]'.format(time.time() - time_start))
```

Listing 1.1. Example of part of traditional LSTM model code

```
model = Sequential()
model.add(tf.keras.layers.InputLayer(input_shape=[X_train.shape[1], \
    X_train.shape[2]]))
config = {'decay': 0.9}
model.add(aitf.layers.SNU(100, return_sequences=False,activation=tf.sigmoid))

model.add(Dense(1))
time_start = time.time()
model.compile(loss='mean_squared_error', optimizer='adam')
history2 = model.fit(X_train, Y_train, epochs=30, batch_size=2000, \
    validation_data=(X_test, Y_test), \
    callbacks=[EarlyStopping(monitor='val_loss', patience=10)], verbose=1, \
    shuffle=False)

print('Finished.␣Total␣time:␣{0:.1f}␣[s]'.format(time.time() - time_start))
```

Listing 1.2. Example of part of NeuroAI model code

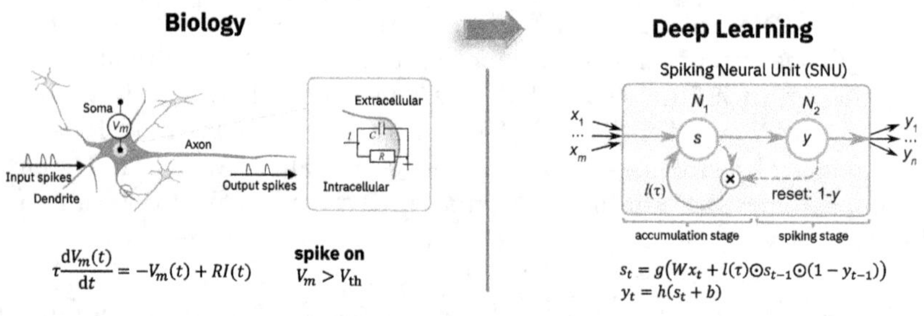

Fig. 1. Biological neuron's inspired the SNU implementation [22].

4 Results

For this work, two different models were trained, one is a traditional LSTM model written in standard TensorFlow, and the second is a NeuroAI modified LSTM model. This second NeuroAI model has identical hyperparameters to the original model and is also written in TensorFlow but utilises the NeuralAI library, so it builds the model using SNU's. To investigate the difference between these models, multiple datasets were used, the first a time series data set that had previously been used successfully in other experiments [20] is selected. This time series data set is a collection of CPU utilisation and fan power of an IBM

POWER8^{TM} system, and the models are then required to predict the rack power. This enables a usable baseline of the performance of the models in terms of training time and accuracy. To further gauge how the models compare, we will also test them using the Etth1 and Etthm1 data sets cite haoyietal-informer-2021, which are standard datasets for benchmarking in the timeseries space. Both models are trained with the same time series data sets, previously mentioned in the paper, on the same hardware. The results presented in this paper are an average over 50 runs.

4.1 Rack Power Dataset

Each model was set to train for 30 epochs and a batch size of 2000 with EarlyStopping monitoring set to 'val_loss'. This meant that the training typically stopped at around 19–24 epochs. In Table 1, we show the training time, Root Mean Squared Error (RMSE) and Mean Absolute Error (MAE) for the test and train runs of both models. As can be clearly seen, the NeuroAI trains much faster, in fact 75% faster than the traditional LSTM model; and outperforms it in the RMSE and MAE scores. In Table 2 we show the CPU time and memory required in training both models on a server where threading was possible. This was performed on separate hardware that allowed specific monitoring of CPU and memory resources. Here again, it is clear that there is a vast reduction in the CPU time required and also a reduction in the amount of memory used. Thereby reduces the amount of resources and power required to train the model.

In Fig. 2 we present the training loss and test loss of training a traditional LSTM and NeuroAI model. We can see that the loss of the NeuroAI model fig (2) reduces much more quickly and ends much closer to zero.

Table 1. Training time and accuracy results for each model using the rack power dataset

	Traditional LSTM	NeuroAI SNN	delta	percentage improvement
Training time (s)	56	14	42	75%
Test RMSE	7.407	0.971	6.436	87%
Test MAE	6.132	0.486	5.646	92%
Train RMSE	8.683	0.408	8.275	95%
Train MAE	8.094	0.213	7.881	97%

4.2 Etth1 and Ettm1 Datasets

These datasets are made of the data of 2 Electricity Transformers at 2 stations, including load, oil temperature. They are for use in long sequence time-series problems. Etth1 has 17421 data points, and Ettm1 has 69681 data points. For these datasets, each model was set to run for 30 epochs and a batch size of 2000

Table 2. CPU and memory usage for each type of model

	Traditional LSTM	NeuroAI SNN	delta	percentage improvement
CPU time (s)	4228	3625	603	14%
Max Memory (MB)	2368	2181	187	8%

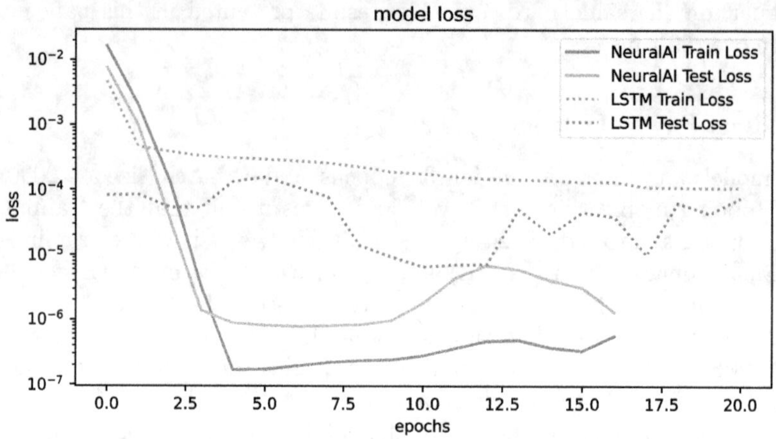

Fig. 2. Model train and test loss over epochs for Traditional LSTM model and NeuroAI model using LSTM.

with EarlyStopping monitoring set to 'val_loss' while training them. In Tables 3 and 4, we show the training time, Root Mean Squared Error (RMSE) and mean absolute error (MAE) of both models. As can be clearly seen, similar to the tests performed on the rack power dataset. The NeuroAI trains 62% faster on the etth1 dataset and 55% on the ettm1 dataset compared with the traditional LSTM model, and outperforms it in the RMSE and MAE scores. This can easily be seen in the Fig. 3 showing the training time, where there is a very visible difference between the traditional LSTM model and the NeuroAI one. Also, in Fig. 4, it can be easily seen that the RMSE and MAE scores are improved by the use of the NeuroAI model on both datasets. This is especially true on the larger ettm1 dataset, where significant improvements can be seen.

4.3 Site-Level Solar Power Forecasting on US NREL UPV/DPV

We used 5-minute AC power measurements (MW) from utility-scale (UPV) and distributed (DPV) photovoltaic sites across Arizona, provided by the National Renewable Energy Laboratory [13]. We used 5-minute AC power measurements (MW). Predictions are generated for each site using the actual AC power series; reported metrics are the across-site mean ± standard deviation. Each model was trained for 30 epochs with a batch size of 300 and EarlyStopping monitoring `val_loss`.

Tables 5 and 6 report wall-clock training time, RMSE, and MAE. NeuroAI SNN trains faster than a standard LSTM—45% on UPV (25.13,s vs. 13.71,s) and 47% on DPV (26.88,s vs. 14.38,s)—but with higher error. On UPV, test RMSE increases by 4% and test MAE by 35%; on DPV, test RMSE increases by 12% and test MAE by 63%. This can be seen in Fig. 6. Training-set metrics show the same pattern (3–11% higher RMSE; 26–53% higher MAE), also shown in Fig. 5. These results indicate a speed–accuracy trade-off on solar power forecasting, with the penalty more pronounced on DPV.

Table 3. Training time and accuracy results along with delta for each using Etth1 dataset

	Traditional LSTM Etth1	NeuroAI SNN Etth1	Etth1 delta	percentage improvement
Training time (s)	4.376	1.675	2.700	62%
Test RMSE	0.755	0.738	0.016	2%
Test MAE	0.538	0.516	0.020	2%
Train RMSE	1.130	1.109	0.023	4%
Train MAE	0.809	0.778	0.030	4%

Table 4. Training time and accuracy results along with delta for each model using Ettm1 dataset

	Traditional LSTM Ettm1	NeuroAI SNN Ettm1	Ettm1 delta	percentage improvement
Training time (s)	10.360	4.617	5.74	55%
Test RMSE	0.598	0.386	0.212	35%
Test MAE	0.500	0.269	0.101	13%
Train RMSE	0.798	0.697	0.231	46%
Train MAE	0.647	0.500	0.147	23%

Table 5. Training time and accuracy results along with delta for each using UPV dataset

	Traditional LSTM UPV	NeuroAI SNN UPV	UPV delta	percentage improvement
Training time (s)	25.133	13.710	11.423	45%
Test RMSE	2.333	2.426	−0.094	−4%
Test MAE	0.818	1.101	−0.284	−35%
Train RMSE	2.826	2.911	−0.085	−3%
Train MAE	1.050	1.327	−0.277	−26%

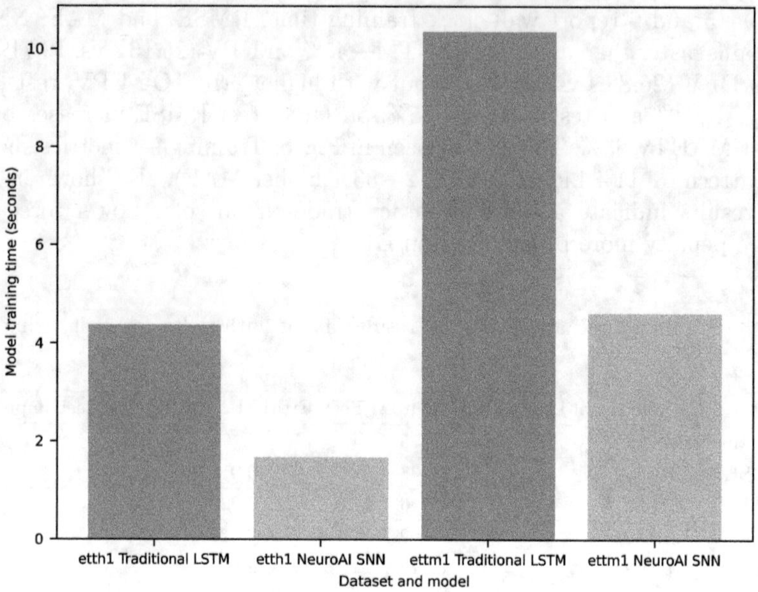

Fig. 3. Training time for models on each data set.

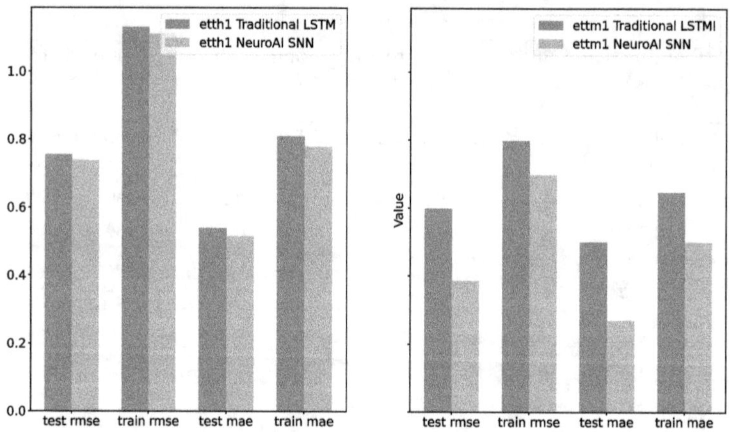

Fig. 4. RMSE and MAE results for etth1 and ettm1 datasets.

Table 6. Training time and accuracy results along with delta for each model using DPV dataset

	Traditional LSTM DPV	NeuroAI SNN DPV	DPV delta	percentage improvement
Training time (s)	26.877	14.377	12.500	47%
Test RMSE	0.836	0.940	−0.104	−12%
Test MAE	0.331	0.539	−0.208	−63%
Train RMSE	0.923	1.023	−0.099	−11%
Train MAE	0.382	0.585	−0.203	−53%

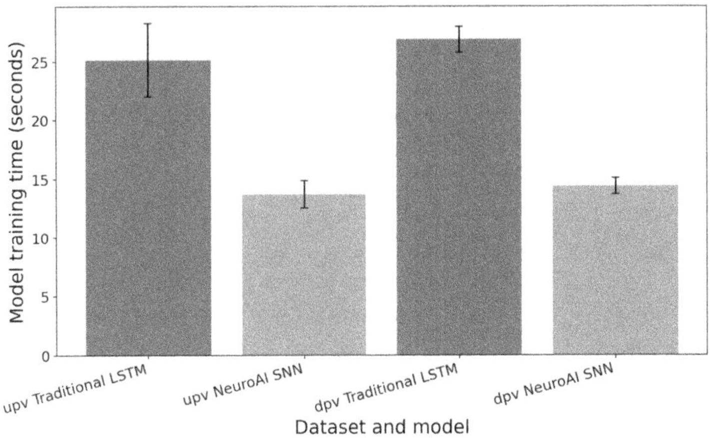

Fig. 5. Training time for models on each data set.

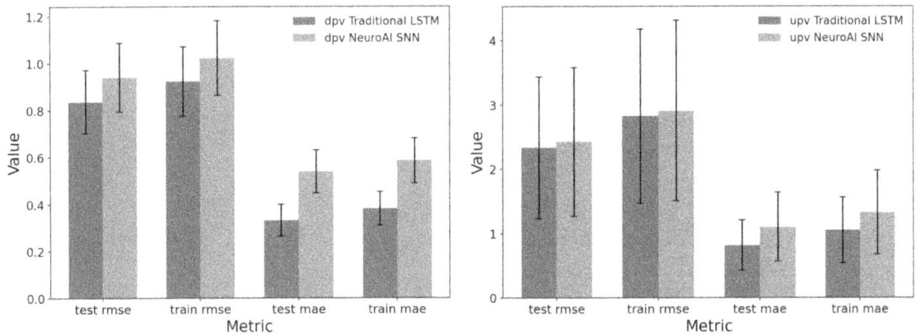

Fig. 6. RMSE and MAE results for etth1 and ettm1 datasets.

5 Conclusion

We have presented a comparison between the traditional LSTM and NeuroAI SNN models running on multiple different datasets and have demonstrated that by utilising NeuroAI, a significant increase in speed, with as much as 75% in some cases, whilst also increasing accuracy. With such big increases in training speed, this will also help in reducing the amount of computing resources required for the calculations, therefore making it a more sustainable method. This was also shown with a reduction of CPU usage of 14% as can be seen in Table 2.

Going forward, we plan to train these models with larger datasets to take advantage of this reduction in training time and look at their potential usage with other types of datasets. Along with using random seeds and subsets of data to compare how the models perform under different conditions. Another task would be taking individual measurements of component power usage during each epoch. This would give us a better understanding of energy savings and from which part of the system, enabling system tuning, potentially reducing power consumption even further.

Acknowledgments. This work was supported by the Hartree National Centre for Digital Innovation, a collaboration between STFC and IBM.

References

1. Alden, R.E., Gong, H., Ababei, C., Ionel, D.M.: LSTM forecasts for smart home electricity usage. In: 2020 9th International Conference on Renewable Energy Research and Application (ICRERA), pp. 434–438 (2020). https://doi.org/10.1109/ICRERA49962.2020.9242804
2. Cassidy, A.S., et al.: 11.4 IBM NorthPole: an architecture for neural network inference with a 12nm chip. In: 2024 IEEE International Solid-State Circuits Conference (ISSCC), vol. 67, pp. 214–215. IEEE (2024)
3. Dalgkitsis, A., Louta, M., Karetsos, G.T.: Traffic forecasting in cellular networks using the LSTM RNN. In: Proceedings of the 22nd Pan-Hellenic Conference on Informatics. PCI 2018, pp. 28–33. Association for Computing Machinery, New York (2018). https://doi.org/10.1145/3291533.3291540
4. Do, J., et al.: Cost-effective, energy-efficient, and scalable storage computing for large-scale ai applications. ACM Trans. Storage **16**(4) (2020). https://doi.org/10.1145/3415580
5. Dollár, P., Singh, M., Girshick, R.: Fast and accurate model scaling. In: Proceedings of the IEEE/CVF Conference on Computer Vision and Pattern Recognition, pp. 924–932 (2021)
6. Dreher, P., Nair, D., Sills, E., Vouk, M.: Cost analysis comparing HPC public versus private cloud computing. In: Helfert, M., Ferguson, D., Méndez Muñoz, V., Cardoso, J. (eds.) Cloud Computing and Services Science, pp. 294–316. Springer, Cham (2017)
7. Hochreiter, S., Schmidhuber, J.: LSTM can solve hard long time lag problems. In: Advances in Neural Information Processing Systems, pp. 473–479 (1997)
8. Hoffmann, J., et al.: Training compute-optimal large language models. CoRR **abs/2203.15556** (2022). https://doi.org/10.48550/ARXIV.2203.15556
9. IBM: Ibm/neuroaikit: Neuro-inspired ai toolkit (2022). https://github.com/IBM/neuroaikit
10. Jackson, A.: Power-hungry data centres put pressure on ireland's grid (2024). https://datacentremagazine.com/critical-environments/power-hungry-data-centres-put-pressure-on-irelands-grid
11. Luo, T., et al.: Achieving green ai with energy-efficient deep learning using neuromorphic computing. Commun. ACM **66**(7), 52–57 (2023). https://doi.org/10.1145/3588591
12. Muir, D.R., Sheik, S.: The road to commercial success for neuromorphic technologies. Nat. Commun. **16**(1), 3586 (2025)
13. National Renewable Energy Laboratory: Solar power data for integration studies (2025). https://www.nrel.gov/grid/solar-power-data. modeled 5-minute PV power and hourly day-ahead forecasts for approximately 6,000 plants representing 2006. Accessed 24 Aug 2025

14. Ouhame, S., Hadi, Y., Ullah, A.: An efficient forecasting approach for resource utilization in cloud data center using CNN-LSTM model. Neural Comput. Appl. **33**(16), 10043–10055 (2021). https://doi.org/10.1007/s00521-021-05770-9
15. Robbins, S., Van Wynsberghe, A.: Our new artificial intelligence infrastructure: becoming locked into an unsustainable future. Sustainability **14**(8), 4829 (2022)
16. Roy, K., Jaiswal, A., Panda, P.: Towards spike-based machine intelligence with neuromorphic computing. Nature **575**(7784), 607–617 (2019)
17. Sardana, N., Portes, J., Doubov, S., Frankle, J.: Beyond chinchilla-optimal: accounting for inference in language model scaling laws. In: Forty-first International Conference on Machine Learning, ICML 2024, Vienna, Austria, July 21–27, 2024. OpenReview.net (2024). https://openreview.net/forum?id=0bmXrtTDUu
18. Sherstinsky, A.: Fundamentals of recurrent neural network (RNN) and long short-term memory (LSTM) network. arXiv preprint arXiv:1808.03314 (2018)
19. Thakur, C.S., et al.: Large-scale neuromorphic spiking array processors: a quest to mimic the brain. Front. Neurosci. **12**, 891 (2018)
20. Tracey, R., Hoang, L., Subelet, F., Elisseev, V.: Ai-driven holistic approach to energy efficient HPC. In: Jagode, H., Anzt, H., Juckeland, G., Ltaief, H. (eds.) High Performance Computing, pp. 267–279. Springer, Cham (2020)
21. Wang, Y., et al.: Benchmarking the performance and energy efficiency of ai accelerators for ai training. In: 2020 20th IEEE/ACM International Symposium on Cluster, Cloud and Internet Computing (CCGRID), pp. 744–751 (2020). https://doi.org/10.1109/CCGrid49817.2020.00-15
22. Woźniak, S., Pantazi, A., Bohnstingl, T., Eleftheriou, E.: Deep learning incorporating biologically inspired neural dynamics and in-memory computing. Nat. Mach. Intell. **2**(6), 325–336 (2020). https://doi.org/10.1038/s42256-020-0187-0
23. van Wynsberghe, A.: Sustainable AI: AI for sustainability and the sustainability of AI. AI and Ethics **1**(3), 213–218 (2021). https://doi.org/10.1007/s43681-021-00043-6
24. Young, A.R., Dean, M.E., Plank, J.S., Rose, G.S.: A review of spiking neuromorphic hardware communication systems. IEEE Access **7**, 135606–135620 (2019)
25. Zhu, S., Ota, K., Dong, M.: Energy-efficient artificial intelligence of things with intelligent edge. IEEE Internet Things J. **9**(10), 7525–7532 (2022). https://doi.org/10.1109/JIOT.2022.3143722

Optimal Planning and Machine Learning for Responsive Tracking and Enhanced Forecasting of Wildfires Using a Spacecraft Constellation

Sreeja Roy-Singh[1,2](✉), Vinay Ravindra[1,2], Richard Levinson[1,3], Mahta Moghaddam[4], Jan Mandel[5], Adam Kochanski[6], Angel Farguell Caus[6], Kurtis Nelson[7], Samira Alkaee Taleghan[5], Archana Kannan[4], and Amer Melebari[4]

[1] NASA Ames Research Center, Moffet Field, CA 94036, USA
sreeja.nag@nasa.gov
[2] Bay Area Environmental Research Institute, Moffet Field, CA 94036, USA
[3] KBR Wyle Services, 601 Jefferson Street, Houston, TX 77002, USA
[4] University of Southern California, 3470 Trousdale Parkway, Los Angeles, CA, USA
[5] University of Colorado, 2055 Regent Drive, Boulder, CO 80309, USA
[6] San Jose State University, 1 Washington Square, San Jose, CA 95192, USA
[7] United States Geological Survey, Sioux Falls, SD 57198, USA

Abstract. We propose a novel concept of operations using optimal planning methods and machine learning (ML) to collect spaceborne data that is unprecedented for monitoring wildfires, process it to create new or enhanced products in the context of wildfire danger or spread monitoring, and assimilate them to improve existing, wildfire decision support tools delivered to firefighters within latency appropriate for time-critical applications. The concept is studied with respect to NASA's CYGNSS Mission – a constellation of passive microwave receivers that measure specular GNSS-R reflections despite clouds and smoke. Our planner uses a Mixed Integer Program formulation to schedule joint observation data collection and downlink for all satellites. Optimal solutions are found quickly that collect 98–100% of available observation opportunities. ML-based fire predictions that drive the planner objective are > 40% more correlated with ground truth than existing state-of-art. The presented case study on the TX Smokehouse Creek fire in 2024 and LA fires in 2025 represents the first high-resolution data collected by CYGNSS of active fires. Creation of Burnt Area Maps (BAM) using ML applied to the data during active fires and BAM assimilation into NASA's Weather Research and Forecasting Model with neural networks to broadcast fire spread are novel outcomes. BAM and CYGNSS-obtained soil moisture are integrated for the first time into USGS fire danger maps. Inclusion of CYGNSS data in ML-based burn predictions boosts accuracy by 13%, and inclusion of high-resolution data boosts ML recall by another 15%. The proposed workflow has an expected latency of 6-30h, improving on the current delivery time of multiple days. All components in the proposed concept are shown to be computationally scalable and globally generalizable, with sustainability considerations such as edge efficiency and low latency on small devices.

Published by Springer Nature Switzerland AG 2026
E. Mercier-Laurent et al. (Eds.): AI4S 2025, IFIP AICT 780, pp. 138–156, 2026.
https://doi.org/10.1007/978-3-032-21184-2_12

Keywords: Artificial intelligence · satellite remote sensing · wildfire predictions

1 Introduction

Wildfires are a serious problem for humans and wildlife, and the impact in the past few decades has grown in terms of area affected and cost to suppress [1]. More than a third of our housing units lie on the wildland-urban interface, thereby increasing the risk of ignition and urban spread [2]. In 2024 alone [3], there were 64,897 wildfires that burned 8.9 million acres in the U.S. Satellite remote sensing is invaluable for tracking the spread of large wildfires owing to its global coverage from a high vantage point and regular revisit of the same ground location. Constellations of low-cost satellites in Low Earth Orbit (LEO) offer unprecedented temporal revisits at 10x closer distances than traditional weather satellites in geostationary orbit. When equipped with passive microwave sensors, they can measure through smoke, clouds, and canopy. The sensor observations, when optimally collected using a novel, proposed AI-based planner and processed using novel, proposed ML-based models, yield fire danger and spread metrics that are functionally novel and/or higher quality. We demonstrate the benefit of the proposed AI/ML technologies by applying the full workflow to pre-fire conditions and active fires over 3 days in the contiguous United States (CONUS). The benefits of active fire ML products are further demonstrated on spacecraft-collected data over the Smoke House Creek fire, which burned more than a million acres in 2024 and was the largest wildfire in Texas history [4], and the Los Angeles fires in 2025 which caused damages over $250 billion, one of the costliest disasters in U.S. history [5]. These geographies (CONUS or 2 wildfires) or time periods (3 days) are examples to demonstrate the value of the technologies, which can be applied to any region, number of fires, or time-period.

1.1 Technology Challenges

The challenges associated with time-critical collection and actionable utilization of spaceborne observations are on both the space technology and the ground logistics sides. Artificial Intelligence (AI) can be applied at multiple stages of the data pipeline from space to actions on Earth and we demonstrate the value in doing so at every stage using novel technologies.

1. *Data Collection:* Appropriate satellite observations are constrained by hardware resources (energy, data) and orbital dynamics (access to observation opportunities and uplink/downlink with Earth) [6]. LEO satellites provide limited Earth coverage at any time. Automated planners can prioritize observations of specific locations and times, given constraints such as limited onboard storage capacity. Collection of high-resolution imagery (e.g., rawIF data, see Sect. 2) is therefore resource-limited and requires multiple passes over ground stations to be downlinked to Earth. AI can be used in scheduling to improve collection and downlink quality and efficiency.
2. *Data Processing:* Remote sensing data can be captured at frequencies beyond human vision, and sensors detect intricate phenomena that are not discernible to the human eye. Machine Learning (ML) is a powerful tool for extracting meaningful information

from complex sensory data, such as Delay-Doppler maps (DDMs, as seen in Fig. 1 inset) from passive microwave signals [7], and creating useful geospatial products such as Burnt Area Maps (BAMs) and Soil Moisture (SM) maps from DDMs.
3. *Data Application to forecasts:* High performance computing (HPC) is necessary for assimilating ancillary, real-time parameters into forecast models such as the WRFx forecasting system, an open-source framework that uses WRF-SFIRE, a high resolution, multi-scale weather forecasting model, coupled with a semi-empirical fire-spread model and a prognostic dead fuel moisture model [8]. ML supports improved predictions of active fire pixels and Fire Arrival Time (FAT) by fusing data from various sensor sources within WRFx and using a multi-head neural network toward classification and regression. This supports firefighters and may provide feedback for the next iteration of intelligent planning to gather responsive data.

1.2 Multidisciplinary, Multilateral Approach

Our goal is to monitor from space, areas that are known to be experiencing severe active fire activity as well as regions at risk of imminent large wildfires. For the proposed concept of operations (ConOps) to provide benefit for on-ground firefighters requires collaborative effort from researchers of several disciplines and institutions. Our collaboration for the presented case study includes:

- Computer scientists for spacecraft planning and scheduling in NASA Ames Research Center (ARC)
- Instrument scientists from NASA Jet Propulsion Laboratory and universities
- Astrodynamics and spacecraft designers from the Bay Area Environmental Research Institute in NASA ARC
- Atmospheric Scientists from the Wildfire Interdisciplinary Research Center, San Jose State University
- HPC Computer scientists in universities for integration into operational wildfire models
- Physical scientists and Incident Management Team (IMT) support from U.S. Geological Survey (USGS)

IMT and USGS also use our data products on ground and can provide actionable feedback to improve the technologies.

2 Concept of Operations

As illustrated in Fig. 1, a distributed spacecraft network can make observations that are periodically downlinked to ground stations (see grey box on "Operations"). All computation to plan those observations and process their data (outside the grey box) runs offboard on Earth and resultant plans are uplinked to the spacecraft when it accesses ground stations. The Planner generates a sequence of feasible satellite observations and downlink operations. The Planner is guided by (A) location-based priorities quantified by the "Priority Framework.", and (B) constraints imposed by physics and hardware limitations, quantified by the "Satellite Orbital Simulator" which predicts observation and downlink opportunities. Collected data is first converted to geophysical products in the "Data Processing" block, then integrated into fire models (pre-fire danger in orange and active-fire in red) to improve wildfire forecasts. These enhanced forecasts support stakeholders in making critical fire management decisions and input to the Planner for the next horizon of scheduling. This feedback loop enables the planner to assess gaps, re-forecasts, and intelligently plan the next iteration. More details on the concept are found in Ref [9].

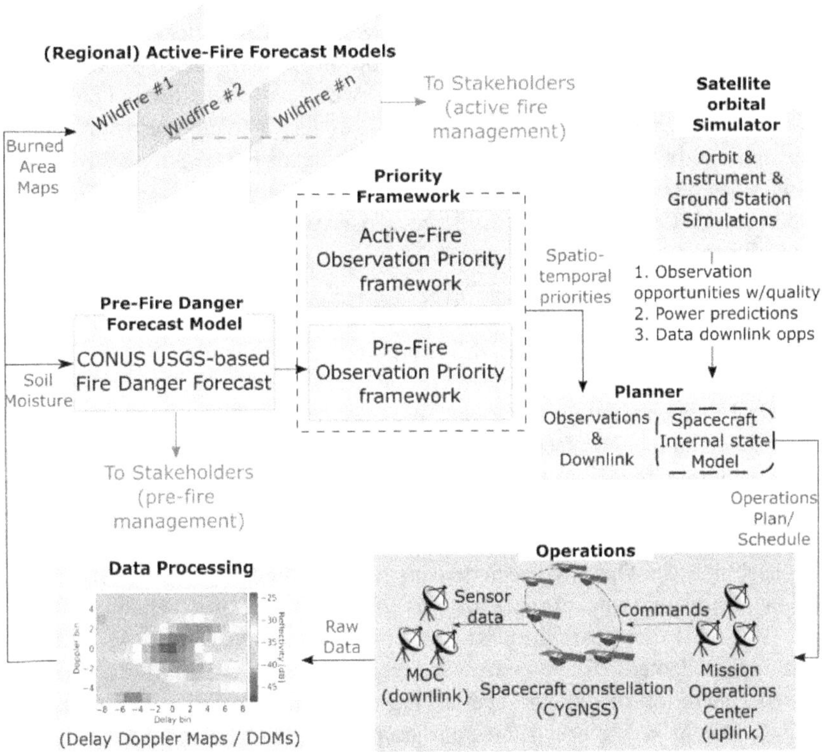

Fig. 1. System Architecture and Concept of Operations (ConOps).

We apply the ConOps to a constellation of 8 designed satellites (of which 7 are active) carrying Global Navigation Satellite Systems Reflectometry (GNSS-R) sensors. GNSS-R measures properties of surface specular points using reflected GNSS signals. It has demonstrated enormous scientific value via NASA's Cyclone Global Navigation Satellite System (CYGNSS) [7] and commercialized by companies like Spire and Muon. The low size, weight, power (SWaP) needs for GNSS-R sensors allows them to be accommodated on small satellites, which enables deployment of numerous copies and enables frequent sampling (observations) globally. While having a small SWaP reduces the overall cost (development, manufacturing, launch) of the constellation, it constrains onboard resources.

For example, CYGNSS usually gathers data in a 'nominal' mode - the raw observation signals are processed onboard and downlinked with information loss but substantially saving onboard data storage and downlink. It is possible to command a 'RawIF' mode of observation, in which the raw signals are downlinked for more informative retrievals on ground, albeit at the cost of 1680x more storage and downlink time required. Only 60 s worth of rawIF images can be stored onboard, which takes ~ 20 min to downlink spread over several hours of access gaps. We seek to command RawIF signals over locations of higher wildfire priority within tight resource and access constraints, therefore the necessity of an AI-planner. The novel geophysical products created after ML-based processing are BAM for active fires and SM of pre-fire unburnt areas. BAM and SM are ingested by a neural network, that simultaneously detects active fires and predicts their arrival times at improved accuracy, and a pre-fire danger predictor, respectively.

Our case study focuses on three dates of significant fire activity over the United States with listed peak fire locations:

A.8 Jan 2025: Los Angeles (LA) fires, comprised of the Palisades, Eaton and Hughes fires in California (CA).

B.26 Feb 2024: Smokehouse Creek fire in Texas (TX).

C.5 July 2023: Day with the highest number of wildfires in CONUS at latitudes less than 35 deg (relevant because CYGNSS orbits are low inclination).

Sections 3–5 describe the three parts of the ConOps, components within them, and the novel AI elements being developed. We describe results from each as would be executed sequentially. The feedback cycle in Fig. 1 is expected to execute every 24 h and all planning/processing to take a few hours as validated in our results. Quicker cycles are possible at the cost of more computation, space-to-ground access contacts, and execution resources.

While similar ConOps have been demonstrated to monitor volcanos [10] and urban floods [11], we present a novel application as enabled by the novel technology described in Sect. 3–5. This represents the ***first*** ever collection of spaceborne GNSS-R raw-IF data for unprecedented passive-microwave resolution of active wildfires, and the ***first*** ML-based generation of medium-spatial, high-temporal resolution BAMs as the fire progresses. Our Mixed Integer Linear Programming (MILP) planner formulation and plan represents the ***first*** optimal solution for a constellation of active, agile satellites where observation and downlink are optimized simultaneously using real-world data-driven prediction models. This ConOps also demonstrates the ***first*** assimilation of any GNSS-R dependent product into ***both*** pre-fire and active fire decision tools: [i] USGS danger maps released daily over CONUS, and [ii] WRFx and ML-based active fire forecasts on fire area extent and progression (outputs are available publicly via Ref [12]).

3 Observation Prioritization Framework

Appropriate quantification of observational priorities is the backbone for *gathering spaceborne data efficiently* for either active fires or pre-fire risk conditions. AI-driven improvements in priorities and uncertainties significantly impacts the quality and responsiveness of subsequent observations.

3.1 Active Fire Prioritization (WRFx Fire Danger)

The state of art (Fig. 2 top half) for evaluating active wildfire danger is non-standardized and varies between companies, applications, users. One of the most advanced products, although experimental, is WRFx and its Fuel Moisture Data Assimilation (FMDA) system, which updates hourly and forecasts *dead fuel* moisture for the next 48 h. FDMA integrates measurements from ground-based Remote Automated Weather Stations (RAWS) and High-Resolution Rapid Refresh (HRRR) of weather data into a fuel moisture model. We propose extending this system by developing a new active fire danger quantification (purple box) that uses multiple Fire Weather Indices (FWIs) *in addition* to HRRR weather data and FMDA fuel moisture estimates. Using this new product and real-time wildfire ignition location data from the National Interagency Fire Center (red), active fire event priority (pink) will be published at hourly cadence.

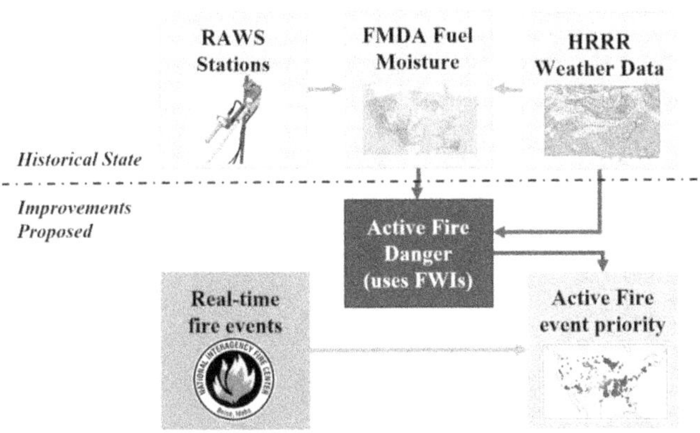

Fig. 2. Algorithmic workflow for Active Fire prioritization

An ML model utilizing XGBoost [13] is being developed for active fire event prioritization, which integrates several state-of-the-art FWIs listed in Fig. 3's X-axis. The model trains on historical wildfire activity data and learns the characteristics of wildfires that grow large and become difficult to control. Model validation shows accuracy at 0.86, recall at 0.87, precision at 0.73. While there is much scope to improve quality by extending to a more diverse set of FWIs and improving XGboost features, the model already outperforms all major FWIs used for fire danger predictions in terms of matching predictions with ground truth. Figure 3 shows the number of large and small fires predicted by the model compared to many FWIs (KBDI, FFWI, LFP, Haines, DMC, FFMC, ISI, SAWTI) and ground truth. The Matthew's Correlation Coefficient (MCC) of the ML model output with respect to ground truth at 0.75 is better than the MCC of all other FWIs. The MCC is >40% higher than the next highest MCC associated with an FWI. The model is also computationally efficient: 1-h processing time for every set of 64 parallel active fires.

The model outputs CONUS location priorities for our case study (Fig. 4). The most destructive wildfires on that day were confirmed to be associated with higher priority values. For example, the 2025 Palisades fire recorded a value of 0.72, the 2025 Eaton fire at 0.77, the 2024 Smokehouse Creek fire at 0.87, and the 2023 Lincoln fire at 0.74 represent the highest danger. As model quality improves, we expect to quantify wildfire severity and behavior more robustly.

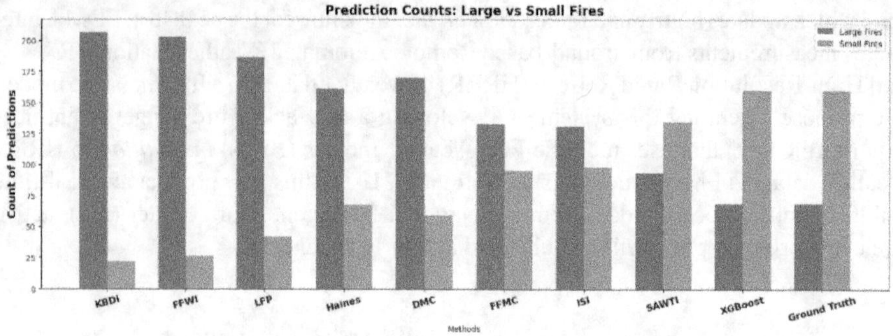

Fig. 3. Comparison of the number of large and small figures predicted by the WRFx Fire Danger ML model (labeled 'XGBoost') compared to 8 state-of-art FWIs and to ground truth.

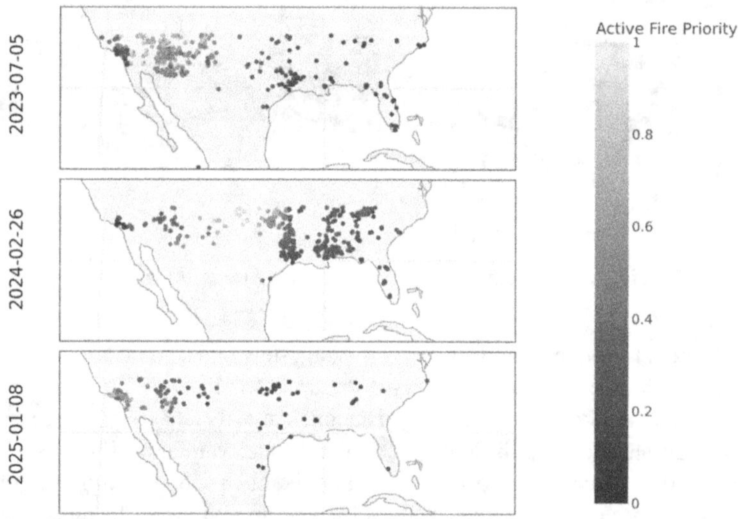

Fig. 4. Active Fire event priority across CONUS for three selected dates

3.2 Pre-Fire Prioritization (USGS Fire Danger)

The USGS produces daily 7-day forecasts of potential large fire activity for CONUS based on satellite-derived pre-fire products such as vegetation greenness, fuel moisture, and predicted weather conditions [14]. These data are used to produce the Wildland Fire Potential Index (WFPI), a unitless value that is correlated with increased incidence of large (>500 acres) wildland fires [15]. WFPI-derived Large Fire Probability (WLFP) and WFPI-derived Fire Spread Potential (WFSP) [16, 17] are predictive indices derived from daily WFPI forecasts. WLFP is the chance that a large fire will occur in a specific location at a specific time based on historical fire activity patterns adjusted to account for the current WPFI values. WFSP is the chance that an existing wildland fire ignition will escape initial attack and grow to be large based on relationships between historical ignitions and WFPI. These indices are used to inform the pre-fire priority of any point in CONUS to optimally plan CYGNSS operations.

4 Planning Satellite Operations

An automated, intelligent planner is used to schedule observations over active fires and pre-fire conditions. Previous studies [11, 20] have shown that even in the lack of optimality guarantees, intelligent planners significantly improve performance compared to manual, static scheduling.

The *planner's input* includes timepoints (seconds) when each satellite may observe geographic locations or downlink data to a ground station (GS). The open-source Earth Observation Simulator (EO-Sim) software is used to predict observation opportunities for target coordinates on Earth, downlinks to GS, and periods of eclipse which is used in the calculations of the battery state [18, 19]. The orbit propagator provides satellite states at discrete time points in 1 s interval over a pre-defined planning horizon (24 h for the case study). EO-Sim calculates observation opportunities based on the computed specular (reflected) locations, i.e., reflection points from pairs of GNSS and CYGNSS satellites, for any selected geography.

Spacecraft Storage and Energy State Models: Each RawIF image spans 1 s of data collection and consumes 1/60 of data buffer capacity. The buffer is a First-In-First-Out queue and can store a maximum of 60 s worth of RawIF data. It takes ~ 20 s to downlink a single image, and ~ 20 min to download a full buffer. Power is produced continuously by solar panels, except during eclipse (~30 of ~ 90 min). Power is consumed continuously by nominal operations, and additionally for data collection and downlink. RawIF and nominal mode images are stored in different buffers.

The *planner's output* is a command schedule for each satellite, each second when it should collect or downlink data.

Large Search Space: The raw inputs identify commands at 1-s granularity, related to the CYGNSS satellite's speed in LEO (~7 km/s). For each satellite, there are ~ 23,000 s when it may collect over CONUS or downlink data or remain idle. This sequence of ~ 23k binary choices per satellite creates a huge search space. Prior time-indexed approaches to this problem modeled each second scaled poorly or formulated the problem non-optimally [20].

Data Cycle Abstraction: We developed a novel interval-based abstraction which improves scaling and enables optimal solutions to be found quickly. Instead of decision variables indexed at the 1-s granularity, we model temporal intervals called Data Cycles, constructed from the raw orbital access times during preprocessing. Each data cycle is a sequence of 2 phases: First the observation phase fills up storage, followed by the downlink phase which frees up storage. The observation time window is associated with a unique set of images available during that interval. An image can cover multiple targets or geo-locations. Rewards are associated with targets and there are many targets per image, while storage is consumed by images. Cycles are repeated over the plan horizon. In our case study, satellites have an average of 6.5 data cycles every 24 h.

4.1 Planning Model for Observation & Downlink

The planner has been implemented as a MILP formulation, generalized for any geography, spacecraft, payload, ground station, with agility options. When applied to our case study, optimal solutions are found within an hour. We highlight the key elements below due to lack of space to describe the full formulation. The planner's *objective* is to maximize the aggregate science rewards collected for all observed targets across all satellites. The planner chooses which satellites will observe which targets, and when to observe them. Each observation collects one *image*, which covers multiple *targets*. The same target may be covered by multiple images.

Parameters:

A = {set of active fire targets}.

P = {set of pre-fire targets}.

T = {set of all targets: $A \cup P$}, $m \in T$ is a target.

I = {set of unique image IDs for each image opportunity}.

$I_{s,k}$ = images visible to satellite s during cycle k.

I_m = images which cover target $m \in T$.

r_a^A = reward for active fire target $a \in A$. This is the active fire target prioritization value described in Sect. 3.1.

r_b^P = reward for pre-fire target $b \in P$. This is the pre-fire prioritization value described in Sect. 3.2.

Binary Decision Variables:

$x_m = 1$ if target $m \in T$ is observed, otherwise $= 0$.

$w_i = 1$ if image $i \in I$ is collected, otherwise $= 0$.

Continuous Decision Variables:

$sa_{s,k}$ = storage available on sat s at the beginning of cycle k.

$sc_{s,k}$ = storage consumed (used) on sat s during cycle k.

$sp_{s,k}$ = storage produced (freed up) on sat s during cycle k.

$ea_{s,k}$ = energy available on sat s at the beginning of cycle.

$eNet_{s,k}$ = net change in energy on sat s during cycle k.

Objective:

$$\text{Maximize}: \ (10 * \Sigma_{a \in A} r_a^A x_a) + \Sigma_{b \in P} r_b^P x_b \tag{1}$$

Constraints:

$$x_m \leq \sum\nolimits_{i \in I_m} w_i \forall m \in T \tag{2}$$

$$x_a = 1 \forall a \in A \tag{3}$$

Our goal is to schedule the active fire targets first and then use remaining resources for pre-fire observations. Equation (1) shows the objective is a weighted sum of the rewards for the active and pre-fire targets observed which favors active fire targets by a factor of 10. Equation (1) ensures target rewards are counted only once in the objective, even when observed by multiple satellites. This produces coordinated observations between satellites. Constraints (2) ensure that if target m is observed, then at least one image

i covering target m must be planned. Constraints (3) say all active fire targets must be scheduled. This hard constraint treats active fire targets as "mandatory", but the constraint itself is optional. If it is not possible to observe all targets, the solver will immediately return "*infeasible*", in which case we can remove this hard constraint and fall back on the weighted sum of active and pre-fire target values in the objective (1). In our case study, it was always possible to observe all active fire targets hence constraint (3) was always included.

The planner enforces these additional constraints:

1. No data collection allowed when storage is full
2. No downlinks allowed when storage is empty
3. Battery charge can never dip below a minimum level
4. Only one satellite at a time can downlink to each GS

Our novel *data cycle* abstraction makes the problem more tractable. The planner tracks aggregate resource (energy, storage) production and consumption during each cycle interval, without modeling individual seconds, as a function of time spent collecting and downlinking data. Resource constraints are enforced at the end of each data cycle.

4.2 Planner Evaluation

All experiments were run on a 2023 MacBook Pro, with an M3 Max chip, 36 GB Memory, using Python 3.8. The MILP solver was Gurobi version 11.0.3 [21]. MILP solvers may be stopped at any time and produce a feasible solution (hence responsive architecture), and solution quality monotonically improves with planning time.

Experiments using the planner and evaluating it are summarized in Table 1. For each date in the case study, we ran three tests. Active & Pre tests include both active fire and pre-fire targets, and use the Gurobi solver's default MIP gap tolerance, i.e. difference between the current best objective score and the solver's estimate of the optimal score, of 1e-4. The solver terminates with an optimal solution when the gap dips below the tolerance. The *Active and Pre* (0.01) tests relax the MIP gap tolerance to 1e-2 instead of 1e-4. The Active Only tests include only the Active Fire targets. Table 1 shows the MIP gap when the solver terminated, either when it timed out after 3-h or when an optimal solution was found. Optimal cases are marked by (*) and colored text in blue.

Table 1. Planner test cases for the case study on LA, TX, CONUS fire dates

			Science Rewards			Active Fire Targets			Pre-Fire Targets		
Test:	MIP Gap	Solve Time	Plan (objective)	Avail-able	Rwd %	Plan	Avail-able	Target %	Plan	Avail-able	Target %
LA Fires, CA: 1/8/2025											
1: Active & Pre	0.005	3 h	47993	48345	99.3	136	136	100	15209	19034	79.9
2: Act & Pre (0.01)*	0.0086	41 m	47863	48345	99.0	136	136	100	14434	19034	75.8
3: Active Only *	0.0001	0.01 s	46741	46741	100	136	136	100	N/A	N/A	N/A
Smokehouse, TX: 2/26/2024											
4: Active & Pre	0.007	3 h	156294	157968	98.9	359	359	100	19702	22119	89.1
5: Act & Pre (0.01)*	0.0099	22.6 m	155856	157968	98.7	359	359	100	19427	22119	87.8
6: Active Only *	0.0001	0.01 s	108046	108046	100	359	359	100	N/A	N/A	N/A
Most fires in 1 day: 7/5/2023											
7: Active & Pre	0.0023	3 h	208727	209617	99.6	367	367	100	19871	23216	85.6
8: Act & Pre (0.01)*	0.0056	22.9 m	208089	209617	99.3	367	367	100	19198	23216	82.7
9: Active Only *	0.0001	0.02 s	157153	157153	100	367	367	100	N/A	N/A	N/A

The science rewards section shows the total rewards collected in the plan using the objective Eq. (1), the total available rewards, and percentage of available rewards collected in the plan. Key performance metrics show that 98–100% of all available rewards are collected. 100% of active fire targets and ~ 80–90% of pre-fire targets are covered. The solver schedules active fire targets immediately (since mandatory), then spends the remaining solve time squeezing in the pre-fire targets to the storage capacity allowed. When the MIP solver is applied to pre-fire targets only using WLFP-based rewards, as a use case not shown in Table 1 [22], it captures > 99% of the targets since there are no active fires to compete for storage. Table 1 also confirms that solutions with a relaxed gap tolerance of 1e-2 collected nearly the same science reward % in ~ 30 min as the 1e-4 cases that timed out at 3 h. As evident from the near maximum rewards collected, the uncaptured targets correspond to very low rewards. Based on cleared buffer, we expect observation-to-downlink latency to be < 24 h. When added to pre- and post-processing expected in Sect. 3 and 5 respectively, the end-to-end workflow execution is expected to fit within 30 h.

5 Data Retrievals and Assimilation

To model plan execution, CYGNSS Mission Operations Center (MOC) commanded rawIF mode over satellite tracks that best matched the geographic polygons the Planner output would have observed, and downlinked the data when GS was available. This step was manual, non-optimal, and caused lengthy delays to collection and downlink. Automation engineering and operationalization are expected to improve execution. Regardless, the AI/ML technologies proposed create novel products like BAMs for active fires, SM for pre-fire conditions, and ML-based fire forecasts and show quantifiable impact to state of art. These improved forecasts support IMTs and may be used toward Sect. 3.1 to inform responsive observations in the next planning horizon.

5.1 Burned Area Maps (BAMs)

BAMs with high temporal, medium spatial resolution and low latency are being developed to be ingested by the wildfire simulation model, per the framework in Fig. 5, from GNSS-R data using the CYGNSS DDMs and an ML model called XGBoost [13]. The training features for the XGBoost binary classification model are extracted from the nominal mode data, consisting of 17 × 11 delay-Doppler bins. The ancillary data layers are collected from Soil Moisture Active Passive (SMAP) [23] and Shuttle Radar Topography Mission (SRTM) [24]. In addition to the features considered in [25], the novel addition of second-order texture information computed from DDMs as training features is hypothesized (and proven) to improve the XGBoost model. The texture analysis includes calculations of dissimilarity, entropy, and correlation of the gray-level co-occurrence matrix [26] formed from DDMs. The outputs of the ML model are the burn area binary classification, and a confidence level associated with the prediction. The model trains on 200k coordinates in < 10 min, predicts 100 coordinates in < 1min and uses memory ~ 1 MiB.

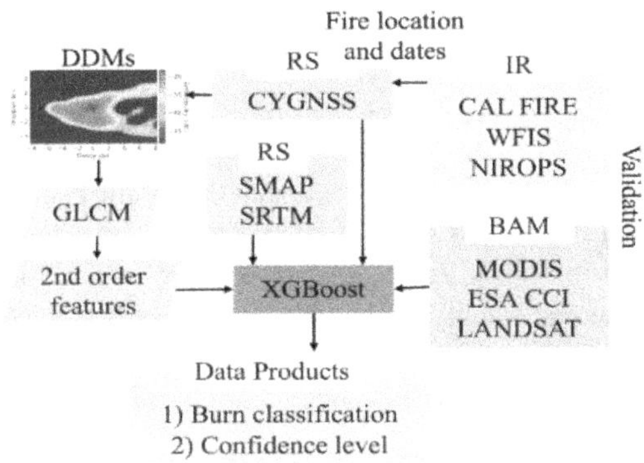

Fig. 5. Overall framework for burned area mapping. RS: Remotely sensed data; IR: Infrared data; GLCM: Grey-level co-occurrence matrix.

The ML model is validated using the set-theoretic union of three different burned area products found in the literature [27–29]. We also validated our ML approach in three different study areas, including California and Texas wildfires from CONUS and a region in northern Angola prone to periodic agricultural burns. See Fig. 6 for examples. For our case study, rawIF mode data were available for the 2024 TX and 2025 LA fires in addition to the nominal mode [30]. With the higher resolution of raw IF DDMs (111 × 69 delay-Doppler bins), we saw a 24% improvement in results from texture analysis when compared to results with just nominal mode data [31]. We saw a 15% boost in the recall of ML model classifications when rawIF was included in training along with nominal mode data. Execution of the optimal plans in a timely manner is expected to improve the collection of raw IF data and BAM classifications even further.

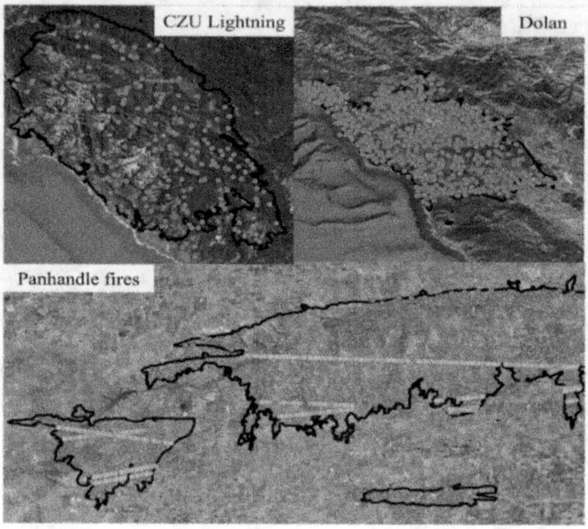

Fig. 6. Examples of BAM from XGBoost predictions for CA and TX fires. Red dots are nominal predictions. Yellow dots are raw IF mode predictions.

5.2 Active Fire Predictions (WRFx Simulation)

Fire Arrival Time (FAT) can be used as an encapsulation of the history and state of a wildland fire simulation in the WRF-SFIRE model [32] to spin up a coupled atmosphere-fire model gradually when initialized from a developed fire or modified by data assimilation [33]. FAT has been historically retrieved using active fire detections from MODIS and VIIRS, NASA's traditional single satellites with visible or near-infrared sensors at latencies less than 6 h, by separating burning and not burning pixels in time-space using a Support Vector Machine (SVM) [34].

The availability of BAMs from CYGNSS allows the addition of complementary fire detection data to a novel, ML-based fusion framework to improve state-of-the-art estimates. The Neural Network (NN) architecture in Fig. 7 employs a multi-head design with two branches of output: a classification head for fire detection and a regression head for FAT prediction. The model's backbone consists of densely connected layers with batch normalization and dropout layers for regularization. The fusion approach concatenates features from both data types with source identifiers at the input level, allowing the network to learn joint representations from these complementary fire detection methods. The network processes normalized input features and implements a custom loss function that combines binary cross-entropy for fire classification and weighted consistency-based penalty for FAT prediction. The FAT loss component penalizes predictions where the predicted arrival time is inconsistent with an observation. For fire-detected points, the loss penalizes predictions where fire is estimated to have arrived later than observed and penalizes early predictions in no-fire areas. There is no conventional label for the FAT. The overall loss is the weighted sum of fire and no-fire penalties.

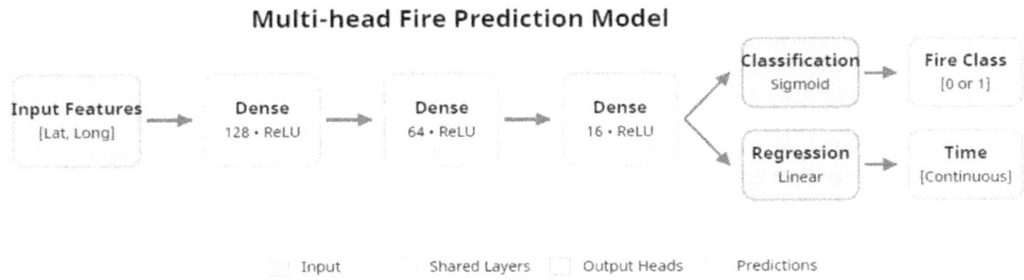

Fig. 7. Neural Network architecture to estimate Fire Arrival Time

The training pipeline implements under-sampling to address the imbalance of a small number of fire or burned points vs. a very large number of no-fire or not burned points, along with confidence-based filtering (threshold > 70) to enhance data quality. The framework is trained through a custom training loop that supports early stopping with a patience of 50 epochs, utilizing the Adam optimizer with a learning rate of 1e-4. The data fusion and FAT estimates from the NN are seen in Fig. 8 the 2024 Smokehouse Creek fire. MODIS/VIIRS data shows the active fire concentrated during the first two days of the fire, complemented by CYGNSS burned area detections starting from day two. The higher versatility of our NN, particularly with Rectified Linear Unit (ReLU) activation, is expected to enable a more accurate representation of FAT in the presence of sharp changes, such as fire suppression, compared to the SVM solution [34], which is fundamentally limited by the properties of the kernel. These improved forecasts will be integrated into data assimilation cycles within WRFx, as described in Ref [35].

CYGNSS microwave data, which can measure through smoke and clouds (unlike MODIS/VIIRS), add further value to the model output. Initial results from the NN classification of burnt area (Table 2) show that training on CYGNSS improves estimates from VIIRS/MODIS Active Fires (AF) alone, and adding RawIF data improves it further. For FAT estimation, the addition of CYGNSS data did not significantly improve results, because burned area detections appear more than a day after the fire starts. If the MILP planner's schedule is executed in a timely way as proposed in our ConOps instead of approximated by the MOC with resource-constrained delays, we expect the collected CYGNSS data to further improve predictions. Future metrics include Fire Arrival Time Prediction Accuracy as a function of clouds and reduction in model forecast error when using CYGNSS data in WRFx.

Table 2. NN performance for the 2024 TX fire. Input features were geographic coordinates, output label was burned at any time during the fire episode. Accuracy is the fraction of test locations classified correctly as Fire/No Fire.

Training Set	Test Set	Accuracy
AF	AF	72.08%
AF + CYGNSS Nominal	AF	84.5%

(*continued*)

Table 2. (*continued*)

Training Set	Test Set	Accuracy
AF + CYGNSS RawIF	AF	86.04%

The model demonstrates exceptional efficiency with only ~ 10k trainable parameters and model size of 0.08 MB, allowing it to achieve impressive inference speeds of 0.17 ms/sample on a single core of EPYC 7502 while maintaining low latency percentiles across various hardware platforms. The minimal resource requirements (3.39 MB memory usage, 2.90 W power consumption) make this model also suitable for low-SWaP deployments on edge devices.

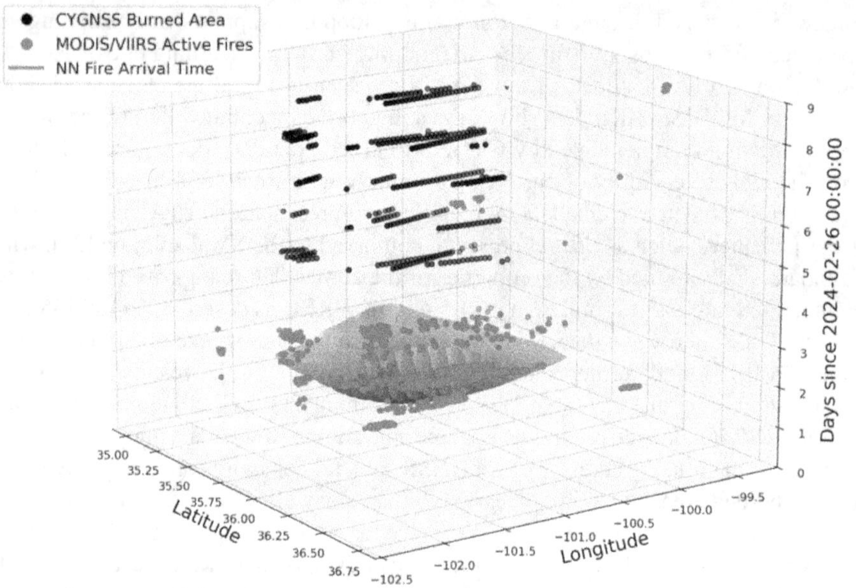

Fig. 8. Spatio-temporal evolution inferred for the 2024 TX Panhandle Smokehouse Creek Fire. CYGNSS burned area detections on specular tracks come in later than the MODIS and VIRRS Active Fires detections. Fire Arrival Time contour, estimated by the proposed NN with a bias correction applied to all data is shown. No-fire detections on CYGNSS specular point tracks and in MODIS/VIIRS granules are not shown for visibility.

5.3 Pre-fire Predictions (USGS Fire Danger)

Multiple algorithms have been developed to estimate surface SM from CYGNSS observations [30, 36]. The nominal mode produces a daily SM product with 9 km and 36 km resolution, estimated using linear regression [37], and delivers it publicly [38]. SM products from rawIF data use ML methods [39], specifically NGBoost/XGBoost, and are experimentally available in our workflow. Previous research has indicated a significant correlation between SM and fuel moisture which affects the estimation of wildland fire potential [40], however neither have been incorporated into USGS' pre-fire danger predictions. We have developed an enhanced WFPI product (WFPI-x) by integrating SM estimates from CYGNSS into WFPI and subsequently derived WLFP-x and WFSP-x from WFPI-x (defined in Sect. 3.2). Details on all models are described in [41], and validation shows expected trends in all *-x products. Since rawIF mode improves the accuracy of retrievals [39], execution of the optimal planner schedules for rawIF observations is expected to improve *-x products, leading to improved priorities set by Sect. 3.1, thereby improving our understanding of pre-fire conditions with every cycle.

6 Discussion and Future Work

We demonstrated an AI-driven spaceborne ConOps that allows the intelligent and responsive collection of novel sensor data that can be processed, and used to track, and forecast the spread of active wildfires and pre-fire danger. Novel AI elements are used in many parts of the concept – ML-based fire predictors, MILP-based planner, ML-based post-processed data products, NN-based assimilations and improved predictions. In addition to enabling wildfire monitoring with novel data and improved performance due to responsive feedback loop, these AI elements demonstrate benefit over current state of art, individually and together. Many sustainability considerations, such as low SWaP for sensors and inference, energy-efficient AI deployment, efficient onboard planning, and low latency of feedback, have been included in the system architecture and ConOps. We also found that, based on runtimes and computational requirements, the concept and its components are feasible to scale with heavier wildfire loads.

While we demonstrate the concept and models on CONUS fires, it is generalizable geospatially with differing considerations for various components. The spacecraft simulator (EO-Sim) and planner models are global. SM and BAM products use global spacecraft sensor and ancillary data, therefore applicable worldwide (as evident in validation outcomes in Angola, Africa). Active fire simulations for initial forecasts depend on the quality and resolution of fire weather and fuels data, e.g. HRRR and RAWS are local to CONUS, but alternative sources are available in other parts of the world. Active fire assimilation needs appropriate sensor data (CYGNSS, MODIS, VIIRS products are global) and WRFx, which can be used internationally if reliable and high-resolution fire behavior fuel model map is available for the region of interest. A global MODIS-based "landuse" product has been used as an alternative in fuel behavior models with a cross-walk of classes. WRF-SFIRE has been demonstrated internationally, e.g. in Greece [42], India [43], Chile, Canada. For pre-fire danger predictions, WFPI, and therefore all dependencies and enhancements described, is a CONUS product. However, similar products are feasible given the availability of fuels and weather data as with active fire cases.

Future work includes ML model improvements for both pre/active fire danger predictors, adding uncertainty quantification to the predictions, and post-assimilation FAT forecasts and using them toward improvement of the AI planner objectives and constraints. We plan to use more GNSS-R sensors beyond CYGNSS, e.g. commercial providers such as Spire Global and Muon Space, which may provide more onboard resources and further improve model outcomes. Engineering improvements to workflow automation will also ensure maximal benefit of the ConOps execution and products.

Ethics Statement and Acknowledgments. There are no ethical issues regarding the research being reported and the broader ethical impact of the work.

All the code and data described will be updated on https://github.com/dshield-proj.

This manuscript is based on work supported by the National Aeronautics and Space Administration (NASA) using primary grant 80NSSC23K1118, and partially on 80NSSC23K1344, 80NSSC22K1717, and 80NSSC22K1405.

References

1. Doerr, Stefan H., Santín, C.: Global trends in wildfire and its impacts: perceptions versus realities in a changing world. Philos. Trans. Royal Soc. B: Biol. Sci. **371**(1696), 20150345 (2016)
2. Radeloff, V.C., et al.: Rapid growth of the US wildland-urban interface raises wildfire risk, Proc. Natl. Acad. Sci. USA 115(13), 3314–3319
3. National Interagency Fire Center Statistics website https://www.nifc.gov/fire-information/statistics and open data website https://data-nifc.opendata.arcgis.com/pages/wfigs-page; Retrieved 9 Feb 2025
4. Bogel-Burroughs, N., Fortin, J., Betts, A.: Texas Fires Span Over 1.2 Million Acres. Here Is What We Know. In: The New York Times (2024). https://www.nytimes.com/article/texas-smokehouse-creek-fire.html. Accessed 31 Jan 2025
5. Karlamangla, S., Jiménez, J., Zhuang, Y., Selig, K., Nostrant, R.: What We Know About the Wildfires in Southern California. In The New York Times (2025). https://www.nytimes.com/2025/01/08/us/wildfires-los-angeles-california.html, Accessed 31 Jan 2025
6. Wang, X., Wu, G., Xing, L., Pedrycz, W.: Agile earth observation satellite scheduling over 20 years: formulations, methods, and future directions. IEEE Syst. J. **15**(3), 3881–3892 (2021)
7. Ruf, C.S., et al.: A new paradigm in earth environmental monitoring with the CYGNSS small satellite constellation. Sci. Rep. **8**(1), 8782 (2018)
8. Mandel, J., et al.: An interactive data-driven HPC system for forecasting weather, wildland fire, and smoke. In: 2019 IEEE/ACM HPC for Urgent Decision Making (UrgentHPC), Denver, CO, USA, pp. 35–44 (2019)
9. Nag, S., et al.: Distributed Spacecraft with heuristic intelligence to monitor wildfire spread for responsive control. In: IEEE International Geoscience and Remote Sensing Symposium 2024, pp. 699–703 (2024)
10. Chien, S.A., et al.: Automated volcano monitoring using multiple space and ground sensors. J. Aerosp. Inf. Syst. **17**(4), 214–228 (2020)
11. Roy-Singh, S., Li, A.P., Ravindra, V., Lammers, R. : Agile, Autonomous Spacecraft Constellations with Disruption Tolerant Networking to Monitor Precipitation and Urban Floods. In: Robotics Science and Systems – Space Robotics Workshop (2025)
12. WRFx simulation database portal, San Jose State University's Wildfire Interdisciplinary Center. Retrieved Feb 2025 https://wrfx.org/ or https://demo.openwfm.org/sj/
13. Chen, T., Guestrin. C.: Xgboost: a scalable tree boosting system. In: Proceedings of the 22nd ACM SIGKDD International Conference on Knowledge Discovery and Data Mining, pp. 785–794 (2016)
14. USGS Fire Danger Forecast. https://www.usgs.gov/fire-danger-forecast, Retrieved 7 Feb 2025
15. Burgan, R., Klaver, R., Klaver, J.: Fuel models and fire potential from satellite and surface observations. Int. J. Wildland Fire **8**(3), 159–170
16. Preisler, H., Brillinger, D., Burgan, R., Benoit, J.W.: Probability based models for estimation of wildfire risk. Int. J. Wildland Fire **13**, 133–142

17. Preisler, H., Burgan, R., Eidenshink, J., Klaver, J., Klaver, R.: Forecasting distributions of large federal-lands fires utilizing satellite and gridded weather information. Int. J. Wildland Fire **18**, 508–516
18. Ravindra, V., Ketzner, R., Nag, S.: Earth observation simulator (EO-Sim): an open-source software for observation systems design. In: 2021 IEEE International Geoscience and Remote Sensing Symposium IGARSS, pp. 7682–7685. IEEE (2021)
19. Earth Observation Simulator (EO-Sim) https://github.com/EarthObservationSimulator, Retrieved 31 Jan 2025
20. Levinson, R., Niemoeller, S., Nag, S., Ravindra, V.: Planning satellite swarm measurements for earth science models: comparing constraint processing and MILP method. In: Proceedings of the International Conference on Automated Planning and Scheduling, ICAPS 2022, Vol. 32, pp. 471–479 (2022)
21. Gurobi Optimization and Decision-making Technology. https://www.gurobi.com/, Accessed 8 Jan 2025
22. Levinson, R., Ravindra, V., Roy-Singh, S.: Optimal planning to coordinate science data collection and downlink for a constellation of agile satellites with limited storage. In: Proceedings of the International Joint Conference on Artificial Intelligence, IJCAI (2025)
23. Entekhabi, D., et al.: The soil moisture active passive (SMAP) mission. Proc. IEEE **98**(5), 704–716 (2010)
24. Earth Resources Observation and Science (EROS) Center. Shuttle Radar Topography Mission 1 Arc-Second Global (2018)
25. Kannan, A., et al.: mapping wildfire burned area using gnss-reflectometry in densely vegetated regions with complex topography: a machine learning approach. In: IEEE International Geoscience and Remote Sensing Symposium (2024)
26. Haralick, R.M., Shanmugam, K., Dinstein, H.: Textural features for image classification. IEEE Trans. Syst. Man Cybern **6**, 610–621 (1973)
27. Giglio, L., Boschetti, L., Roy, D.P., Humber, M.L., Justice, C.O.: The Collection 6 MODIS burned area mapping algorithm and product. Rem. Sens. Environ. **217**, 72–85 (2018)
28. Chuvieco, E.M., Pettinari, L., Lizundia-Loiola, J., Storm, T., Padilla Parellada, M.: ESA fire climate change initiative (Fire_cci): MODIS Fire_cci burned area pixel product, version 5.1. (2018)
29. Hawbaker, T.J., et al.: The Landsat Burned Area algorithm and products for the conterminous United States. Rem. Sens. Environ. **244**, 111801 (2020)
30. Carreno-Luengo, H., Ruf, C.S., Gleason, S., Russel, A.: A new multi-resolution cygnss data product for fully and partially coherent scattering. IEEE Trans. Geosci. Rem. Sens. (2023)
31. Kannan, A., et al.: Enhancements to mapping wildfire burned areas using GNSS-Reflectometry raw intermediate frequency data. IEEE Int. Geosci. Rem. Sens. Symp. (2025)
32. Mandel, J., et al.: Recent advances and applications of WRF-SFIRE. Nat. Hazards Earth Syst. Sci. **14**(10), 2829–2845 (2014)
33. Mandel, J., Kochanski, A.K., Vejmelka, M., Beezley, J.D.: Data assimilation of satellite fire detection in coupled atmosphere-fire simulations by WRF-SFIRE. In: Advances in Forest Fire Research, pp. 716–724. Coimbra University Press (2014)
34. Farguell, A., Mandel, J., Haley, J., Mallia, D.V., Kochanski, A., Hilburn, K.: Machine learning estimation of fire arrival time from level-2 active fires satellite data. Rem. Sens. **13**(11), 2203 (2021)
35. Kochanski, A.K., Clough. K., Farguell, A., Mallia, D.V., Mandel, J., Hilburn, K.: Analysis of methods for assimilating fire perimeters into a coupled fire-atmosphere model. Front. Forests Global Change **6**, 1203578 (2023)
36. Wu, X., Ma, W., Xia, J., Bai, W., Jin, S., Calabia, A.: Spaceborne GNSS-R soil moisture retrieval: status, development opportunities, and challenges. Rem. Sens. **13**(1), 45 (2020)

37. Chew, C., Small, E.: Description of the UCAR/CU soil moisture product. Rem. Sens. **12**(10), 1558 (2020)
38. CYGNSS. 2024. CYGNSS Level 3 Soil Moisture Version 3.2. NASA Physical Oceanography Distributed Active Archive Center. https://doi.org/10.5067/CYGNU-L3S32
39. Tsagkatakis, G., Melebari, A., Akbar, R., Campbell, J.D., Hodges, E., Moghaddam, M.: Uncertainty quantification in machine learning based retrieval of soil moisture from GNSS-R observations. In: IEEE International Geoscience and Remote Sensing Symposium, Athens, Greece (2024)
40. Krueger, E., Levi, M., Achieng, K., Bolten, J., Carlson, J., Coops, N., et al.: Using soil moisture information to better understand and predict wildfire danger: a review of recent developments and outstanding questions. Int. J. Wildland Fire **32**, 111–132 (2023)
41. Nelson, K., Roy-Singh, S., Ravindra, V., Moghaddam, M., Kannan, A., Melebari, A.: Integration of GNSS-R derived Soil Moisture into the USGS Wildland Fire Potential Index. In: IEEE International Geoscience and Remote Sensing Symposium (2025)
42. Kartsios, S., Karacostas, T., Pytharoulis, I., Dimitrakopoulos, A.P.: Numerical investigation of atmosphere-fire interactions during high-impact wildland fire events in Greece. Atmos. Res. **247**, 105253 (2021)
43. Kale, M.P., et al.: Operational Forest-Fire Spread Forecasting Using the WRF-SFIRE Model. Remote Sensing **16**(13), 2480 (2024)

DDHC: Domain-Driven Hierarchical Classification Framework for Stream Type Identification

Golnaz Mesbahi[1](✉), Shadan Golestan[1], Buddy Brown[2], Joey Ton[1], and Jerome Cranston[2]

[1] Alberta Machine Intelligence Institute, Edmonton, Canada
golnaz.mesbahi@amii.ca
[2] Carson Integrated, Edmonton, Canada

Abstract. Land Surface Water (LSW) such as streams is vital for human and ecological systems. Efficient stream type classification is crucial for sustainable resource management. However, it traditionally relies on costly field surveys. This paper proposes a framework that leverages diverse geospatial features and domain knowledge to curate representative datasets and design hierarchical classification models for stream type identification. Empirical results demonstrate that our framework outperforms standard multi-class models, offering great promise for exploring data representation and structure in environmental classification tasks.

Keywords: Water Stream Classification · Hierarchical Classification · Supervised Learning

1 Introduction

Land Surface Water (LSW), such as streams and rivers, is the most accessible water resource to the human population, and can be a critical source of drinking water, agricultural water, and water used for various ecosystems [26]. The impact of LSW spans multiple domains of the United Nations Sustainable Development Goals [25]. LSW plays a crucial role in preserving the ecological equilibrium of ecosystem function and supporting the sustainable urban development [6].

Sustainable resource management requires efficient evaluation and classification of streams (see for example [14,19,27]) and including information about their flow characteristics, i.e., stream types, in order to make decisions around construction projects that may include infrastructure. Traditional stream type classification relies on manual field surveys and expert judgment. This process is extremely time-consuming and costly when covering vast or remote areas. Timely and accurate stream classification enables both resource-efficient and evidence-based ecological planning. Recent advances in AI-driven remote sensing promise automated, interpretable, and ecologically sensitive stream-type mapping across entire river systems.

Published by Springer Nature Switzerland AG 2026
E. Mercier-Laurent et al. (Eds.): AI4S 2025, IFIP AICT 780, pp. 157–168, 2026.
https://doi.org/10.1007/978-3-032-21184-2_13

A growing body of literature have applied rule-based or machine/deep learning algorithms for related tasks such as water presence detection [9], water body extraction [5,24], and monitoring changes in surface water area [8], significantly advancing LSW monitoring, and offering scalable and cost-effective alternatives to traditional methods. However, most existing methods focus on high-level tasks where coarse-grained satellite imagery data is often sufficient. In contrast, several related works rely on pixel-level information for higher-granularity tasks, such as river surface elevation estimation [10]. Similarly, stream type classification requires high-resolution datasets and tailored machine learning frameworks to capture subtle hydrological and morphological features, which sits at the intersection of environmental science and AI-driven remote sensing, demanding hybrid approaches that can embed domain expertise while remaining adaptable to regional variability.

In this paper, we investigate whether diverse features associated with water stream data can be used to classify streams. To investigate, we design the Domain-Driven Hierarchical Classification (DDHC) framework specifically tailored for stream type classification tasks. This framework utilizes the Geographic Information System (GIS) technology that stores and analyze the spatial information contained in raster images. To design a hierarchical classification method, we develop a set of domain-driven heuristic approaches to construct different hierarchy structures. Each hierarchy consists of multiple sub-classification tasks, whose aggregation collectively approximates the original multi-class stream type classification problem. Our hypothesis is that this reformulation of the original task with curating a high-granularity dataset restructures the problem in a way that can lead to improved performance and better alignment with domain knowledge. In addition, we incorporate interpretability features such as SHAP-based importance analysis at each node of the hierarchy, allowing practitioners to trace which geospatial or hydrological features drive each sub-classification decision.

Our main contributions are: 1) We curate a high-granularity labeled dataset from diverse sources, suitable for stream type classification 2) We design a hierarchical classification framework guided by domain-driven heuristics 3) We empirically validate the superior performance of our framework compared to conventional multi-class classification methods. To ensure reproducibility, we have made our code publicly available at https://github.com/golnaz-amii/DDHC.

2 Related Work

At the high-granularity end of the spectrum, some studies leverage high-resolution imagery and deep learning methods to identify LSW. For instance, studies such as [20,22] demonstrated that applying convolutional neural networks to commercial satellite imagery can successfully detect LSW. Moortgat et al. [20] specifically illustrated that these models can learn the morphological features of rivers. One challenge for this line of research is the limited availability of high-granularity datasets. To address this, Li et al. [17] proposed a

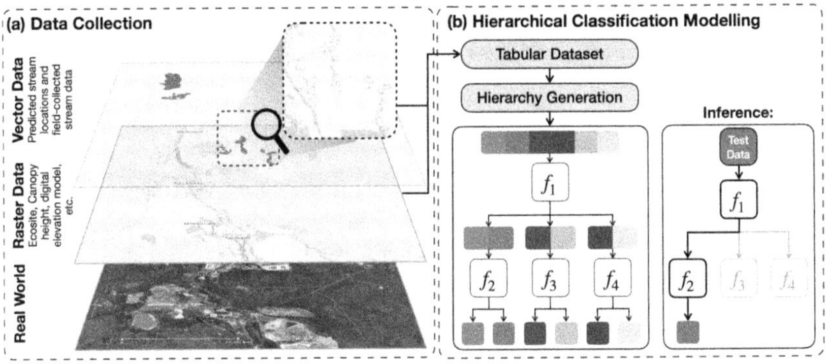

Fig. 1. Overview of our proposed DDHC framework.

large-scale dataset (GLH-Water) that offers sub-meter imagery containing 40.96 billion labelled water pixels to support fine-grained flood mapping tasks.

On the other hand, several studies [9,11,13,21] use lower-granularity tabular datasets (10–30 m). Specifically, Ko et al. [13] trained two boosted random-forest classifiers on 30 m Landsat 8 reflectance[1] and water indices for water-body classification.

Several other studies adopt rule-based algorithms or exploratory data analysis. For instance, Xu et al. [27] introduced a modified normalized difference water index that uses a simple band ratio to highlight open water and suppress built-up areas, while Kuenzer et al. [12] have compared the performance of different radar sensors for flood monitoring. Other studies leverage machine learning models for improving the quality of the data. Hao et al. [12] utilized a self-supervised learning method to fill missing observations in the JRC Global Surface Water Dataset.

In this paper, we address a more fine-grained problem: distinguishing among multiple stream types. Instead of relying solely on pixel-level imagery datasets, we curate a highly granular tabular dataset and use domain-driven heuristics to define a spatial buffer around each stream segment to collect and compute multi-source geospatial features. Similar to Ko et al. [13], our DDHC framework leverages tree-based models (XGBoost). Unlike the related work, DDHC utilizes domain-driven heuristics to restructure the multiclass LSW classification problem into a hierarchy of classification tasks, enabling the framework to incorporate expert knowledge and outperform the standard multi-class approach.

3 Methodology

We present our proposed Domain-Driven Hierarchical Classification (DDHC) framework in Fig. 1, specifically designed for the stream type classification using

[1] see https://www.usgs.gov/landsat-missions/landsat-8 for details.

raster and vector data sources. The DDHC framework consists of two main parts: (a) Data Collection, and (b) Hierarchical Classification Modeling.

3.1 Data Collection

The Data Collection involves integrating both raster and vector geospatial data, forming the foundation for constructing a high-granularity, feature-rich dataset for downstream tasks. Raster data is a grid of pixels encoding continuous geographical information. Vector data consists of predicted stream and river locations generated using the Wet Areas Mapping (WAM) algorithm, along with ground-truth stream type labels.

The dataset curated for this work consists of 11547 samples and 88 features, collected from multiple geospatial data sources. Each sample represents a water stream segment in a high-dimensional feature space, capturing its characteristics obtained from multiple sources: the Derived Ecosite Phase (DEP) dataset [3]; a Digital Elevation Model (DEM) with 1 m resolution, generated from LiDAR data used in WAM algorithm, with a point cloud density specification of 2.5-4 points per square meter [1]; and a Canopy Height Model with a 3 m resolution, also derived from LiDAR. In addition, terrain-related features such as depth to water, terrain ruggedness index, topographic position index, and convergence index, were incorporated to further enrich the environmental representation. Note that raster-based features are derived by aggregating values within a specified surrounding area, defined by a buffer size B, to compute the final feature values. The samples are labeled with one of six potential watercourse classes: dry, ephemeral, intermittent, transitional, small permanent, and large permanent. These labels are assigned by field crews based on on-site assessments, following definitions established by the Government of Alberta and documented in Forest Harvesting Operating Ground Rules [2]. The resulting dataset is compiled into a unified tabular format for downstream classification tasks.

3.2 Hierarchical Classification Modelling

In this part, the tabular dataset is used to design various hierarchical classification structures. Rather than solving an imbalanced 6-class classification, we decompose it into a hierarchy of simpler classification sub-tasks. Specifically, a high-level classifier initially makes a high-level decision between broader, merged categories. Then, lower-level classifiers are used to refine the predictions within each merged category. This process progressively narrows down the classification, ultimately assigning each stream segment to one of the six classes.

We propose two approaches for the Hierarchy Generation module: a *domain-knowledge-based* method, and a *greedy-based* method. In the former, the hierarchy is manually designed using prior expert insights and intuitions into stream type characteristics and their relationships. The latter begins by grouping classes into pairs based on their proximity in the feature space, using predefined domain-informed similarity metrics (i.e., Canberra, Euclidean, Pearson correlation, and

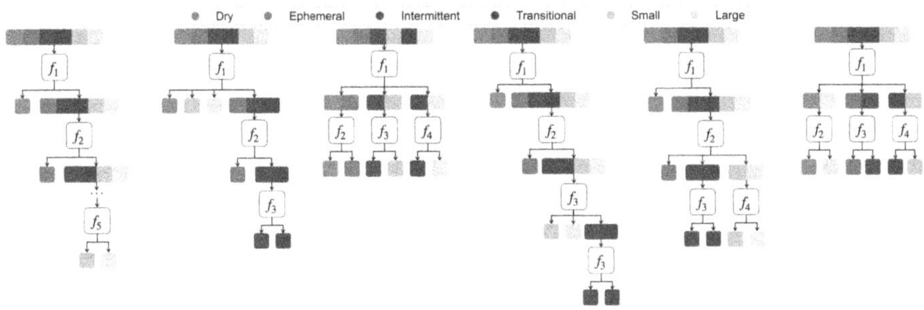

(a) HC1 (b) HC2 (c) HC3 (d) HC4 (e) HC5 (f) HC6

Fig. 2. Visualization of six different hierarchy configurations. HC1, HC2, HC4, and HC5 are domain-knowledge-based hierarchies, while HC3 and HC6 are generated using the greedy-based approach.

cosine similarity). This results in the first level of the hierarchy, where a high-level classifier (f_1) is trained to distinguish between these superclasses. In the next level of the hierarchy, separate classifier models (f_2, f_3, and f_4 in our case) are trained within each superclass, with each model responsible for classifying the two original classes that comprise its corresponding superclass. During inference, a test sample is first processed by the higher-level classifier and, based on its prediction, routed to the appropriate lower-level classifier for final classification.

We compare hierarchies obtained from both approaches with a standard multi-class classifier model that directly predicts among all six classes. Figure 2 illustrates the hierarchies generated by DDHC framework. According to the figure, `HC1`, `HC2`, `HC4` and `HC5` are obtained via domain-knowledge-based method: `HC1` adopts a one-vs-rest strategy; `HC2` is motivated by the observation that standard XGBoost struggles to distinguish ephemeral, intermittent, and transitional streams; `HC4` and `HC5` are variants of `HC1` to isolate larger stream types (small and large) earlier in the hierarchy. In addition, `HC3` and `HC6` are generated via the greedy-based method: `HC3` can be obtained using Euclidean and Canberra distances, while `HC6` results from Pearson correlation and cosine similarity. We choose Extreme Gradient boosting (XGBoost) [7] for all of the $f_1, ..., f_5$ classifiers, as it is robust and effective in many applications ranging from predictive modeling in healthcare [15] to chemical and environmental settings [23]. It has also demonstrated robust performance on imbalanced datasets [16].

4 Experiments

We evaluate hierarchy models using *balanced accuracy* given the imbalanced nature of the dataset.

We assemble multiple datasets, each calculated with a particular buffer size $B \in \{5, 10, 20, 30, 40, 50, 60\}$ meters. We also curate a dataset with variable buffer sizes, tailored to each stream type. This setting, which we refer to as

Table 1. Average balanced accuracy (± 95% CI) for each HC and XGBoost for different buffer sizes. A star (*) indicates methods that are significantly outperformed by the best-performing HC (shown in bold) at that buffer size.

Method	5 m	10 m	20 m	30 m	40 m	50 m	60 m	`variable-size`
XGBoost	67.09±0.74*	67.40±0.69*	67.08±0.77*	67.77±0.86*	68.04±0.79*	67.86±0.77*	67.71±0.73*	74.83±0.77*
`HC1`	68.57±0.53*	68.55±0.62*	68.99±0.98	68.82±0.66*	68.92±0.84*	69.96±0.72	69.73±0.78*	75.88±0.65*
`HC2`	**69.83±0.62**	**69.94±0.78**	**69.83±0.84**	**69.92±0.69**	69.94±0.69	**70.56±0.66**	70.39±0.88	76.96±0.63
`HC3`	68.61±0.53*	69.06±0.76	69.28±0.70	69.60±0.71	69.72±0.73	69.75±0.58	69.77±0.80*	76.99±0.53
`HC4`	69.24±0.57	69.63±0.67	69.76±0.75	69.80±0.64	**70.05±0.61**	70.46±0.78	**71.08±0.81**	76.63±0.58
`HC5`	69.00±0.55*	69.28±0.72	69.01±0.94	69.04±0.53*	69.48±0.88	69.54±0.75*	69.81±0.97*	**77.07±0.59**
`HC6`	68.36±0.77*	67.63±0.83*	68.02±0.81∗	68.64±0.77*	68.12±0.72*	68.72±0.57*	68.26±0.58*	74.70±0.70*

"`variable-buffer`", accounts for the class-specific spatial characteristics. Specifically, we use B=10 for dry, ephemeral, intermittent, and transitional streams, B=30 for small streams, and B=60 for large streams, following the intuition that smaller classes exert less spatial influence on their surroundings.

To test our hypotheses empirically, we train each HC across all buffer sizes, as shown in Table 1. Each experiment is repeated with 10 random seeds, and we report the mean balanced accuracy with ±95% confidence interval (CI). We perform pairwise Welch's t-test between the methods, for each B value separately, and consider results statistically significant at $p < 0.05$. We identify the best-performing method in each buffer-size in bold, and mark methods with a star ($*$) if the best-performing HC significantly outperforms at the corresponding buffer size. We make the following observations: 1) Across all the buffer sizes, all HCs significantly outperform XGBoost, except `HC6`, which is significantly better than XGBoost only at $B = 5$; 2) `HC2`, `HC4` and `HC5` are among the best-performing methods across multiple buffer sizes; 3) `HC5` has the highest performance for the `variable-size` buffer and its performance is significantly better than `HC1` and XGBoost; 4) for each HC, performance improves with larger fixed B, the `variable-size` buffer consistently attains the best balanced accuracy.

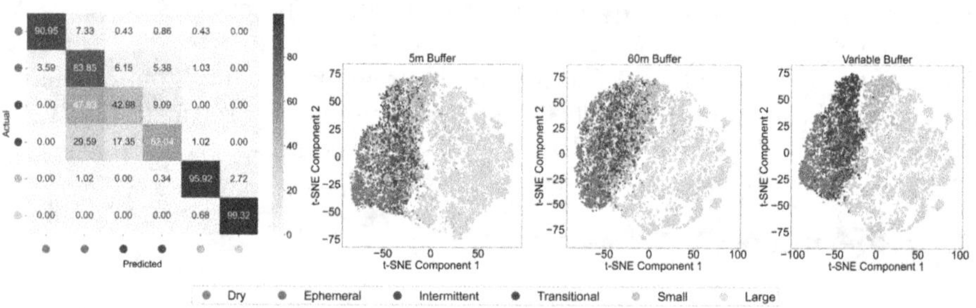

Fig. 3. Normalized confusion matrix for `HC5` for B=`variable-size` and t-SNE visualizations for datasets with B=5m, B=60m and B=`variable-size`.

To investigate observation 3 and 4 further, we calculate the normalized confusion matrix for `HC5` in `variable-size` buffer (for one run) in Fig. 3. It can be seen that there is confusion in classifying the intermediate classes, i.e., ephemeral, intermittent, and transitional. However, the model outperforms XGBoost on intermittent and matches its accuracy on ephemeral and transitional. We believe this is because the combination of specific buffer sizes in `variable-size` buffer setting with appropriate HCs can effectively capture stream characteristics and improve classification performance. To examine this, we calculate t-distributed stochastic neighbor embedding (t-SNE) to project the original high-dimensional data into two dimensions for $B \in \{5, 60, \texttt{variable-size}\}$, as illustrated in Fig. 3. It can be seen that classes are more separable for B=`variable-size`. By comparing the dataset's t-SNE representation with the structures of `HC5`, we observe that the large and small classes are clearly separated from the rest. `HC5` exploits this structure by grouping these two classes early via f_2, and easily separating them with f_4. Also, by grouping intermittent and transitional in f_2 and classifying them with a dedicated binary classifier (f_5), `HC5` reduces confusion between these overlapping classes. Nevertheless, the model can still struggle at f_2 and f_3 when classifying ephemeral from the tightly clustered intermittent and transitional classes. Note that `HC2`, `HC3`, and `HC4` share relatively similar structures, hence their comparable performance.

4.1 Interpretability of the Experiments

We analyze how individual features contribute to the models' predictions, with the features and their descriptions listed in Table 2. For features denoted with a trailing –, we compute multiple statistics within a buffer, including mean, maximum, minimum, range, and standard deviation. In contrast, standalone features such as length and location are used without additional statistics. As a first step, we examine the feature importance rankings of XGBoost, used as our baseline, and identify the top ten features contributing most significantly to the predictions, as shown in Fig. 4.

The importance score is defined as the average information gain from splits involving a given feature. Specifically, it is computed as the reduction in entropy between parent and child nodes, averaged across all trees in the ensemble. This metric highlights the features that play the greatest role in reducing impurity in the decision trees. The results are reported on the variable-buffer dataset. Observing the importance plot can help us understand and validate the assumptions coming from domain knowledge about the contribution of features. For instance, two features of interest in Fig. 4 are caatinlet and caatoutlet, which refer to catchment inlet and outlet, respectively. These features are defined as the number of pixels at the inlet and outlet of a stream, indicating how wide the stream is. As expected, these features play an important role, as the original definition of the stream classes comes from their width, defined by the Government of Alberta Harvesting Operating Ground Rules [2].

Using the Hierarchical structures, we can gain insights into which features were more influential in separating a particular class from others in each hier-

Table 2. Features and their descriptions

Feature	Description
aspect−	Direction of the slope at a point
chm−	Canopy Height Model
cnvgi−	Convergence index, indicating terrain slopes convergence or divergence
curve−	curvature of the terrain surface
dem−	Digital Elevation Model
dtw−	Depth-to-water index
tri−	Terrain Ruggedness Index, a measure of surface roughness
tpi−	Topographic Position Index
location	Spatial coordinates, indicating the start and end of a stream
length	Length of the stream
slope	Steepness of terrain
edge_flow	Flow accumulation
z	Elevation value at the point
order	Strahler Stream Order
magnitude	Number of Strahler order 1 stream segments flowing into the segment
wsno	Watershed number
caatinlet	number of pixels at the catchment inlet
caatoutlet	number of pixels at the catchment outlet

archy. This can help understand how the hierarchies are built. As an example, we look into the most important features based on XGBoost feature importance (information gain metric), for the `HC1` model, where at each level, one class is separated from the others (one-vs-rest strategy). We find that features related to catchment play an important role, constantly being at the top for classifiers at multiple hierarchies. Other important features are magnitude, and curvature statistics.

Finally, we examine the SHAP (SHapley Additive exPlanations) plots for `HC5`, the best-performing model. SHAP is a game-theoryâĂŞinspired method for model interpretability [18], quantifying how each feature value contributes to a prediction. For example, Fig. 5a shows the summary plot for the f_1 classifier in `HC5`, which is responsible for distinguishing the No class from the rest. The x-axis represents the magnitude of feature contributions. Positive values indicate a push toward predicting the sample as part of the rest class, while negative values indicate contributions toward predicting it as No. The color scale reflects feature values; for instance, for the caatinlet feature, higher values tend to push predictions toward the rest class. By examining the SHAP plots for each classifier separately, we can gain more detailed insights into how different features influence positive or negative predictions depending on their values.

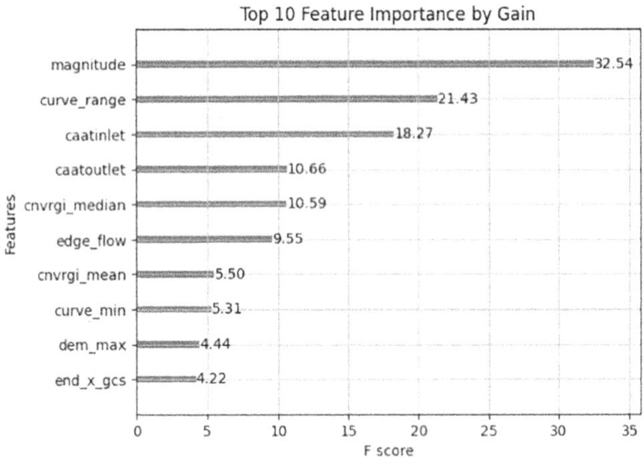

Fig. 4. XGBoost Feature Importance List, Based on Gain.

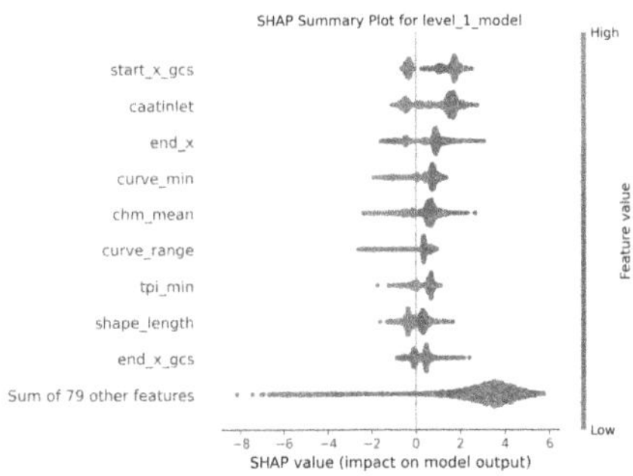

(a) Summary plot for f_1 classifier in HC5

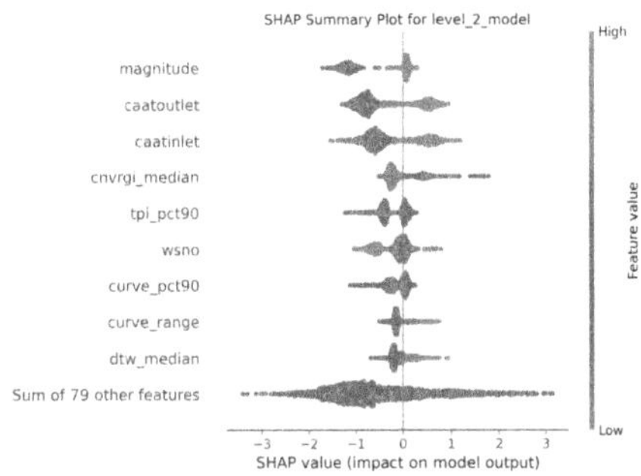

(b) Summary plot for f_2 classifier in HC5

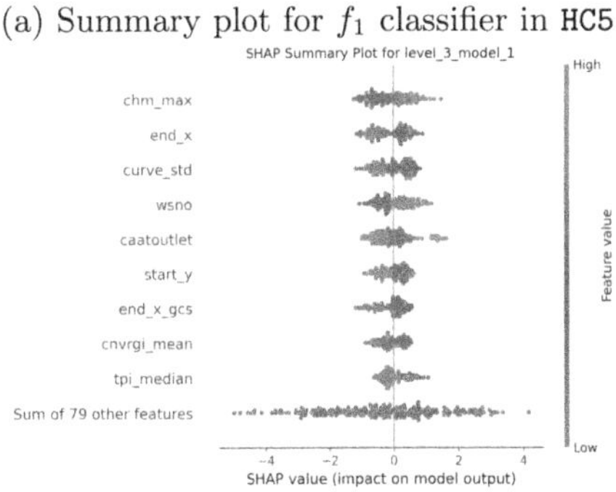

(c) Summary plot for f_3 classifier in HC5

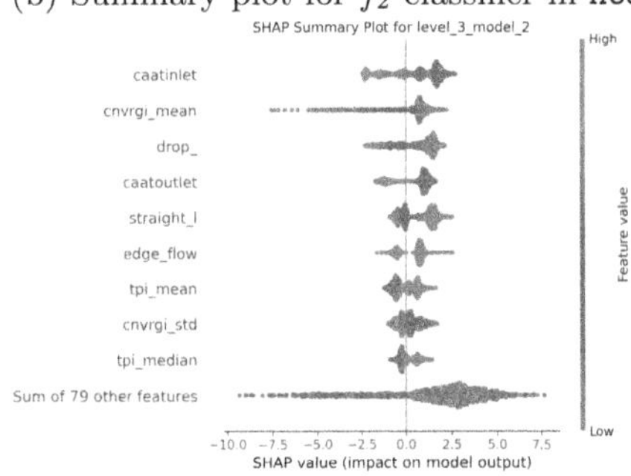

(d) Summary plot for f_4 classifier in HC5

Fig. 5. SHAP summary plots for HC5.

5 Discussion and Conclusion

In this work, we proposed a domain-driven hierarchical framework for stream-type classification. First, we curated a multi-source dataset using various buffer sizes to capture spatial context at varying scales. Using the framework, we built hierarchical models guided by domain heuristics. Results showed that leveraging stream structure through hierarchies yielded notable performance gains and that class-specific buffer sizes were critical for accurate feature representation.

We emphasize that intermediate stream classes, i.e., ephemeral, intermittent, and transitional, were harder to classify than others, likely due to subjective field definitions and strong visualâĂŞspatial overlap. This ambiguity appeared in the t-SNE plot and confusion matrix (Fig. 3). These challenges have direct implications for regulatory permitting, habitat assessment, and flood management, where accurate differentiation between streams can affect legal protections and planning outcomes.

We argue that several promising research directions remain for improving DDHC. Our implementation used XGBoost at every node; second, the greedy hierarchy generator permitted only binary merges. More effective hierarchies could mix classifiers and allow flexible merges. Note that DDHC's computational cost rises only modestly, as each additional hierarchy level trains its classifier on a progressively smaller subset of data. Expanding the search space of possible hierarchy configurations also encourages using optimization algorithms, e.g., evolutionary techniques, Bayesian optimization, and Reinforcement Learning. Automating buffer-size selection with heuristics or unsupervised clustering also appears promising, given its influence on feature quality. Finally, we note the need to test DDHC across diverse regions, since region-specific training data could bias results and stream characteristics vary geographically.

Beyond stream classification, we propose extending DDHC to wetland and flood-zone mapping and habitat classification, where hydrological hierarchies help capture nested patterns. Key ethical risks in deploying DDHC include ecological harm if misclassifications guide development into sensitive habitats.

Manual labeling by field experts remains the costliest step. To reduce dependency on labeled data, self-supervised learning approaches, such as [12], could be employed to extract useful representations from the roughly 34,000 unlabeled samples. These methods utilize the structure of the unlabeled data by creating pseudo-labels and using them to teach the model to predict withheld portion of the input. Once these general representations are learned, the performance of the downstream stream and river classification task can be improved by using the labeled set. In parallel, active learning algorithms can complement this by ranking unlabeled samples according to their information content and querying only the most uncertain or representative examples for manual labeling. For an example of integrating active learning into a self-supervised framework, see [4], which introduces an oracle-driven approach that queries semantic relationships between samples to build positive pairs for self-supervised training. Such strategies have been shown to be sample-efficient, substantially reducing the need for manual labeling

Our current formulation treats stream type classification as a discrete multi-class prediction problem. An alternative method is to formulate this problem as a regression task, predicting continuous scores representing flow persistence. These scores could then be clustered to drive type categories, potentially capturing gradual transitions in stream characteristics, that may be lost in categorical approaches.

Finally, the DDHC framework could benefit from integration with remote-sensing resources, such as the GLH-Water dataset [17], enabling pixel-level identification of streams and rivers. Combining such fine-grained segmentation outputs with our hierarchical classifiers could enrich the feature space and improve the classification of ambiguous classes.

References

1. Alberta Provincial Digital Elevation Model - Open Government — open.alberta.ca. https://open.alberta.ca/opendata/gda-c16469a2-5541-455c-bba0-63a24c0ff08a, Accessed 14 Apr 2025
2. Alberta timber harvest planning and operating ground rules (2024) - Open Government. https://open.alberta.ca/publications/timber-harvest-planning-and-ogr-2024
3. Derived ecosite phase. version 2, alberta government (2020). https://open.alberta.ca/publications/derived-ecosite-phase-version-2
4. Cabannes, V., Bottou, L., Lecun, Y., Balestriero, R.: Active self-supervised learning: A few low-cost relationships are all you need. In: Proceedings of the IEEE/CVF International Conference on Computer Vision, pp. 16274–16283 (2023)
5. Cao, H., Tian, Y., Liu, Y., Wang, R.: Water body extraction from high spatial resolution remote sensing images based on enhanced u-net and multi-scale information fusion. Sci. Rep. (2024)
6. Chen, F., Chen, X., Van de Voorde, T., Roberts, D., Jiang, H., Xu, W.: Open water detection in urban environments using high spatial resolution remote sensing imagery. Rem. Sens. Environ. (2020)
7. Chen, T., Guestrin, C.: Xgboost: a scalable tree boosting system. In: Proceedings of the ACM SIGKDD International Conference on Knowledge Discovery and Data Mining (2016)
8. Composto, R.W., Tulbure, M.G., Tiwari, V., Gaines, M.D., Caineta, J.: Quantifying urban flood extent using satellite imagery and machine learning. Nat. Hazards (2025)
9. Cordeiro, M.C., Martinez, J.M., Peña-Luque, S.: Automatic water detection from multidimensional hierarchical clustering for sentinel-2 images and a comparison with level 2a processors. Rem. Sens. Environ. **253** (2021)
10. Dai, C., Durand, M., Howat, I.M., Altenau, E.H., Pavelsky, T.M.: Estimating river surface elevation from arcticdem. Geophys. Res. Lett. (2018)
11. Du, Z., et al.: Estimating surface water area changes using time-series landsat data in the qingjiang river basin, China. J. Appl. Rem. Sens. (2012)
12. Hao, Z., et al..: Resolving data gaps in global surface water monthly records through a self-supervised deep learning strategy. J. Hydrol. (2024)
13. Ko, B.C., Kim, H.H., Nam, J.Y.: Classification of potential water bodies using landsat 8 oli and a combination of two boosted random forest classifiers. Sensors (2015)

14. Kuenzer, C., Guo, H., Huth, J., Leinenkugel, P., Li, X., Dech, S.: Flood mapping and flood dynamics of the mekong delta: Envisat-asar-wsm based time series analyses. Rem. Sens. (2013)
15. Kumar, P., Pandi, S.S., Karthikeyan, N., Balakrishnan, V.P.: Utilizing cloud computing to deliver accurate and scalable diabetes predictions with xgboost. 2nd IEEE International Conference on Advances in Information Technology, ICAIT 2024 - Proceedings (2024)
16. Le, T.T.H., Oktian, Y.E., Kim, H.: Xgboost for imbalanced multiclass classification-based industrial internet of things intrusion detection systems. Sustainability **14**, 8707 (2022)
17. Li, Y., Dang, B., Li, W., Zhang, Y.: Glh-water: a large-scale dataset for global surface water detection in large-size very-high-resolution satellite imagery. In: Proceedings of the AAAI Conference on Artificial Intelligence (2024)
18. Lundberg, S.M., Lee, S.I.: A unified approach to interpreting model predictions. Adv. Neural Inf. Process. Syst. (2017)
19. Mason, D.C., Speck, R., Devereux, B., Schumann, G.J.P., Neal, J.C., Bates, P.D.: Flood detection in urban areas using terrasar-x. IEEE Trans. Geosci. Rem. Sens. (2009)
20. Moortgat, J., Li, Z., Durand, M., Howat, I., Yadav, B., Dai, C.: Deep learning models for river classification at sub-meter resolutions from multispectral and panchromatic commercial satellite imagery. Rem. Sens. Environ. (2022)
21. Pekel, J.F., Cottam, A., Gorelick, N., Belward, A.S.: High-resolution mapping of global surface water and its long-term changes. Nature (2016)
22. Popandopulo, G., Illarionova, S., Shadrin, D., Evteeva, K., Sotiriadi, N., Burnaev, E.: Flood extent and volume estimation using remote sensing data. Rem. Sens. **15**(18), 4463 (2023)
23. Shaik, N.B., Jongkittinarukorn, K., Bingi, K.: Xgboost based enhanced predictive model for handling missing input parameters: a case study on gas turbine. Case Stud. Chem. Environ. Eng. (2024)
24. Song, Y., Rui, X., Li, J.: Aednet: An attention-based encoder–decoder network for urban water extraction from high spatial resolution remote sensing images. IEEE J. Selected Top. Appl. Earth Observations Rem. Sens. (2023)
25. United Nations: the 17 goals: Sustainable development (2023). https://sdgs.un.org/goals, Accessed 17 Apr 2025
26. World Health Organization: Drinking-water (2023). https://www.who.int/news-room/fact-sheets/detail/drinking-water, Accessed 17 Apr 2025
27. Xu, H.: Modification of normalised difference water index (ndwi) to enhance open water features in remotely sensed imagery. Int. J. Rem. Sens. (2006)

Explainable AI for Wildfire Management and Ecological Sustainability: A Short Survey of Techniques, Case Studies, and Operational Pathways

Akshata Kishore Moharir[1], Jay Prakash Thakur[2], Divya Arvainda Nayak[3(✉)], and Sameera S Kashyap[4]

[1] Portland, OR, USA
[2] Los Angeles, CA, USA
[3] Chesapeake, VA, USA
nayakdivya01@gmail.com
[4] University of California, Santa Cruz, Santa Cruz, CA, USA

Abstract. The growing destructiveness of wildfires, amplified by climate change, demands transparent predictive tools. AI models have advanced wildfire prediction, but their "black-box" nature hinders operational adoption. This short survey investigates explainable AI (XAI) methods for wildfire management and ecological sustainability, focusing on bridging the experimental to operational gap. We systematize XAI techniques (model-agnostic, model specific, hybrid/emerging) into a unified taxonomy, evaluating their interpretability, computational efficiency, and fidelity. Deployed case studies show that XAI integration yields substantial benefits but also reveals critical implementation barriers such as resource constraints and legacy system integration. Synthesizing fragmented literature and empirical insights, this work provides actionable guidelines for selecting the right XAI methods that balance technical stability with operational practicality. It also highlights research priorities to align the advances of XAI with the needs of firefighters, policymakers, and ecosystem sustainability managers to promote climate resilience and biodiversity conservation.

Keywords: Explainable AI · Wildfire Prediction · Climate Change · Ecological Sustainability · Model Interpretability · Operational Deployment · XAI Taxonomy · Climate Resilience · Biodiversity Conservation · Black-box Models · Wildfire Management · Model-agnostic Methods

1 Introduction

Climate change has impacted the frequency of severe wildfires, increasing the number of such occurrences each year. This has repeatedly caused damage to

A. K. Moharir, J. P. Thakur and D. A. Nayak—Independent Researcher.

Published by Springer Nature Switzerland AG 2026
E. Mercier-Laurent et al. (Eds.): AI4S 2025, IFIP AICT 780, pp. 169–189, 2026.
https://doi.org/10.1007/978-3-032-21184-2_14

infrastructure and ecosystems and is a great threat to surrounding populations and the economy.

Although several advances in deep learning have enabled AI models to improve wildfire prediction, their "black-box" nature often limits operational trust and adoption in critical environments. Explainable AI (XAI) aims to bridge this gap by making model decisions transparent, enabling stakeholders to understand, trust, and act upon predictions in high stakes scenarios. This understanding is crucial, as wildfire management decisions have immediate safety consequences and profound long-term implications for biodiversity, community resilience, and broader goals of sustainable development.

Artificial intelligence and machine learning methods are considered powerful tools for predicting wildfires. These approaches use diverse data sets such as satellite imagery, weather data, topography, and vegetation indices data to predict the origin of the fire, its behavior, propagation, and preventive actions that can be applied [49]. However, the opacity of many AI models significantly challenges their operational adoption. Deep learning complexity, for example, often obscures predictive reasoning, creating a "black box" effect that undermines stakeholder trust [61]. Emergency responders, policymakers, AI developers, and the public need transparent and interpretable AI insights to make confident decisions under pressure. These decisions impact the immediate safety of hazardous incidents, further setting the stage for the degradation of long-term ecological health and biodiversity, which are essential for a sustainable ecosystem [20].

The importance of accountability where lives, property and the ecosystem are endangered calls for predictive fire management in high-stakes environments [26].

XAI techniques aim to focus on explaining the key drivers driving AI predictions, allowing business users to understand why a particular outcome was predicted and how the model arrived at its conclusion [45]. XAI can foster trust, transparency, and enable stakeholders to critically evaluate system output by providing insight into the model's reasoning. The ultimate goal is to create AI and machine learning systems that are not only accurate but also transparent, reliable, and understandable, enabling informed decision making and promoting responsible use of AI in critical applications [42]. Explainability is also valuable for evaluating and improving AI, ML model performance, identifying weaknesses, and supporting targeted adaptive fire management strategies. Such strategies, informed by transparent and explainable AI, are essential to move beyond reactive suppression to proactive measures such as prescribed burns, sustainable fuel load management, and community preparedness, the cornerstones of long-term resilience to wildfires and ecosystem health [23,27,54].

Although deep learning techniques show promise for the fight against wildfires [25], improving their explainability, interpretability, and operational deployment in wildfire management remains limited and fragmented. This implementation gap is a critical challenge. This survey critically assesses the current XAI landscape for wildfire prediction, moving beyond a descriptive account. We delve into the nuances of XAI methods, scrutinize their operational viability through case studies, and synthesize insights on their practical strengths, limitations, and actual impact on decision making. A central aim is to illuminate pathways towards realizing AI-driven systems' full potential not just for improved prediction but for fostering genuinely sustainable and resilient wildfire management practices globally.

Given the breadth of both XAI and wildfire science, this "short survey" necessarily adopts a focused scope. We prioritize XAI techniques and case studies that demonstrate significant progress toward operational deployment or highlight critical challenges and opportunities for sustainable wildfire management. Although not exhaustive, this paper aims to provide a representative, critically evaluated overview of the current trajectory and future needs at the intersection of these rapidly evolving fields.

1.1 Background

Although several advances in deep learning have enabled AI models to improve wildfire prediction, their "black-box" nature often limits operational trust and adoption in critical environments. Explainable AI (XAI) aims to bridge this gap by making model decisions transparent, enabling stakeholders to understand, trust, and act upon predictions in high stakes scenarios. This understanding is important, as wildfire management decisions have immediate safety consequences and profound long-term implications for biodiversity, community resilience, and broader goals of sustainable development.

Current wildfire prediction tools analyze the ignition potential, the dynamics of the spread of the fire, and assess the impact caused at the incident site [57]. Inflammation models assess fire start conditions, while spread models simulate fire movement using weather and satellite data [36]. Impact models calculate expected harm to infrastructure, ecosystems, and human settlements [9]. The Canadian Fire Weather Index and the US National Fire Danger Rating System use traditional models that utilize deterministic equations with meteorological data [30].

These systems are capable of generating understandable results from proven scientific theories, but lack the predictive power of modern AI models for complex spatio-temporal patterns or unusual fire behavior that often arises due to climate change [25]. Wildfire management faces challenges from uncertainties such as unpredictable weather, making deterministic fire growth prediction impossible, and incomplete fire-spread theory [59]. Existing limitations in risk assessment and planning create opportunities to improve wildfire management and policy by better understanding the interactions between climate and fire [20].

AI-based systems, including random forests, gradient boosting, deep learning, and hybrid architectures, show better predictive performance in operational metrics [49]. Recent deep learning techniques achieve better detection lead times and reduced false alarm rates compared to conventional methods [51]. AI models excel at identifying complex patterns in extensive datasets and are therefore suitable for the use case of satellite imagery and meteorological data [25]. Machine learning has improved the prediction, management and detection systems of wildfires [26].

1.2 Types of Explainable AI Techniques

XAI techniques address interpretability deficits using feature attribution methods (quantifying the importance of input variables), surrogate models (approximating complex systems with simpler and interpretable ones), and visualization techniques (highlighting influential regions in spatial/temporal data) [45].

Despite rapid advancement, most of the XAI research has focused on technical capabilities over operational deployment. Transitioning research prototypes to operational systems faces challenges from computational constraints, diverse stakeholder needs, integration complexities, and time-sensitive emergency requirements [42].

1.3 Contributions and Survey Paper Structure

This survey paper focuses on the critical examination of XAI techniques for sustainable wildfire prediction, synthesizing operational insights, and charting future research. The key contributions of the paper include the following.

- **Operational Perspective:** In this paper, we analyze XAI techniques for their practical effectiveness and integration difficulties in real-world wildfire management with a focus on sustainable results.
- **Contextualized XAI Taxonomy:** The survey introduces a systematic XAI taxonomy which includes model-agnostic and model-specific and hybrid/emerging categories to evaluate their effectiveness and constraints for wildfire prediction.
- **Comparative Evaluation Focused on Operational Impact:** This paper provides a detailed comparison of XAI methods, showing how different methods trade off between interpretability and computational cost and their suitability for various wildfire prediction tasks.
- **Real-World Deployment Lessons:** The survey paper draws on deployed case studies to combine essential findings with persistent obstacles that occur during the translation of XAI research into operational tools for decision-making and resource allocation assessment.
- **Analysis of Implementation paths and Barriers:** The survey examines major implementation barriers that include technical and computational limitations, as well as human-AI interaction and trust issues, and presents new approaches to achieve operational success.
- **Highlighting XAI for Sustainable Ecosystem Management:** The main theme of the survey document demonstrates how XAI enables sustainable ecosystem management through explainable predictions that lead to proactive strategies to reduce climate and preserve biodiversity.

- **Forward-Looking Research Direction:** The paper outlines a research direction that identifies essential gaps while suggesting paths to match the developments of XAI with the operational requirements to manage wildfires and the goals of sustainable development.

The paper proceeds as follows. Section 2 presents XAI techniques. Section 3: This section of the paper critically examines deployed case studies, followed by a detailed comparative analysis of the XAI methods in Sect. 4. The highlights of operational challenges, emerging solutions, and evaluation metrics are detailed in Sects. 5 and 6. Section 7: discusses and focuses on an XAI effectiveness matrix and future research. Implementation strategies, barriers, and ethical considerations are in Sects. 8 and 9. Section 10 The survey paper concludes with synthesized guidelines and overarching remarks.

2 Role of XAI Techniques in Wildfire Prediction

XAI methods such as SHAP and LIME used for wildfire prediction, are generally model-agnostic. On the other hand, model-specific methods like Grad-CAM are tailored to architectures such as Convolutional Neural Networks for visual explanations. Hybrid/emerging methods include graph neural networks for complex regional modeling and Edge XAI for real-time field alerts. These categories offer various strategies to improve interpretability. The choice depends on the AI model, the specific wildfire task (risk assessment, spread modeling), the data modalities, and operational needs such as computational efficiency and the desired type of explanation. Sections 2.1 and 2.2 detail the mathematical concepts of SHAP and LIME, illustrating the core XAI principles. In Sect. 4 a comprehensive comparative analysis of more XAI techniques is provided, evaluating their strengths, limitations, and operational fit for various wildfire prediction tasks (Table 1). This overview frames the XAI landscape before that deeper comparison (Fig. 1).

2.1 Mathematical Foundation for Widely Used Technique: SHAP

SHAP (SHapley Additive exPlanations) [43] is a widely used XAI technique for wildfire prediction. For a prediction function f, an input x', and a feature i, the SHAP value is:

$$\phi_i(f, x) = \sum_{S \subseteq N \setminus \{i\}} \frac{|S|!(|N| - |S| - 1)!}{|N|!} [f_x(S \cup \{i\}) - f_x(S)] \tag{1}$$

where N are all features, S represents feature coalitions and $f_x(S)$ is the expected model output with only the features in S known. This allows SHAP to fairly distribute the prediction among features, which is valuable to identify key drivers of the risk of wildfires.

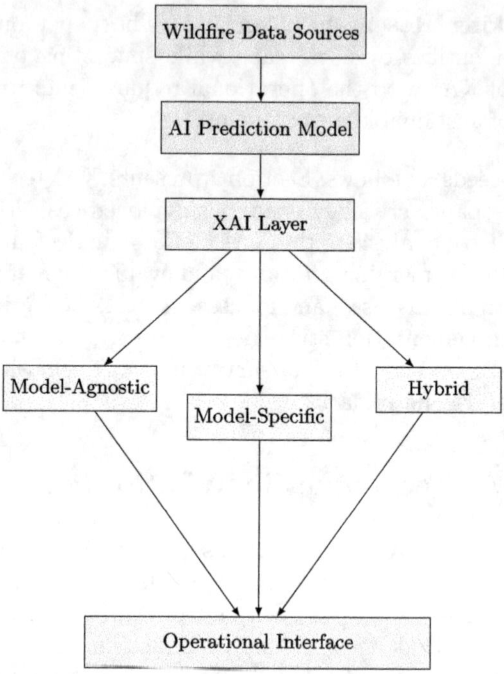

Fig. 1. Conceptual diagram of an XAI-enhanced wildfire prediction system. The XAI layer enables interpretability of model predictions through various techniques, ultimately translating complex model outputs into actionable insights for stakeholders.

2.2 Mathematical Foundation for Widely Used Technique: LIME for Wildfire Alerts

[53] While SHAP offers comprehensive feature attribution, LIME (Local Interpretable Model-agnostic Explanations) has complementary strengths for real-time wildfire alerting. LIME's mathematical foundation is:

$$\xi(x) = \arg\min_{g \in G} \mathcal{L}(f, g, \pi_x) + \Omega(g) \tag{2}$$

where:

- f is the original wildfire prediction model
- g is an interpretable model from class G (e.g., linear regression)
- π_x is a proximity measure around instance x
- $\mathcal{L}$ is a loss function that measures how well model g approximates f
- $\Omega(g)$ is a complexity penalty for g

LIME creates locally faithful explanations by minimizing approximation error and model complexity. In wildfire applications, LIME generates explanations like, for example, "humidity $< 40\% \rightarrow$ high risk," which are actionable for emergency managers and less computationally intensive than SHAP. This makes LIME valuable for time-sensitive operations which require rapid decision making.

3 Deployed XAI Case Studies in Wildfire Prediction

This section critically examines selected real-world XAI deployments in wildfire prediction. The aim is to dissect their operational utility, the impact of XAI on decision making, and the contributions of sustainable fire management. These case studies illustrate how various XAI approaches address operational challenges and stakeholder needs in various settings.

Global Lightning-Ignited Wildfire Prediction [52]

- **Use Case:** Predict and distinguish global lightning-ignited wildfires.
- **Applied XAI/AI Technique:** XGBoost [11], which is a boosting-based ensemble learning technique with SHAP, using MODIS/VIIRS burned area data, lightning detection, and fire weather index data was used in this use case.
- **Deployment Strategy:** A Cloud-based API was deployed for national wildfire agencies.
- **Limitations:** In this case study, it required regional calibration; it was not fully integrated into real-time alerts.

Critical Analysis: While SHAP clarified global risk factors (e.g., lightning, fuel dryness), translating these into actionable regional alerts proved challenging, limiting tactical use. Its main operational impact was strategic, improving the understanding of ignition patterns.

California Wildfire Spread Forecasting [56]

- **Use Case:** Real-time wildfire spread forecasting in California.
- **Applied XAI/AI technique:** Conditional Wasserstein GANs (cWGANs) with Grad-CAM were used, using satellite imagery, weather forecasts, and fuel moisture data.
- **Deployment Strategy:** this use case was used in real-time by emergency agencies for dynamic resource allocation.
- **Limitations:** Required substantial GPU clusters; faced generalization challenges from simulated to real events.

Critical Analysis: Grad-CAM's visuals helped managers assess the plausibility of cWGAN's forecast for resource allocation. However, ensuring consistent interpretation of these visual cues, especially given simulation-to-reality gaps, was a key XAI hurdle (the 'explanation-to-action' gap). More analysis is needed to determine whether these XAI-informed allocations enhanced long-term ecological resilience and to underscore the need for XAI to also explain the forecast uncertainties.

Next-Day Wildfire Danger Prediction in the Mediterranean [37]

- **Use Case:** Prediction of Next-day wildfire danger in the Eastern Mediterranean.
- **XAI/AI Technique Applied:** Deep learning with SHAP and LIME was applied to interpret meteorological/environmental features in this use case.
- **Deployment Strategy:** the solution was integrated with local weather station networks providing risk maps to fire agencies.
- **Limitations:** Explanations often require expert interpretation, while performance may be sensitive to the quality or density of local input data.

Critical Analysis: SHAP and LIME explainability approaches were applied to provide the importance of the characteristics of meteorological factors, in order to build trust in the next-day danger predictions of the deep learning model. However, the need for expert interpretation of these explanations (an "expertise bottleneck") limited their utility for rapid operational use. To effectively guide sustainable practices such as targeted warnings or fuel treatments, XAI insights must be reliable even with sparse data typical of Mediterranean regions.

Australian Explainable Wildfire Risk Mapping [17]

- **Use Case:** Regional wildfire risk mapping and early warning in Australia.
- **XAI/AI Technique Applied:** XGBoost with SHAP and partial dependence plots (PDPs) were applied as techniques in this paper.
- **Deployment Strategy:** This use case was deployed for regional risk mapping/early warning. The results of the model were integrated into public dashboards and agency platforms.
- **Limitations:** Required high-quality, real-time data; there were model transferability challenges to regions with different vegetation/climate indexes.

Critical Analysis: SHAP and PDPs successfully created accessible public dashboards by translating complex XGBoost model outputs into understandable risk drivers. The key challenges of XAI included maintaining the fidelity of the explanation during rapid fire escalations and in various ecological zones. For true long-term sustainability, the XAI needed to move beyond the current risk explanation to offer insights for adaptive land management and region-specific community preparedness.

The deployed case studies from literature demonstrate that basic interpretability methods are not enough. True operational value and sustainability emerge when XAI becomes deeply embedded in workflows to deliver strong, actionable explanations that users understand within their boundaries. Success hinges on clear use cases and user co-design. The main obstacles in implementing XAI include creating connections between explanations and actions while scaling up the system and validating its results and making it useful for different ecological and data environments.

Comparative Insights from Deployed WildFire XAI Case Studies: Key cross-cutting insights consist of the persistent 'explanation-to-action' gap, where translating XAI transparency (e.g., from SHAP, Grad-CAM) into effective decisions often demands significant expertise or else is faced with regional and computational hurdles. Successful XAI integration relies on specific use cases and end-user inputs and codesign (e.g., Mediterranean alerts, Australian dashboards). Scalability and transferability challenges remain crucial, with performance varying between data-rich and data-sparse environments. Although model-agnostic tools like SHAP aid in strategic understanding, model-specific or hybrid approaches have been observed to better suit specialized tasks. No single XAI solution is universally optimal and hence effective implementation is a sociotechnical task requiring contextual adaptation and integration with local knowledge.

4 Comparative Analysis of XAI Techniques for Wildfire Prediction

This section of the survey paper focuses on systematically comparing XAI techniques to guide researchers and domain experts in selecting appropriate methods to achieve the specific goal in a particular operational context, highlighting strengths, limitations, and suitability. Table 1 comparatively evaluates the leading XAI methods to predict wildfires (type, application, mechanism, strengths, limitations) to help assess the suitability of the interpretability.

Interpreting the Comparative Landscape of XAI Techniques: The table analysis 1 reveals essential strategic information. The trade-off between explanation depth and fidelity exists primarily in model-agnostic methods such as SHAP, which produces more detailed results at a higher computational cost. The high computational expense of SHAP for dynamic models makes us select LIME for real-time alerts, even though it produces local approximations. Model-specific techniques (e.g. Grad-CAM for CNNs/imagery) provide visual clarity for spatial risk but lack flexibility.

The combination of GNN+XAI and ConvLSTM+attention works well for complex spatio-temporal tasks, but requires significant computational resources. [58] [55] XSFL and Edge XAI meet the requirements for low-latency and privacy-preserving field explanations, which are essential in high-stakes environments, but interpretability depends on the quality of local models and networks.

No single technique is universally optimal. The selection of XAI methods depends on the requirements of wildfire task and data modalities, available resources and speed, and the level of user expertise. The explanation methods for tabular data (Random Forest/XGBoost + SHAP) operate differently from those for image data (Grad-CAM). PDPs suit planning, not tactical response. A multifaceted XAI toolkit with complementary methods is vital, as also shown in Fig. 2. Practitioners must orchestrate the XAI tools for a comprehensive, reliable, and actionable understanding.

The table functions as a useful selection guide for explanation methods during different phases of the assessment of wildfire risk.

5 Barriers and Breakthroughs for XAI in Wildfire Prediction

Transitioning XAI from research to robust operational tools in wildfire prediction requires facing multifaceted challenges. Beyond technical hurdles (Table 2), critical questions of explanation quality and usability must be addressed [31,32,44]. This section of the survey paper highlights these, including explanation fidelity nuances, Human-XAI Interaction (HCI) complexities, and outlines emerging solutions to bridge research innovations with operational demands for trustworthy, actionable XAI.

A significant challenge is making sure the quality, fidelity, and robustness of the explanation are. Explanations are not infallible, that is, their utility depends on accurately reflecting model reasoning (fidelity) and stability under minor input perturbations (robustness) [53]. Misleading explanations can erode trust more than a recognized black-box, potentially leading to misguided actions. This survey paper explores methods to detect/mitigate explanations susceptible to adversarial manipulation and develop XAI with formal guarantees of explanation properties. End users, especially in high-stakes wildfire scenarios, need mechanisms to assess the reliability of the explanation (e.g. confidence scores, indicators of trustworthiness due to uncertainties in the data/model) [39]. This affects bridging the "Black-Box Trust Gap," as trust must extend to the explanation.

Table 1. Comparison of XAI Techniques for Wildfire Prediction

Technique	Type	Application	Explanation Mechanism	Strengths	Limitations
SHAP [5]	Model-Agnostic	Global wildfire risk attribution using climate and human factors (e.g., vegetation dryness, wind speed)	Shapley values from game theory assign contribution scores to input features; identifies nonlinear interactions (e.g., drought × wind)	Global + local interpretability	Computationally expensive for large models
LIME [23]	Model-Agnostic	Real-time fire alerts in high-risk regions (e.g., EU FireTrust framework)	Trains local surrogate models (e.g., linear regression) around specific predictions to generate human-readable rules (e.g., "humidity < 40% → high risk")	Lightweight, adaptable to dynamic conditions	Sensitive to perturbation radius
Grad-CAM(Gradient-weighted Class Activation Mapping [2]	Model-Specific (CNN)	Satellite imagery classification for burn area mapping (e.g., Yosemite wildfires)	Generates heatmaps by weighting CNN activation maps with gradient information; highlights dry vegetation patches in satellite images	Visual clarity for spatial risk zones	Limited to CNN architectures
Random Forest [8] + SHAP [54]	Model-Agnostic	Wildfire hotspot detection in Mediterranean regions	Feature importance ranks drivers (e.g., Fire Weather Index); SHAP values quantify marginal contributions of soil moisture and wind direction	Robust to outliers, handles tabular data	Poor scalability for global datasets

(continued)

Table 1. (*continued*)

Technique	Type	Application	Explanation Mechanism	Strengths	Limitations
XGBoost + SHAP [29]	Model-Agnostic	Drought severity prediction and ignition risk in Australia	SHAP ranks drought indices (SPI/SPEI); PDPs reveal nonlinear effects of fuel moisture on burn probability	Handles large datasets, fast predictions	Black-box without SHAP integration
ConvLSTM [56]	Hybrid	Hourly wildfire spread forecasting in California	Attention mechanisms prioritize temporal sequences (e.g., humidity drops); combines CNN for spatial features and LSTM for temporal dynamics	Captures spatio-temporal dependencies	High computational cost
Attention Mechanisms in U2-NET [64]	Model-Specific	Wildfire spread prediction using remote sensing data	Spatial attention highlights critical fire zones; improves feature localization in complex landscapes	High-resolution fire spread prediction; suitable for complex terrains	Requires extensive labeled data; limited temporal generalization
GNN + XAI [10]	Hybrid	Global wildfire prediction	Integrated Gradients for feature importance; community detection for correlation clusters	Captures spatial correlations globally	Training complexity
Explainable Semantic Federated Learning (XSFL) [18]	Hybrid	Real-time fire surveillance using IIoT edge devices	Combines federated learning with semantic-aware communication and interpretable local model updates using ReLU-based attribution	Maintains privacy, reduces communication overhead, and adapts to heterogeneous devices	High complexity in coordination; interpretability depends on local model quality; limited scalability in unstable networks
Partial Dependence Plots (PDP) [19]	Model-Agnostic	Fuel moisture and biomass-climate analysis for wildfire risk prediction	Visualizes marginal effects of fuel moisture, biomass, and precipitation on wildfire probability; assumes feature independence	Captures nonlinear trends; highlights thresholds; supports fire planning across different ecosystems	Limited to low-dimensional data; assumes feature independence; sensitive to variable correlation
Symbolic Neural Networks [38]	Hybrid	Wildfire damage classification and driver identification	Integration of neural networks with symbolic reasoning; explanations via SHAP for feature importance and model decisions	Combines learning capabilities of neural networks with the interpretability of symbolic reasoning	May require extensive domain knowledge for symbolic component; potential complexity in integration

Human-in-the-Loop Explainability and operational usability are equally important challenges. However, an XAI insight is valuable only if it is effectively understood, interpreted, and acted upon by diverse stakeholders, such as field crews, incident commanders, and policy makers. Designing intuitive interfaces conveying complex information (SHAP plots, LIME rules, Grad-CAM heatmaps) without overwhelming users or requiring extensive XAI knowledge is paramount. What a data scientist or machine learning developer finds insightful might be opaque or black-box to a firefighter. This necessitates an 'actionable' explanation that is participatory, co-designed with end-users, and is iteratively tested for usability in realistic operational contexts, ensuring production of an adaptive explanation tailored to user expertise, task, and cognitive bandwidth. Without addressing Human in loop, sophisticated XAI risks remaining academic curiosities [47].

Beyond the questions of explanation quality and usability, deploying XAI in wildfire prediction also involves persistent technical and resource bottlenecks. Table 2 of the survey paper details key challenges, data limitations, and emerging technical solutions.

Table 2. Challenges in Existing XAI Techniques and Emerging Solutions for Wildfire Prediction

Challenge	Description	Emerging Technique	How It Helps
Computational Cost [12,16]	Running SHAP/LIME are computationally expensive for large-scale models.	Federated XAI	Distributes computation across agencies, reducing local resource load.
Black-Box Trust Gap [33,66]	Stakeholders do not trust non-transparent models like deep learning.	Symbolic Neural Networks	Provides human-readable rules for transparency.
Data Scarcity [35]	Labeled data is limited in remote regions.	Physics-Guided XAI	Uses domain knowledge to reduce data dependency.
Spatial Scalability [10]	CNNs fail to capture fire relations across distant regions.	GNN + XAI	Learns long-range dependencies using graph networks.
Real-Time Explainability [48,50]	Post-hoc methods such as LIME are too slow.	Edge XAI	Deploys lightweight explanation models on IoT devices for fast response.
Data Quality and Uncertainty [7]	Incomplete, noisy, or varied resolution operational data (weather, fuel, satellite); impacts model/explanation reliability.	Uncertainty-Aware XAI/Robust XAI	Quantifies prediction/explanation uncertainty; highlights reliance on poor inputs or model sensitivity.

6 Critical XAI Evaluation Metrics

This section of the survey paper presents evaluation metrics for consistent XAI assessment across different techniques and contexts, based on previous XAI analysis.

Table 3 organizes key metrics (interpretability, fidelity, robustness, user trust) by category, purpose, and importance, assisting the selection and validation of XAI strategies for wildfire risk.

Table 3. Key Evaluation Metrics for XAI Techniques in Wildfire Prediction

Metric	Description	Importance
Fidelity [23,34]	Explanation matches model logic.	Builds user trust in predictions.
Stability [1,23]	Similar inputs yield similar explanations.	Ensures operational consistency.
Sparsity [53,54]	Focus on few key features.	Quick, clear interpretation.
Runtime [23]	Time to generate explanations.	Enables real-time response.
Usability [34,54]	Actionable for end-users.	Supports practical decisions.

6.1 Evaluating Explanation Quality Beyond AI/ML Model Performance

Beyond Table 3, operational XAI needs methods to assess the quality of the explanation itself, which significantly affects the results.

Evidence based Evaluation Methods. Agencies use several approaches: **Outcome Matching Evaluation** (e.g., EFFIS: improved decision consistency with XAI [22]); **Counterfactual Completeness** (e.g., Australian CRC: faster adaptation with high-scoring techniques [17]); **Black-Box Usability Trials** (e.g., CAL FIRE: higher adoption if AI/expert explanations indistinguishable [67]); and **Trust Lifecycle Monitoring** (e.g., Colorado: consistent fidelity metrics built stable trust in 4–6 months [24]).

Quantifying Impact. High-quality explanations improve operations with fewer false evacuations (Australia [17]); better resource allocation efficiency (Mediterranean [37]); faster operator response (California [56]).

Evaluating explanation quality, not just model performance, is therefore critical for operational effectiveness.

7 XAI Effectiveness Matrix and Future Research Directions for Wild Fire Predictions

The evaluation of XAI's wildfire prediction system in its current state leads to a methodical assessment of its operational effectiveness and future research possibilities. This section of the survey paper helps practitioners select XAI methods for specific wildfire management objectives and highlights future development pathways.

The effectiveness matrix (Fig. 2) is a practical decision support tool for selecting XAI techniques based on specific wildfire management tasks, reflecting characteristics (computational requirements, explanation types, data modalities) discussed above (e.g. Table 1). For instance, LIME's suitability for real-time alerts versus the more intensive SHAP, or Grad-CAM's visual strengths for imagery-based tasks, are apparent. No single technique excels in all operational needs, highlighting the importance of a complementary, multitechnique approach.

The comparative analysis of this survey, juxtaposing XAI capabilities with operational wildfire management realities and sustainability goals, illuminates critical future research areas. The following section and Table 4 map these emerging paths, linking identified operational needs and current XAI gaps to promising technical and methodological solutions. These directions aim to foster XAI systems that are more powerful, interpretable, ethically robust, user-centric, and effectively integrated into the sociotechnical fabric of wildfire response and long-term ecosystem management.

	Ignition Risk	Spread Modeling	Resource Allocation	Real-time Alert
SHAP	High	High	Medium	Low
LIME	High	Medium	High	High
Grad-CAM	Medium	High	Medium	High
GNN+XAI	High	High	Medium	Low
Edge XAI	Medium	Low	High	High

Fig. 2. XAI Effectiveness Matrix for Wildfire Prediction. Green indicates high effectiveness, yellow indicates medium, and red indicates low or limited effectiveness.

Table 4. Mapping Future XAI Directions to Operational Needs in Wildfire Prediction

Research Direction	Operational Need	Example Technique
Real-Time XAI [23]	Faster evacuation, dynamic resource allocation	Edge-compatible LIME
Causal XAI [65]	Climate-informed policy design	Causal SHAP
Human-AI Interfaces [23]	Improved field crew decision-making	3D Grad-CAM + VR
Ethical XAI [3,46,62]	Equitable resource distribution	Fairness-aware SHAP
Federated XAI [13,14]	Global risk modeling without data sharing	Federated SHAP
Multimodal Explanations [23]	Stakeholder-specific communication	GPT-4 + Grad-CAM
Community-Driven XAI [42]	Integration of traditional knowledge	Explanation Dialogues
XAI for Climate Adaptation [6]	Resilient infrastructure planning	CF Analysis, SHAP, Participatory

8 Operational Implementation Strategies and Barriers

Successful XAI transition to operational wildfire management requires overcoming implementation challenges. Key pathways and barriers emerge from the analysis of the systems deployed, highlighting XAI as a sociotechnical challenge that requires attention to human factors, workflow integration, and trust [67].

8.1 Integration with Existing Infrastructure

Effective integration of XAI techniques involves phased approaches. For example, by integrating LIME-based explanations, Colorado's WildCAD system ended up reducing resource allocation decision time [23]. Legacy systems with limited computation for intensive XAI (e.g. full SHAP) without hardware upgrades act as a common barrier. Strategies to overcome this challenge include starting with lightweight XAI (LIME) on current hardware, then adding intensive techniques (SHAP) as infrastructure allows [24].

8.2 User Interface Design for Different Stakeholders

User interface design must cater to diverse stakeholders, which include policymakers, field crews, and scientists needing varied explanation formats with the granularity of different data points, as a one-size-fits-all solution is ineffective. The Fire Incident Dashboard, which was successfully used, uses role-specific interfaces with tailored explanations, improving situational awareness [21]. Interfaces must support cognitive offloading, enhancing judgment of the user under pressure, while minimizing misinterpretation. These solutions require deep co-design, iterative prototyping, and usability testing. Future adaptive interfaces may tailor explanations to user expertise, task, load, and uncertainty [40].

8.3 Computational Efficiency for Real-Time Operation

Real-time XAI demands computational efficiency. The complete SHAP for high-dimensional imagery can take minutes, which is not suitable for rapid decisions. EFFIS, for example, uses a progressive SHAP computation, with explanations calculated incrementally as resources allow [22]. Other strategies include precomputed explanation templates, dynamic resource allocation according to urgency, and approximate methods that trade low fidelity for speed [63].

8.4 Training and Trust Building with Domain Experts

- **Barrier:** Wildfire personnel often lack AI/ML and explainable AI familiarity, leading to skepticism or trust problems if explanations are not critically evaluated. Building trust and transparency requires demonstrating model performance and educating users on AI/XAI capabilities and limitations, including potential explanation infidelity and non-robustness [4].
- **Implementation strategy:** Successful strategies emphasize ongoing comprehensive training beyond basic software operation, covering: (1) AI/XAI principles, (2) interpreting and critically evaluating explanation types and weaknesses, (3) understanding model/explanation competence boundaries, and (4) simulated XAI decision-making exercises. Implementing gradual rollouts, side-by-side operation with legacy systems, and transparent communication about XAI updates are vital to producing the desired results. Feedback loops must be implemented where users can question, validate and correct explanations, as these help build trust and improve the XAI system and the HCI design [28,41].

9 Ethical Considerations and Societal Impact

The deployment of XAI in high-stakes wildfire prediction requires a review of its ethical and societal impacts. Proactive risk mitigation during operational integration is vital for responsible and equitable use.

9.1 Ensuring Fair AI Predictions

XAI systems can inherit or amplify historical data biases (for example, in inequitable resource allocation), perpetuating them in models and explanations [15].

Empirical Evidence of Bias Risks. Audits show lower precision in underserved areas due to data and monitoring gaps; XAI sometimes rationalized these gaps with problematic proxies, risking unfair resource distribution [67].

Bias Mitigation Strategies. Key operational approaches include:

- **Fairness-Aware XAI**: Constraint models to reduce accuracy disparities between demographic groups [37].
- **Community-Based Equity Assessment**: Multi-stakeholder boards review and correct biases, improving trust between diverse communities [17].
- **Transparent Bias Reporting**: Regular audit reports on model biases and corrective measures to guide equitable decision making [67].

9.2 Establishing Accountability in Explainable AI Systems

Clear accountability for XAI-informed critical decisions (for example, in evacuations) is essential, as adverse AI-guided outcomes raise complex responsibility issues. Effective governance clarifies AI-human decision authority, mandates audit trails, uses multi-stakeholder oversight, and provides accessible avenues for challenge and appeal. The "tiered accountability" mechanism of EFFIS has been shown to improve the clarity of responsibility [22].

9.3 User-Centered Transparency in XAI Systems

Poorly designed explanations can lead to overtrust or cognitive overload in emergencies [24,67]. Interfaces must clearly present limitations and uncertainties, supported by user training and trust calibration exercises.

9.4 Balancing Transparency with Security

Highly transparent systems may reveal vulnerabilities [24]. Tiered access controls and adversarial testing help balance transparency with security concerns.

10 Conclusion

The survey paper integrates essential findings into practical recommendations while providing general observations on the future direction of the wildfire prediction. The main objective of this survey article is to extract valuable information that will help practitioners and researchers choose and develop appropriate XAI methods for sustainable wildfire management.

The selection and adoption of XAI techniques for practitioners and domain experts depends on the prediction task and operational constraints together to generate meaningful explanations for sustainable wildfires. The survey shows that XAI stands as an essential element for the advancement of operational wildfire prediction systems. The traditional lack of interpretability in AI/ML (Machine Learning) models, together with their real-time operational requirements, has historically limited their adoption in critical wildfire management. The comparative analysis showed that XAI techniques exist in various forms that produce different operational consequences.

The implementation of XAI will lead to substantial sustainability benefits. The combination of precise predictions with transparent, reliable, and trustworthy systems leads to better resource management and less ecological harm, and extended ecosystem preservation that supports the Sustainable Development Goals of the United Nations (13, 15, 11) [60]. The future development of XAI through ethical guidance will create substantial opportunities to enhance resilience to wildfires among communities in a changing climate.

References

1. Alvarez-Melis, D., Jaakkola, T.S.: On the robustness of interpretability methods. arXiv preprint arXiv:1806.08049 (2018). https://arxiv.org/abs/1806.08049
2. Apostolopoulos, I.D., Athanasoula, I., Tzani, M., Groumpos, P.P.: An explainable deep learning framework for detecting and localising smoke and fire incidents: evaluation of Grad-CAM++ and LIME. Mach. Learn. Knowl. Extr. **4**(4), 1124–1135 (2022). https://doi.org/10.3390/make4040057
3. Balasubramaniam, N., Kauppinen, M., Rannisto, A., Hiekkanen, K., Kujala, S.: Transparency and explainability of ai systems: from ethical guidelines to requirements. Inf. Softw. Technol. **159**, 107197 (2023). https://doi.org/10.1016/j.infsof.2023.107197, https://www.sciencedirect.com/science/article/pii/S0950584923000514
4. Balayan, V., Saleiro, P., Belém, C., Krippahl, L., Bizarro, P.: Teaching the machine to explain itself using domain knowledge. arXiv preprint arXiv:2012.01932 (2020). https://arxiv.org/abs/2012.01932
5. Ballard, T., Erinjippurath, G., Cooper, M.W., Lowrie, C.: Wildfire risk to global forest carbon offset projects revealed by explainable ai (2023). https://arxiv.org/abs/2305.02397
6. Barutcu, H.C., Çelik, S., Gezer, M.: Explainable artificial intelligence-based approaches for climate change: a review. Int. J. Global Warming **35**(2/3/4) (2025). https://doi.org/10.1504/IJGW.2025.145102, https://www.inderscienceonline.com/doi/10.1504/IJGW.2025.145102
7. Benali, A., Sá, A.C.L., Ervilha, A.R., Trigo, R.M., Fernandes, P.M., Pereira, J.M.C.: Deciphering the impact of uncertainty on the accuracy of large wildfire spread simulations. Sci. Total Environ. **569–570**, 73–85 (2016). https://doi.org/10.1016/j.scitotenv.2016.06.112, https://www.sciencedirect.com/science/article/abs/pii/S0048969716312852

8. Breiman, L.: Random forests. Mach. Learn. **45**(1), 5–32 (2001). https://doi.org/10.1023/A:1010933404324
9. Brillinger, D.R., Autrey, B.S., Cattaneo, M.D.: Probabilistic risk modeling at the wildland urban interface: the 2003 cedar fire. Environmetrics **20**(6), 607–620 (2008). https://doi.org/10.1002/env.959
10. Chen, D., Cheng, S., Hu, J., Kasoar, M., Arcucci, R.: Explainable global wildfire prediction models using graph neural networks. arXiv preprint arXiv:2402.07152 (2024). https://doi.org/10.48550/arXiv.2402.07152, https://arxiv.org/abs/2402.07152
11. Chen, T., Guestrin, C.: Xgboost: a scalable tree boosting system. In: Proceedings of the 22nd ACM SIGKDD International Conference on Knowledge Discovery and Data Mining, pp. 785–794. ACM (2016).https://doi.org/10.1145/2939672.2939785, https://doi.org/10.1145/2939672.2939785
12. Chiarani, M., Roy, S., et al.: Xai-driven client selection for federated learning in scalable 6g network slicing. arXiv preprint arXiv:2503.12435 (2025). https://arxiv.org/abs/2503.12435
13. Corcuera Bárcena, J.L., et al.: Fed-xai: Federated learning of explainable artificial intelligence models. In: Proceedings of the 3rd Italian Workshop on Explainable Artificial Intelligence (XAI.it 2022) (2022). https://ceur-ws.org/Vol-3277/paper8.pdf
14. Daole, M., Schiavo, A., Corcuera Bárcena, J.L., Ducange, P., Marcelloni, F., Renda, A.: Openfl-xai: Federated learning of explainable artificial intelligence models in python. SoftwareX **23**, 101505 (2023). https://doi.org/10.1016/j.softx.2023.101505
15. Deck, L., Schoeffer, J., De-Arteaga, M., Kühl, N.: A critical survey on fairness benefits of explainable ai. In: Proceedings of the 2024 ACM Conference on Fairness, Accountability, and Transparency, pp. 1579–1595 (2024). https://doi.org/10.1145/3630106.3658990, https://arxiv.org/abs/2310.13007
16. Desai, H.A., Hilal, A., Eldardiry, H.: Reft: Resource-efficient federated training framework for heterogeneous and resource-constrained environments (2023). https://arxiv.org/abs/2308.13662
17. Di Giuseppe, F., McNorton, J., Lombardi, A., Wetterhall, F.: Global data-driven prediction of fire activity. Nat. Commun. **16**(2918) (2025). https://www.nature.com/articles/s41467-025-58097-7
18. Dong, L., Peng, Y., Jiang, F., Wang, K., Yang, K.: Explainable semantic federated learning enabled industrial edge network for fire surveillance. arXiv preprint arXiv:2412.19979 (2024). https://arxiv.org/abs/2412.19979
19. Dorph, A., Marshall, E., Parkins, K.A., Penman, T.D.: Modeling ignition probability for human- and lightning-caused wildfires in Victoria, Australia. Nat. Hazard. **22**(11), 3487–3503 (2022). https://doi.org/10.5194/nhess-22-3487-2022
20. Dunn, C.J., Thompson, M.P., Calkin, D.E.: A framework for developing safe and effective large-fire response in a new fire management paradigm. For. Ecol. Manage. **404**, 184–196 (2017). https://doi.org/10.1016/j.foreco.2017.08.039
21. Esri: Introduction to fire incident dashboard, https://doc.arcgis.com/en/arcgis-solutions/latest/reference/introduction-to-fire-incident-dashboard.htm
22. Essenfelder, A.H., Toreti, A.: Seguini: Expert-driven explainable artificial intelligence models can detect multiple climate hazards relevant for agriculture. Commun. Earth Environ. **6**(1), 124 (2025). https://doi.org/10.1038/s43247-024-01987-3
23. Fan, D., Biswas, A., Ahrens, J.P.: Explainable AI integrated feature engineering for wildfire prediction (2024). https://arxiv.org/abs/2404.01487

24. of Fire Prevention Control, D.: 2023 wildfire preparedness plan. Technical report, Colorado Department of Public Safety (2023). https://spl.cde.state.co.us/artemis/psserials/ps912internet/ps9122023internet.pdf
25. Ghali, R., Akhloufi, M.A.: Deep learning approaches for wildland fires using satellite remote sensing data: Detection, mapping, and prediction. Fire **6**(5) (2023). https://doi.org/10.3390/fire6050192, https://www.mdpi.com/2571-6255/6/5/192
26. Guede-Fernández, F., Martins, L., de Almeida, R.V., Gamboa, H., Vieira, P.: A deep learning based object identification system for forest fire detection. Fire **4**(4), 75 (2021). https://doi.org/10.3390/fire4040075
27. Hamida, S.U., Chowdhury, M.J.M., Chakraborty, N.R., Biswas, K., Sami, S.K.: Exploring the landscape of explainable artificial intelligence (XAI): a systematic review of techniques and applications. Big Data Cogn. Comput. **8**(11), 149 (2024). https://doi.org/10.3390/bdcc8110149
28. Happer, C.: Creating feedback loops between human experts and AI systems. ResearchGate (2025). https://www.researchgate.net/publication/391398367
29. He, K., Shen, X., Merow, C., Nikolopoulos, E., Gallagher, R.V., Yang, F., Anagnostou, E.N.: Improving fire severity prediction in south-eastern Australia using vegetation-specific information. Nat. Hazard. **24**(10), 3337–3355 (2024). https://doi.org/10.5194/nhess-24-3337-2024
30. Hoffman, C.M., Canfield, J., Linn, R.R., Mell, W., Sieg, C.H., Pimont, F., Ziegler, J.P.: Evaluating crown fire rate of spread predictions from physics-based models. Fire Technol. **52**(1), 221–237 (2016). https://doi.org/10.1007/s10694-015-0500-3
31. Hoffman, R.R., Mueller, S.T., Klein, G., Litman, J.: Metrics for explainable AI: Challenges and prospects. arXiv preprint arXiv:1812.04608 (2018). https://arxiv.org/abs/1812.04608
32. Holzinger, A., Carrington, A., Müller, H.: Measuring the quality of explanations: The system causability scale (SCS). Comparing human and machine explanations. arXiv preprint arXiv:1912.09024 (2019). https://arxiv.org/abs/1912.09024
33. Iqbal, J., Eldred, A.: Symbolic ai meets deep learning: a hybrid approach to improving explainability and predictive accuracy (2025). https://www.researchgate.net/publication/389249740_Symbolic_AI_Meets_Deep_Learning_A_Hybrid_Approach_to_Improving_Explainability_and_Predictive_Accuracy
34. Jacovi, A., Goldberg, Y.: Towards faithfully interpretable NLP systems: How should we define and evaluate faithfulness? (2020). https://arxiv.org/abs/2004.03685
35. Jadouli, A., El Amrani, C.: Enhancing wildfire forecasting through multisource spatio-temporal data, deep learning, ensemble models and transfer learning. Adv. Artif. Intell. Mach. Learn. **4**(3), 2614–2628 (2024). https://doi.org/10.54364/AAIML.2024.43152
36. James, G.L., Ansaf, R.B., Al-Samahi, S.S.A., Parker, R.D., Cutler, J.M., Gachette, R.V., Ansaf, B.: An efficient wildfire detection system for ai-embedded applications using satellite imagery. Fire **6**(4), 169 (2023). https://doi.org/10.3390/fire6040169
37. Ji, Y., Wang, D., Li, Q., Liu, T., Bai, Y.: Global wildfire danger predictions based on deep learning taking into account static and dynamic variables. Forests **15**(1), 216 (2024). https://doi.org/10.3390/f15010216
38. Jia, Y., Opabola, E.: Interpretable machine learning insights into wildfire damage drivers in california, USA (2025). https://ssrn.com/abstract=5177407
39. Lakshminarayanan, B., Pritzel, A., Blundell, C.: Simple and scalable predictive uncertainty estimation using deep ensembles. Adv. Neural Inf. Process. Syst. **30** (2017). https://doi.org/10.48550/arXiv.1612.01474

40. Lei, D., He, Y., Zeng, J.: What is the focus of xai in ui design? prioritizing ui design principles for enhancing xai user experience. arXiv preprint arXiv:2402.13939 (2024). https://arxiv.org/abs/2402.13939
41. Liao, Q.V., Gruen, D., Miller, S.: Questioning the ai: Informing design practices for explainable ai user experiences. In: Proceedings of the 2020 CHI Conference on Human Factors in Computing Systems, ACM (2020). https://doi.org/10.1145/3313831.3376590, https://arxiv.org/abs/2001.02478
42. Liao, Q.V., Varshney, K.R.: Human-centered explainable ai (xai): From algorithms to user experiences. arXiv preprint arXiv:2110.10790 (2021). https://doi.org/10.48550/arXiv.2110.10790
43. Lundberg, S.M., Lee, S.I.: A unified approach to interpreting model predictions. arXiv preprint arXiv:1705.07874 (2017). https://arxiv.org/abs/1705.07874
44. Löfström, H., Hammar, K., Johansson, U.: A meta survey of quality evaluation criteria in explanation methods. arXiv preprint arXiv:2203.13929 (2022). https://arxiv.org/abs/2203.13929
45. Muhammad, D., Bendechache, M.: Unveiling the black box: a systematic review of explainable artificial intelligence in medical image analysis. Comput. Struct. Biotechnol. J. **24**, 542–559 (2024). https://doi.org/10.1016/j.csbj.2024.08.005
46. Nannini, L., Manerba, M.M., Beretta, I.: Mapping the landscape of ethical considerations in explainable ai research. Ethics Inf. Technol. (2024) https://doi.org/10.1007/s10676-024-09773-7
47. Nguyen, T., Canossa, A., Zhu, J.: How human-centered explainable ai interfaces are designed and evaluated: a systematic survey. arXiv preprint arXiv:2403.14496 (2024). https://arxiv.org/abs/2403.14496
48. Nguyen, T.T.H., Nguyen, P.T.L., Cao, H.: Xedgeai: A human-centered industrial inspection framework with data-centric explainable edge ai approach. arXiv preprint arXiv:2407.11771 (2024). https://arxiv.org/abs/2407.11771
49. Okoro, S., Lopez, A., Unuriode, A.: A synergistic approach to wildfire prevention and management using ai, ml, and 5g technology in the united states. SSRN (2024). https://doi.org/10.2139/ssrn.4739361
50. Rahmati, M.: Towards explainable and lightweight ai for real-time cyber threat hunting in edge networks. arXiv preprint arXiv:2504.16118 (2025). https://arxiv.org/abs/2504.16118
51. Ranadive, O., et al.: Image-based early detection system for wildfires. arXiv preprint arXiv:2211.01629 (2022). https://arxiv.org/abs/2211.01629
52. Regional Fire Management Resource Center – South East Asia: Global wildland fire early warning. https://rfmrc-sea.org/global-wildland-fire-early-warning/
53. Ribeiro, M.T., Singh, S., Guestrin, C.: Why should i trust you?: Explaining the predictions of any classifier. In: Proceedings of the 22nd ACM SIGKDD International Conference on Knowledge Discovery and Data Mining, pp. 1135–1144. ACM (2016). https://doi.org/10.1145/2939672.2939778
54. Schnoor, E., Vielhaben, J.: Explainable artificial intelligence for improving natural disaster management. TEMA Project (2023). https://tema-project.eu/articles/explainable-artificial-intelligence-improving-natural-disaster-management, Accessed 15 May 2024
55. Sha, N., Wu, X., Wen, J., Li, J., Li, C.: Adaptive spatio-temporal graph convolutional network with attention mechanism for mobile edge network traffic prediction. Clust. Comput. **27**, 13257–13272 (2024). https://doi.org/10.1007/s10586-024-04577-2

56. Shaddy, B., et al.: Generative algorithms for fusion of physics-based wildfire spread models with satellite data for initializing wildfire forecasts. arXiv preprint arXiv:2309.02615 (2023). https://arxiv.org/abs/2309.02615
57. Shmuel, A., Lazebnik, T., Glickman, O., Heifetz, E., Price, C.: Global lightning-ignited wildfires prediction and climate change projections based on explainable machine learning models. Sci. Rep. (2025). https://doi.org/10.1038/s41598-025-92171-w
58. Tang, J., Xia, L., Huang, C.: Explainable spatio-temporal graph neural networks. arXiv preprint arXiv:2310.17149 (2023). https://arxiv.org/abs/2310.17149
59. Thompson, M.P., Calkin, D.E., Scott, J.H., Hand, M.S.: Uncertainty and probability in wildfire management decision support. In: Wildfire Risk: Human Perceptions and Management Implications, pp. 31–52. American Geophysical Union and Wiley (2016). https://doi.org/10.1002/9781119028116.ch4
60. United Nations: The 17 goals - sustainable development goals (2015). https://sdgs.un.org/goals
61. University of Michigan-Dearborn: Ai's mysterious 'black box' problem, explained (2023). https://umdearborn.edu/news/ais-mysterious-black-box-problem-explained
62. Vainio-Pekka, H., Agbese, M.O.O., Jantunen, M., Vakkuri, V., Mikkonen, T., Rousi, R., Abrahamsson, P.: The role of explainable AI in the research field of AI ethics. ACM Trans. Inter. Intell. Syst. **13**(4), 1–39 (2023) https://doi.org/10.1145/3599974
63. Wacker, J.: Explainable AI in practice: Finding the right method to open the black box (2024). https://www.statworx.com/en/content-hub/blog/explainable-ai-in-practice-finding-the-right-method-to-open-the-black-box/
64. Xiao, H., Zhu, Y., Sun, Y., Zhang, G., Gong, Z.: Wildfire spread prediction using attention mechanisms in u2-net. Forests **15**(10), 1711 (2024). https://www.mdpi.com/1999-4907/15/10/1711
65. Xu, S., Ge, Y., Zhang, Y.: Causal explainable ai. In: Machine Learning for Causal Inference, pp. 137–159. Springer International Publishing (2023). https://doi.org/10.1007/978-3-031-35051-1_7
66. Zhang, X., Sheng, V.S.: Neuro-symbolic ai: Explainability, challenges, and future trends. arXiv preprint arXiv:2411.04383 (2024). https://arxiv.org/abs/2411.04383
67. Zhou, Y., Kong, R., Xu, Z., Xu, L., Cheng, S.: Comparative and interpretative analysis of cnn and transformer models in predicting wildfire spread using remote sensing data. arXiv preprint arXiv:2503.14150 (2025). https://arxiv.org/abs/2503.14150

RAG4MuniESG: ESG Assessment for Municipal Sustainability Through Retrieval-Augmented Generation

Dorsaf Sallami[1(✉)] and Juste Rajaonson[2]

[1] Department of Computer Science and Operations Research, Université de Montréal, Montreal, Canada
dorsaf.sallami@umontreal.ca

[2] Department of Urban and Tourism Studies, Université du Québec à Montréal, Montreal, Canada
rajaonson.juste@uqam.ca

Abstract. Climate change impacts, social inequalities, and governance challenges increasingly impact municipal performance, creating challenges for financial institutions, insurers, and asset managers trying to evaluate sustainability-related risks. Unlike corporations, municipalities spread ESG-relevant information across various fragmented sources, including plans, bylaws, reports, and public outreach efforts. We present *RAG4MuniESG*, an AI-driven framework that combines Large Language Models (LLMs) with Retrieval-Augmented Generation (RAG) to facilitate ESG assessments specific to municipalities. Using a public-sector-oriented ESG scheme, the system semantically merges structured indicators with unstructured documents. We propose a hybrid AI-human approach wherein automated extraction and scoring are augmented with transparent and traceable justifications to support human-in-the-loop verification and auditability. *RAG4MuniESG* produces (1) aggregated ESG profiles for overarching risk screening and (2) detailed outputs with page-level references for auditing and oversight. Tested in three Canadian cities, it shows how such tool can convert fragmented municipal data into actionable ESG insights that meet the needs of financial and regulatory stakeholders.

Keywords: Municipal Sustainability · Environmental · Social · Governance · Large Language Models · Retrieval-Augmented Generation

1 Introduction

Climate change impacts, growing social inequalities, and governance instabilities present significant challenges, compelling financial institutions, insurers, asset managers, and oversight bodies to better evaluate how municipalities respond to these critical risks. Unlike corporations, which typically disclose sustainability

Published by Springer Nature Switzerland AG 2026
E. Mercier-Laurent et al. (Eds.): AI4S 2025, IFIP AICT 780, pp. 190–201, 2026.
https://doi.org/10.1007/978-3-032-21184-2_15

information through structured, investor-oriented reports, municipalities distribute relevant environmental, social, and governance (ESG) data across numerous, fragmented sources. Such sources include strategic plans, zoning bylaws, budget documents, policy frameworks, and even public outreach materials aimed at influencing citizen behavior [8]. Consequently, municipal ESG assessment remains complex, resource-intensive, and underdeveloped, often relying on manual analysis or simplified statistical indicators that overlook essential contextual nuances and interdependencies [4,6].

Although integrating ESG factors into municipal practice can enhance accountability and align local policies with global sustainability objectives, systematic ESG assessment methodologies tailored specifically for the public sector are lacking. Recent studies have proposed frameworks incorporating smart city data into municipal ESG evaluation [4], yet practical, scalable implementations remain rare.

As discussed in Sect. 2 on related works, current AI-driven ESG assessment tools, primarily developed for corporate contexts, perform poorly in municipalities due to the highly contextual and less standardized nature of public-sector data [7]. Advancements in Artificial Intelligence (AI), particularly Large Language Models (LLMs), offer new possibilities for processing and analyzing the rich yet fragmented municipal documentation landscape [7,12]. LLMs excel in synthesizing information from lengthy, unstructured texts, enabling the extraction of relevant ESG insights from complex municipal sources such as policy reports, council minutes, and sustainability plans. However, fully automated solutions still struggle to capture local policy contexts and subtleties essential for rigorous ESG analysis, highlighting the need for hybrid systems that integrate AI automation with human oversight.

To address these challenges, we introduce RAG4MuniESG, an AI-powered framework combining LLMs with Retrieval-Augmented Generation (RAG) specifically designed for ESG evaluation at the municipal level. As detailed further in Sect. 3 regarding the methodology, RAG4MuniESG leverage a public-sector-specific ESG scheme to systematically retrieves, analyzes, and synthesizes fragmented municipal documentation and statistical indicators. It operationalizes a hybrid AI-human approach by combining automated ESG data extraction into an aggregated municipal ESG profile suitable for rapid, high-level risk assessment, with a detailed disaggregated output offering transparent, document-level justifications and page-specific references. As shown in Sect. 4 on the experiments and results from Canadian municipalities studied, this dual structure enables scalability and efficiency through automation, while ensuring human interpretability, validation, and regulatory oversight. As a result, financial institutions, policymakers, and oversight bodies gain actionable insights that bridge ESG evaluation requirements with the complex realities of municipal governance.

2 Related Works

ESG assessment has advanced rapidly in the corporate sector, where standardized disclosures and AI-assisted tools increasingly support investment decision-making. Most corporate ESG ratings rely on structured data from sustainability reports, often aggregated into opaque scores by rating agencies [1]. In response to the opacity of these systems, recent studies have proposed more explainable AI-driven frameworks. ChatReport [7] and ESGReveal [12] use Retrieval-Augmented Generation (RAG) to extract verifiable evidence from lengthy corporate ESG disclosures. While ESGReveal focuses on structured data extraction, ChatReport extends this by combining semantic search with grounded summarization to reduce hallucinations. Similarly, ClimRetrieve [9] benchmarks information retrieval for climate disclosures, offering high-precision filtering for climate adaptation content.

Building on these, the Adaptation Alignment Assessment Framework (A3F) [10] integrates LLMs with a custom NLP pipeline to assess firm-level adaptation planning. The A3F translates indicators derived from climate and nature disclosure standards (e.g., TCFD, ISSB, TNFD) into close-ended, binary questions. These are used to evaluate corporate alignment with climate resilience through publicly disclosed reports, with a focus on both climate and nature-related physical risks. Importantly, their framework is implemented using a RAG system grounded in verifiable references and critical prompt engineering, offering a transparent and scalable assessment method.

Despite these advances, these tools remain poorly suited for public-sector contexts, where ESG-relevant information is highly fragmented and narrative-driven. Municipal governance presents distinct challenges for automated ESG evaluation. As introduced in the previous section, information is often embedded across multiple sources, vary widely in format and are rarely unified through standardized reporting templates. As such, applying corporate-grade ESG models directly to the public sector results in poor interpretability, lack of contextualization, and low transferability.

Several recent efforts have attempted to bridge this gap between AI capabilities and municipal realities. Gu *et al.* [4] propose integrating smart city data into ESG frameworks, but highlight the absence of robust, scalable methodologies for implementation. Matviaková [6] stresses the need to capture the narrative nuances and interdependencies that shape local governance responses to climate change. From a methodological perspective, Natural Language Processing (NLP) has been applied to municipal planning documents to support sustainability analysis. Huang *et al.* [5] used topic modeling and semantic similarity analysis to detect sustainability priorities in city plans, but their approach emphasizes theme identification over ESG benchmarking. Zapata*et al.* [2] and Brinkley and Stahmer [3] applied NLP-based classifiers to general plans in California, offering insight into document contents and thematic clustering across cities. These approaches improve our ability to read and compare local plans at scale, yet they often rely on static rule-based or topic modeling methods that fall short on interpretability and explainability.

More recent developments suggest that LLMs, especially when paired with RAG architectures, could significantly improve contextualization and traceability in public-sector applications. Xu *et al.* [11] propose a GenAI-powered multi-agent framework for urban mobility management that uses RAG to integrate structured and unstructured data sources and tailor advisory outputs to diverse public audiences. Although focused on transportation, their approach offers a transferable blueprint for human-centric, explainable AI in urban governance.

Taken together, these studies illustrate a growing recognition of the need for AI systems that are not only scalable and accurate but also transparent, locally contextualized, and aligned with the unique complexities of municipal administration.

3 Methodology

The architecture of *RAG4MuniESG* (Fig. 1)is organized into three principal modules: (1) Data Collection, (2) Multimodal Knowledge Construction, and (3) the LLM Agent Module.

3.1 Data Collection Module

RAG4MuniESG introduces automation in sourcing ESG-relevant information from public municipal resources. This sets it apart from prior systems that rely on manually curated PDFs. The module extracts two primary types of data: unstructured textual documents and structured numerical datasets. For the quantitative data, the system targets authoritative sources such as national and regional statistical agencies. These platforms host valuable Excel and CSV datasets, which we systematically locate and extract using browser automation and web scraping tools. The extracted numerical content is stored in a structured repository for further use in ESG evaluation.

Simultaneously, for qualitative data, we developed a scraping tool specifically designed to retrieve municipal reports in PDF format. The scraper automatically navigates through municipal websites and downloads available PDF files. To enhance relevance, we apply a keyword-based and date-based filtering mechanism, retaining documents that match ESG-related terms (e.g., "plan", "strategie", "finance") and fall within a designated time window. Documents lacking verifiable dates are flagged for manual analyst review. Our pipeline also incorporates refinements such as regular expression matching for PDF URLs, detection of embedded documents within iframe or object tags, and the use of headless browser tools to handle dynamic, JavaScript-loaded content. This comprehensive and automated collection strategy ensures the inclusion of timely and context-rich municipal content, representing a scalable and reproducible alternative to manual data aggregation.

3.2 Multimodal Knowledge Construction Module

This module transforms diverse municipal data into a semantically indexed knowledge base. Beginning with the PDF documents, we developed a parser that extracts not only the main body text but also text embedded within tables

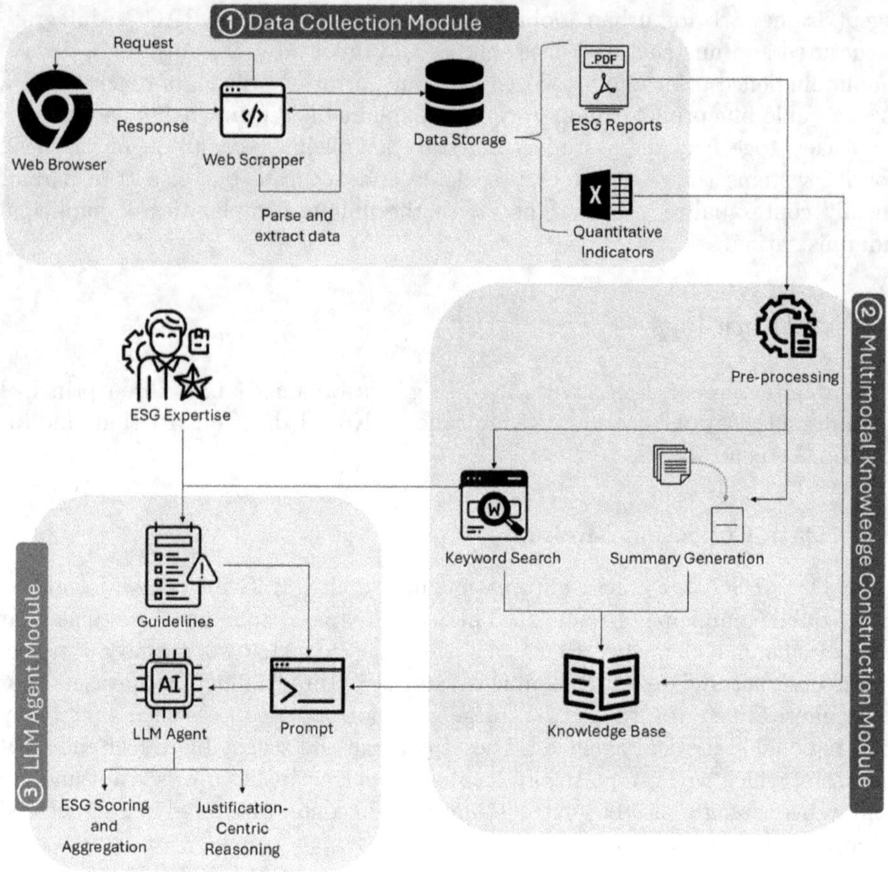

Fig. 1. *RAG4MuniESG* Framework Architecture.

and images, as shown in Fig. 2. This ensures that no relevant information is lost during preprocessing. After text extraction, we perform keyword analysis based on a curated list provided by ESG experts to identify and prioritize meaningful content. Each keyword is counted, and its distribution across the document is recorded.

To further enrich this process, we extract numerical values associated with ESG-related keywords by capturing a context window of 100 words (50 before and 50 after each keyword). This window is used to formulate structured prompts for a LLM, which identifies and returns any quantitative indicators semantically linked to the keyword. This hybrid process en- hances the semantic density of the knowledge base by blend- ing raw numerical evidence with structured qualitative inter- pretation, laying the foundation for targeted, context-aware retrieval in the subsequent module.

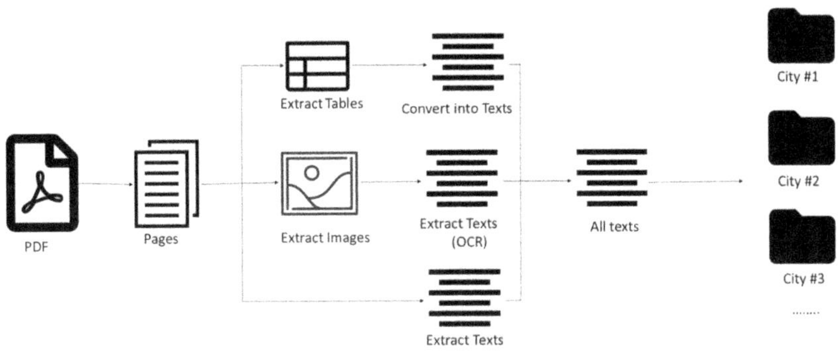

Fig. 2. Overview of the preprocessing pipeline.

3.3 LLM Agent Module

The LLM agent module manages information retrieval and generates structured, evidence-based ESG responses. For each indicator, it formulates a query using the label, expert-provided description, and the multimodal knowledge. Queries are embedded in a vector space to retrieve the top-k relevant document chunks via similarity search. Retrieved texts are reranked and passed to the LLM with a tailored prompt, producing an assessment and justification grounded in source content. Indicator-level results are then aggregated by ESG category, with the LLM generating a summary, a 0–100% performance score, and individual indicator scores with justifications.

We adopt Retrieval-Augmented Generation over conventional prompt engineering approaches because RAG mitigates the limitations of relying solely on a model's pre-trained knowledge. Prompt engineering often yields inconsistent results, constrains customization, and is entirely dependent on the static knowledge encoded in the model. In contrast, RAG integrates a retrieval mechanism with the LLM, enabling the system to dynamically access and incorporate contextually relevant information from the knowledge base. This hybrid design reduces hallucination, improves evidence grounding, and ensures that generated ESG assessments are accurate, explainable, and aligned with source content.

The LLM agent module is responsible for orchestrating information retrieval and generating structured, evidence-based responses to ESG queries. For each indicator, the module synthesizes a targeted query combining the indicator label, its qualitative description, and contextual guidance. These indicators and their interpretive guidelines are provided by ESG domain experts to ensure alignment with relevant policy expectations and thematic framing. The query is embedded into vector space and used to retrieve the top-k most relevant document chunks from the knowledge base via similarity search. The retrieved content is reranked and passed to the LLM along with a carefully designed prompt. The prompt instructs the model to determine whether the indicator is meaningfully addressed in the document, extract and interpret relevant policy actions or statistical evidence, and justify by referencing specific textual or numerical content.

The output is expected to include both an assessment and a justification grounded in the retrieved document context. Following individual indicator evaluation, results are aggregated at the category level (Environmental, Social, Governance). For each category, the LLM is prompted to generate a summary interpretation of municipal performance, a success score on a 0–100% scale for the category, and the individual indicator scores and justifications. The outputs are structured and normalized to facilitate comparative visualization across municipalities and ESG dimensions.

4 Experiments and Results

4.1 Experiment Setup

Municipal Selection and Document Collection

For experimentation purposes, the *RAG4MuniESG* framework was applied to three French-speaking mid-sized municipalities in the province of Québec: Longueuil, Saguenay, and Saint-Jean-sur-Richelieu. Cities were selected based on policy maturity, adequate public documentation, and geographic and socio-economic diversity. The data collection focused on documents published between 2015 and 2024, capturing nearly a decade of municipal sustainability activity. Across the three municipalities, our system retrieved a total of 788 PDF documents after filtering. The automated scraping pipeline significantly reduced manual effort, achieving an average processing time of 15 min per municipality.

Indicator Framework

For this study, we adopted a set of 28 ESG indicators specifically tailored to the municipal context, each defined and validated by domain experts. These indicators, presented in French, are organized across the three ESG pillars. A detailed description accompanies each indicator. The indicators cover critical dimensions such as air quality, income inequality, access to public consultation, biodiversity, and strategic urban planning. Their inclusion ensures a balanced evaluation that reflects both policy intent and implementation maturity, and they serve as the foundation for prompt construction, document analysis, and final scoring within the *RAG4MuniESG* framework.

Prompt Design

To support both qualitative interpretation and quantitative scoring, we employed two complementary prompts within the LLM Agent (GPT-4). The first prompt guided the model to respond in French with a detailed explanation explicitly grounded in the retrieved document context: *"Répondez de manière détaillée en français. Expliquez pourquoi vous donnez cette réponse, en vous basant explicitement sur des éléments du contexte fourni. Mentionnez les phrases clés ou données spécifiques qui justifient votre analyse."* This ensured that each ESG evaluation was not only linguistically aligned with the municipal context but also transparent in its rationale. In addition, a second prompt was designed to generate structured, numerical scoring for each ESG category. It requested the model to estimate an overall success rate for the municipality within a specific ESG dimension and provide a breakdown of scores for each underlying indicator: *"Veuillez fournir une estimation en pourcentage de la réussite globale de la municipalité dans cette catégorie (entre 0 et 100%), suivie d'une liste détaillée de chaque indicateur avec son score estimé. Le format doit être: Pourcentage global: XX% Détails: - Indicateur 1: XX% - Indicateur 2: XX% ..."*. This prompt enabled standardized aggregation of ESG performance, facilitating comparative analysis across municipalities.

Knowledge Base Construction
During preprocessing, each municipal document was segmented into manageable text chunks to facilitate semantic retrieval. Using a sliding window approach, documents were split into overlapping chunks of 3,000 characters with 200-character overlap. This granularity allowed for more accurate vector embeddings and ensured that contextually relevant information could be retrieved at query time. Keyword matching was conducted using the 28 predefined ESG indicators, each associated with domain expert-validated descriptions.

4.2 Results

To illustrate the effectiveness of *RAG4MuniESG* framework, we present results from the municipality of Longueuil. The system begins by generating an aggregated ESG profile at the dimension level, providing a high-level snapshot of municipal sustainability performance. As shown in Fig. 3, Longueuil achieves a high Governance score (75%), a moderate Social score (60%), and a comparatively lower Environmental score (45%). These aggregate ratings offer an immediate understanding of the municipality's relative strengths and weaknesses across ESG domains.

The system then offers a second-level breakdown by listing scores for each indicator within a given dimension. For example, Fig. 4 presents the social indicator scores, which range from 50% to 80%. This level of disaggregation supports more nuanced analysis and actionable insights. The social domain, scoring 60%, reflects moderate municipal efforts across a range of indicators. Figure 4 and the detailed breakdown below show that high scores were recorded for indicators such as "Criminalité" and "Activités sportives (parcs, piscines)", both at 80%, while others like "Création d'emplois" and "Faibles revenus" remained at 50%. This level of disaggregation supports more nuanced analysis and actionable insights.

Finally, at the most granular level, the system generates natural-language justifications for each indicator. As shown in Fig. 5, the indicator "Bruit nocturne" is accompanied by an explanation synthesized by the LLM, which references Longueuil's proactive communication around nighttime construction. This justified a high score by demonstrating municipal sensitivity to quality-of-life concerns through direct evidence from the retrieved documents.

This layered architecture—from ESG dimension scores to indicator-specific diagnostics and evidence-based justification—exemplifies how *RAG4MuniESG*

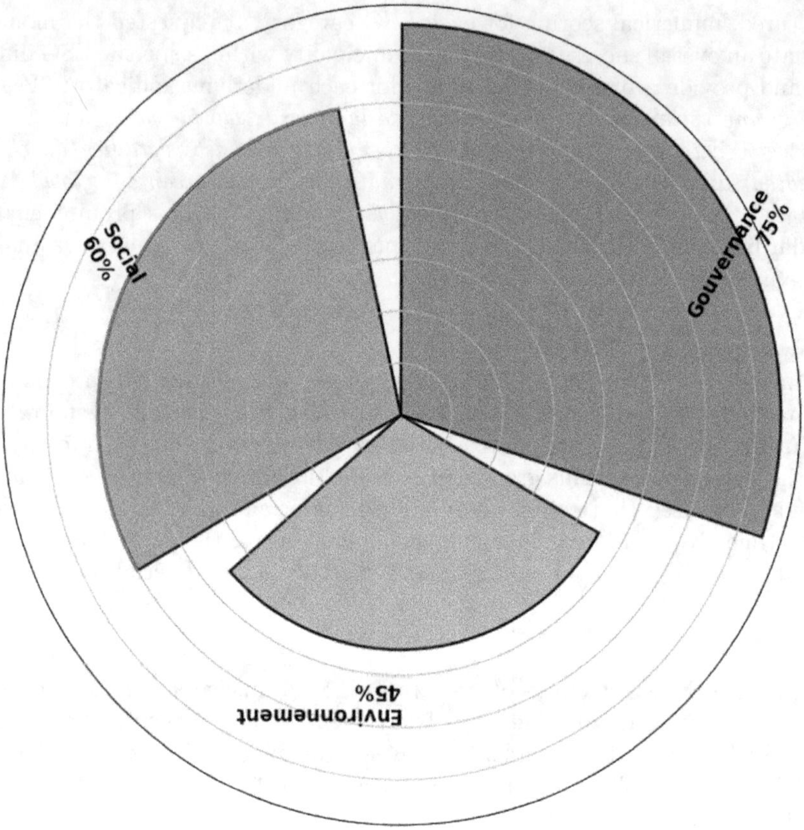

Fig. 3. High-level ESG overview.

balances breadth and depth in municipal sustainability assessment, enabling traceable evaluation processes. *RAG4MuniESG* enables consistent ESG evaluation across multiple municipalities using standardized prompts and indicators. As shown in Fig. 6, the framework can support scalable cross-city comparisons, making it a practical tool for benchmarking and policy monitoring.

5 Discussion

While this study shows the potential of a RAG framework for municipal ESG assessment, it is only a first step, with room for improvement. First, in the absence of standardized ESG benchmarks for cities, the evaluation is qualitative; future work should include external audits or expert comparisons for validation. Second, the current set of 28 expert-validated indicators offers a solid foundation, and further breaking them down could improve the granularity and coverage of sustainability issues. Third, system performance depends heavily on

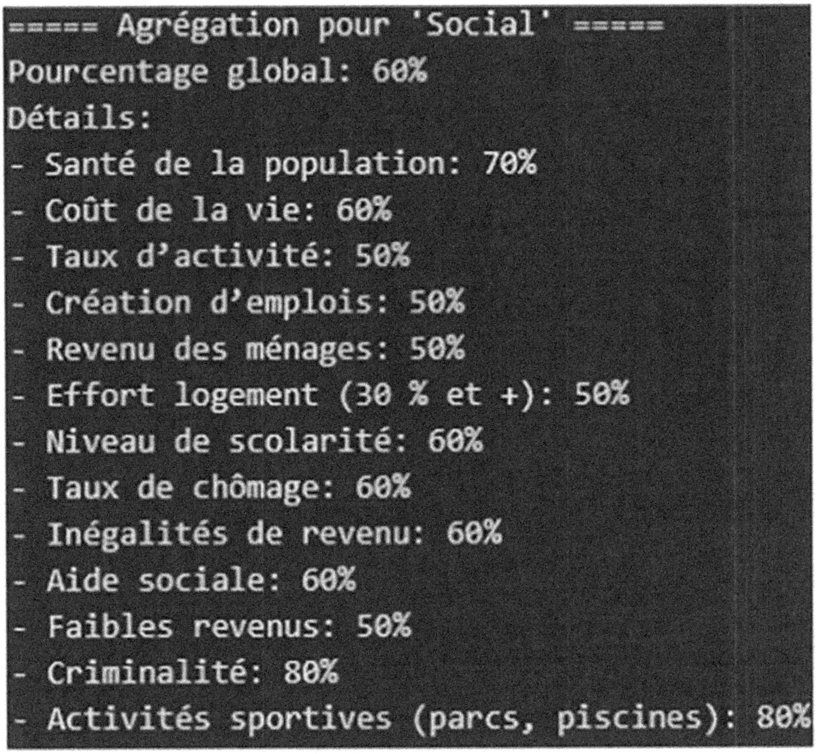

```
===== Agrégation pour 'Social' =====
Pourcentage global: 60%
Détails:
- Santé de la population: 70%
- Coût de la vie: 60%
- Taux d'activité: 50%
- Création d'emplois: 50%
- Revenu des ménages: 50%
- Effort logement (30 % et +): 50%
- Niveau de scolarité: 60%
- Taux de chômage: 60%
- Inégalités de revenu: 60%
- Aide sociale: 60%
- Faibles revenus: 50%
- Criminalité: 80%
- Activités sportives (parcs, piscines): 80%
```

Fig. 4. Mid-level view: aggregated ESG scores by category.

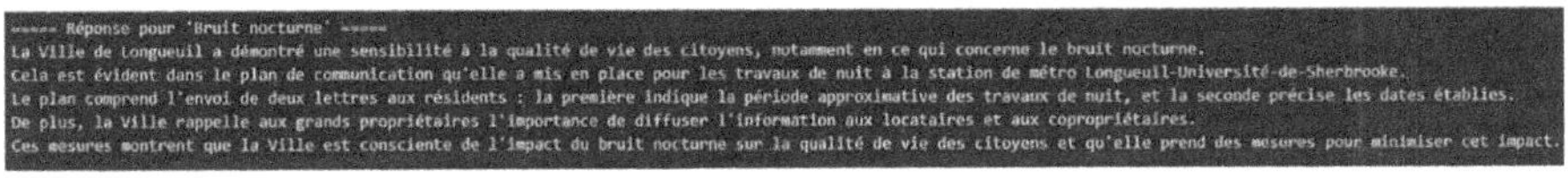

```
===== Réponse pour 'Bruit nocturne' =====
La Ville de Longueuil a démontré une sensibilité à la qualité de vie des citoyens, notamment en ce qui concerne le bruit nocturne.
Cela est évident dans le plan de communication qu'elle a mis en place pour les travaux de nuit à la station de métro Longueuil-Université-de-Sherbrooke.
Le plan comprend l'envoi de deux lettres aux résidents : la première indique la période approximative des travaux de nuit, et la seconde précise les dates établies.
De plus, la Ville rappelle aux grands propriétaires l'importance de diffuser l'information aux locataires et aux copropriétaires.
Ces mesures montrent que la Ville est consciente de l'impact du bruit nocturne sur la qualité de vie des citoyens et qu'elle prend des mesures pour minimiser cet impact.
```

Fig. 5. Detailed view: individual ESG indicator scores.

document parsing and retrieval quality—enhancements in layout-aware extraction and table handling are needed. Fourth, we tested on familiar municipalities to ensure grounded, human-verifiable results; broader generalization will come as experimentation expands across diverse contexts. Finally, ethical use demands human oversight: while the system ensures transparency and traceability, it should serve as a decision-support tool within a hybrid AI-human framework to maintain accountability and contextual relevance.

While this study demonstrates the promise of using a RAG framework for municipal ESG assessment, several limitations should be acknowledged. First, due to the absence of standardized ESG benchmarks for cities, our evaluation remains qualitative. Future work should explore external audits or comparisons with human expert assessments to validate scoring. Second, although the 28

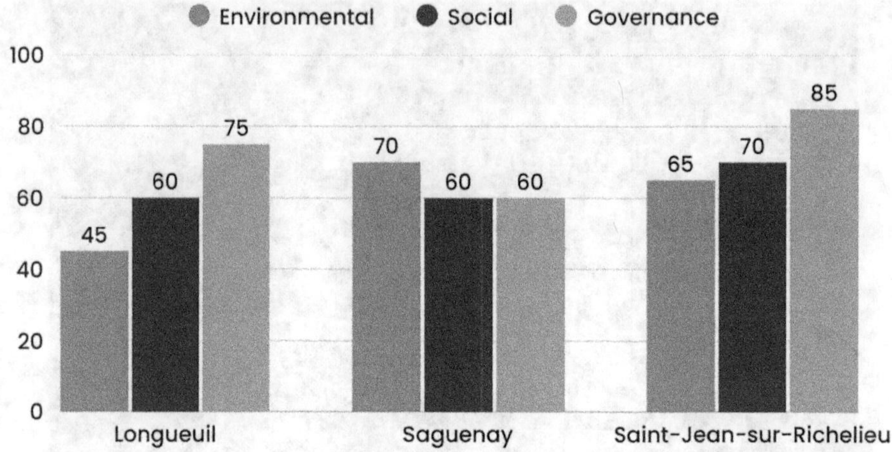

Fig. 6. Comparative ESG scores across three municipalities.

indicators were validated by ESG experts, they are not exhaustive. Expanding the indicator set would allow the framework to capture a broader range of sustainability dimensions and adapt to emerging priorities. Third, the quality of document parsing and retrieval significantly impacts system performance. Enhancing layout-aware extraction and better handling of tables or fragmented content would improve evidence relevance and coverage. Fourth, while the system was tested on three mid-sized Québec municipalities, broader validation across diverse regions, languages, and governance models is necessary to assess generalizability. Finally, ethical deployment requires human oversight. While the system provides transparency and traceability, it should function as a decision-support tool. A hybrid AI-human approach remains critical to ensure accountability, contextual relevance, and trust—beyond indicator design, humans guide interpretation and oversight.

6 Conclusion

This study proposed *RAG4MuniESG*, a RAG framework for ESG assessment in the municipal context. By combining expert-validated indicators, automated document retrieval, and LLM-based analysis, the system produces grounded, transparent evaluations. *RAG4MuniESG* demonstrated its ability to deliver both high-level ESG summaries and detailed justifications. Future work will expand its scope, refine extraction, and validate outputs across diverse municipal contexts.

References

1. Berg, F., Kölbel, J.F., Rigobon, R.: Aggregate confusion: the divergence of esg ratings. Rev. Finan. **26**(6), 1315–1344 (2022)
2. Berrío-Zapata, C., da Silva, E.F., Condurú, M.T.: The technological informavore: Information behavior and digital sustainability in the global platform ecosystem. First Monday (2021)
3. Brinkley, C., Stahmer, C.: What is in a plan? using natural language processing to read 461 California city general plans. J. Plan. Educ. Res. **44**(2), 632–648 (2024)
4. Gu, Y., Katz, S., Wang, X., Vasarhelyi, M., Dai, J.: Government esg reporting in smart cities. Int. J. Account. Inf. Syst. **54**, 100701 (2024)
5. Huang, B., Wang, J.: Big spatial data for urban and environmental sustainability. Geo-Spatial Inf. Sci. **23**(2), 125–140 (2020)
6. Matviaková, N.: Esg concept in the condition of municipalities: the case study of beskydy region. WSEAS Trans. Bus. Econ. **21**, 2513–2527 (2024)
7. Ni, J., et al.: Chatreport: Democratizing sustainability disclosure analysis through llm-based tools. arXiv preprint arXiv:2307.15770 (2023)
8. Rajaonson, J., Régnier-Sakamoto, P., Vivin, C., Guillemette, M.: Addressing sustainability challenges in micro-municipalities: insights from the study of quebec's smallest municipalities. Can. Geogr./Le Géographe Can. **68**(1), 88–100 (2024)
9. Schimanski, T., Reding, A., Reding, N., Bingler, J., Kraus, M., Leippold, M.: Bridging the gap in esg measurement: Using nlp to quantify environmental, social, and governance communication. Financ. Res. Lett. **61**, 104979 (2024)
10. Spacey Martín, R., Ranger, N., Schimanski, T., Leippold, M.: Harnessing ai to assess corporate adaptation plans on alignment with climate adaptation and resilience goals. Available at SSRN 4878341 (2024)
11. Xu, H., et alX.: Genai-powered multi-agent paradigm for smart urban mobility: Opportunities and challenges for integrating large language models (llms) and retrieval-augmented generation (rag) with intelligent transportation systems. arXiv preprint arXiv:2409.00494 (2024)
12. Zou, Y., et al.: Esgreveal: an llm-based approach for extracting structured data from esg reports. J. Clean. Prod. **489**, 144572 (2025)

Changing the Perception of Digital Competences in 2018 and 2025 Among Polish and Slovak Students - Comparative Analysis

Marek Szajt[1](✉), Klaudia Smoląg[1], Katarzyna Kozioł-Nadolna[2], Zuzana Hajduová[3], and Iwona Chomiak-Orsa[4]

[1] Czestochowa University of Technology, Częstochowa, Poland
{marek.szajt,klaudia.smolag}@pcz.pl
[2] University of Szczecin, Szczecin, Poland
katarzyna.koziol-nadolna@usz.edu.pl
[3] University of Economics in Bratislava, Bratislava, Slovakia
zuzana.hajduova@euba.sk
[4] Wroclaw University of Economics and Business, Wrocław, Poland
Iwona.Chomiak-Orsa@ue.wroc.pl

Abstract. The aim of the study was to assess changes in the level of self-esteem of digital competences among Polish and Slovak students conducted and their use of electronic tools and online activities in 2018 and in 2025. Pandemic Covid-19 and the revolutionary development of ICT technology certainly had an impact on the assessment of student competences. The research was conducted in parallel by the survey among students in Poland and Slovakia. The same survey questionnaire was used for research. The analysis of the results was carried out using statistical tests and spatial simulations performed in the STATISTICA 13 program. The obtained research results indicate that students assess their digital competences at a higher but even and more flattened and less diverse level. In addition, students from Slovakia assess their digital competences as significantly higher in the second study compared to the 2018 study.

Keywords: digital competences · information competences · IT competences

1 Introduction

Information and IT competences play an increasingly important role on both the labor market and in everyday life. Their acquisition takes place not only in the course of education at various levels but above all in everyday life. The level of advancement of the technique we have today is disproportionately higher than twenty years ago. The revolution 4.0 defined in the literature has actually happened and we are currently dealing not so much with the revolution but the use of its results [21]. Society currently operates not only in cooperation with a number of electronic and IT tools, information channels, but sometimes in almost two parallel beings. One is the one we know since the dawn of

Published by Springer Nature Switzerland AG 2026
E. Mercier-Laurent et al. (Eds.): AI4S 2025, IFIP AICT 780, pp. 202–217, 2026.
https://doi.org/10.1007/978-3-032-21184-2_16

time. The second, used by both individuals and entire organizations, is the virtual one based on websites, messengers, social media, etc. Functioning in these realities required an increase in competence in society [13]. This applies to us as people in general. Of course, we can assume that in poorer countries, those with limited access to the Internet, authoritarian managed, the level of commitment and competence will stand out, but in a group of countries with a similar level of development it should also be similar. Hargittai and Hinnant emphasize that the differences in digital skills more determine the actions taken by Internet users than access to technology itself [18]. The self-esteem, which in this study was analyzed as in the research of Schmid and Petko [33], is influenced by, among others access to information technologies at home or school and experience in using these technologies [45]. At this point, however, history, culture and everything that is extremely difficult to define [42], while for learning society and influencing its preferences, worth identifying.

2 Theoretical Background

Digital competences are an important factor that shapes the quality of life of individuals and entire organizations. Their importance is constantly growing, which results from the dynamics of the development of information and communication technologies (ICT) and continuous digital transformation [2]. Digital competences are referred to as one of the priority factors for the construction of a digital economy strategy and a digital society [36]. According to European Parliament and the Council, Digital Competence involves the confident and critical use of Information Society Technology (IST) for work, leisure and communication. It is underpinned by basic skills in ICT: the use of computers to retrieve, assess present and exchange information, and to communicate and participate in collaboration networks via the Internet [13]. Another definition says that digital competences are defined as skills, knowledge and attitudes that make the learners use digital media to participate, work and solve problems, independently and in cooperation with others in a critical, responsible and creative way [19]. They define digital competences as the ability to use technology to use and access information. Information competences relate to all processes related to: the ability to search and obtain information from various sources, its assessment (assessment of credibility and usefulness) and its use [38].

It is assumed that digital competences depend on age, professional activity or education of a person [13]. These competences are important because they enable the functioning in the world currently dominated by ICT in almost all areas and are part of the lifelong learning process [11, 28]. It is assumed that people studying in the digital era, who commonly use ICT solutions almost in the 24-hour rhythm, should naturally shape their digital competences or at least certain groups of these competences. However, research to date shows that knowledge and skills play an important role in shaping competences, not access and use itself [10, 17, 41]. Higher education also plays a significant role because it creates the independence of students in the academic and professional environment [30]. That is why the development of digital competences, regardless of what level of education takes place, is targeted at the entity of the respective educational impact and the corresponding level of digital literacy [32].

The technical service of IT devices, which constituted the basis of digital competences, become basic knowledge, are acquired by us from an early age and accompany us basically in every element of functioning [9]. Importantly, apart from a number of intuitive solutions dedicated to individual age groups, it is important to properly process information reaching us. Therefore, the competences related to the conscious acquisition of the right information and its processing for their own needs are becoming more and more important [6]. Effective information management includes the ability to specify information needs, define sources of acquisition, select information and use them to carry out planned tasks [1, 25, 37]. In the area of digital communication, which is the basis of communication of enterprises with the environment in the current reality, the ability to acquire and communicate with a wide audience has gained special significance [43] and the ability to interact, regardless of distance and time, which allows the creation and strengthening of social relationships [22]. Enterprises using this type of solutions reacted much faster in the case of Covid-19 pandemic and suffered clearly lower losses.

Creative Digital Skills are another category of digital skills, more and more often analyzed. These skills are particularly important for employers, so more and more often students are analyzed and more and more courses are created to develop these competences [22, 23, 26].

Currently, activities in the digital space are characterized by a high level of diversity and intensity. Electronic tools such as a desktop computer, laptop or increasingly a smartphone are also becoming fundamental tools in everyday online activities. Students typically use laptops to shop online, comment on cultural content (such as books and videos), and create and share content on social media platforms such as Facebook or Instagram [7]. In addition, it is worth emphasizing that online activity, despite its apparent passivity, contributes to the development of digital competencies, especially in the field of information assessment, communication and online expression [14]. in the process of forming digital competencies, it is also important to take various forms of online activity, including opportunities for interaction [24]. Creating digital content and participating in online discussions enriches the digital experience, as well as allows you to create digital competencies.

Different approaches to defining digital skills focus on two main areas: IT and information skills. These two groups used these skills as a basis for studying digital skills of Polish and Slovak students. People undergoing research are those that in practice are brought up in the digital world and naturally shape their digital competences. On the other hand, it should be assumed that there is a large variety of how they deal with ICT technology and what they know about it [17]. In addition, the development of digital competences is associated with some effort and time [4]. The study seems to be justified in the light of postulates indicating the need for a broader analysis of a group of young people - the most susceptible to the development of these competences [20, 28]. Digital competences are increasingly the subject of many research. Digital skills related to the use of ICT are examined and evaluated, among others pupils, students [5, 15, 35], also in Polish research conditions [26, 33, 40]. This subject is also dedicated research conducted by the authors of this article [8, 16, 39]. The presented study is therefore part of this trend. His goal was to assess changes in the level of self-esteem of Polish and Slovak students of students resulting from their personal verification, which was influenced by the Covid-19 pandemic.

3 Methodology

This study just attempts to compare information and IT competences of Polish and Slovakia students. The research was conducted in two turns. The first took place in 2018 and included 172 students from Poland and 171 from Slovakia. The second round took place in 2025 and covered 435 and 203 students respectively (as of 30.04.2025). Differences in sample sizes are not of much importance from the point of view of the test used, but in the case of correlation analysis, they would play a role. In the case of the presented simulations - made separately for each country - a larger number brings them closer to the real image. Despite the fact that the study is still ongoing, based on flowing surveys, an attempt was made to compare both the levels of self - assessment of competence, as well as changes and equals in these changes in relation to both countries and individual competences. The survey contained 28 substantive questions - including in question 18–38 variables identifying IT competences and 11 describing information competences, and 6 metric questions. Validation with the help of Alfa Cronbach confirmed the proper construction of the questionnaire. Questions on competences were constructed based on a compartmental five-point scale, where the maximum assessment was 5. The study used the U-Mann-Whitney test appropriate for the scale used, as a replacement for the Student's t-test and simulations of exponential equalization weighted at distances.

Table 1. Characteristics of the study

	Subject area	Subject range	Time range	Spatial range
Survey characteristics	Digital skills: Computer science and information for students Online student participation is understood as students' activity in a digital environment Use of electronic tools by students	Students of higher educational institutions Polish and Slovak	Two stages: Stage I-2018 and Phase II-2025	Poland and Slovakia

The study was multi-faceted and multi-faceted (Table 1). The subject area of the study included such issues as digital competence, online activity, and electronic tools. In turn, the field of activity included students of higher educational institutions from two countries-Polish and Slovak (spatial range). The dates are 2018 and 2025.

4 Results

To assess the changes of self-esteem posing in the area of competence, the U-Mann-Whitney test was used, which is the equivalent of the average test at the lower scales.

Table 2. Evaluation of the results of the U-Mann-Whitney test for comparisons of self-assessment of information and IT competences for Polish and Slovak students in 2018 and 2025.

Variable	SK 18/25	PL 18/25	18 PL/SK	25 PL/SK
Write, edit and send text in a text editor	⇑	⇓	PL	-
Verify the text by checking spelling/dictionaries	⇑	⇓	PL	-
Insert images/symbols in a text editor	⇑	⇓	PL	SK
Insert and edit tables in a text editor	⇑	⇓	PL	-
Use drawing/graphics applications such as Power Point	⇑	⇓	PL	-
Move files/folder on the computer	-	⇓	PL	-
Move files/folder on the smartphone	⇑	⇓	PL	-
Move files/folder between devices	⇑	⇓	PL	-
Use copy tools	⇑	⇓	PL	-
Use cutting tools	⇑	-	PL	-
using the basic functions of spreadsheet	⇑	⇓	PL	-
compressing files	⇑	⇓	PL	-
Connect and install new devices, e.g. printer, scanner, etc.?	⇑	⇓	PL	-
Install the program on your computer	⇑	⇓	PL	-
Search for and install a smartphone program/application	⇑	-	PL	-
Uninstall the program on your computer	⇑	⇓	PL	-
Uninstall the program/application yourself on your smartphone	⇑	-	PL	-
Programming in a specialist language	⇑	-	PL	SK
Use internet search engines (e.g. Google, Yahoo etc.)	-	⇓	PL	-
transferring data from a spreadsheet	-	⇓	-	-
Set up/create private email address	⇑	-	PL	-
Send and receive e-mail	⇑	-	PL	-
Send an email with attachments	⇑	-	PL	-
using e-mail/calendar systems	⇑	-	PL	-
using file-sharing programs (P2P)	⇑	-	PL	PL
creating websites	⇑	-	PL	PL
Transfer photos from a digital camera to a computer	⇑	⇓	PL	-
Transfer photos from a smartphone to a computer	⇑	⇓	PL	PL
Make calls via the Internet	-	⇓	PL	PL

(*continued*)

Table 2. (*continued*)

creating an electronic signature	-	⇓	PL	PL
Send/receive SMS/MMS from a mobile phone	⇑	⇓	PL	-
Connect to the Internet using a mobile phone	⇑	-	PL	SK
Order and buy tickets online	⇑	⇓	PL	SK
Buy and sell goods via native websites	⇑	⇓	PL	PL
making commercial transactions using languages other than native	⇑	⇓	PL	PL
Use IP telephony or Skype	⇑	-	PL	PL
Using an electronic signature	⇑	⇑	PL	-
Participate in online communities, e.g. Facebook or Instagram	⇑	-	PL	-
locating pages with the information you need	⇑	-	PL	-
web browsing	-	⇓	PL	PL
Finding the specific information you need on the websites of government agencies	⇓	⇓	PL	PL
searching websites in a language other than Polish	⇑	-	PL	PL
selection of the right information from the Internet	⇑	-	PL	SK
organized file organization on your computer	⇑	-	PL	-
Organize the information found by e.g. placing them in lists and tables	⇑	-	PL	-
Assess the quality of information that can be found on the Internet, for example, whether it is old, biased or unreliable	⇑	⇓	PL	-
Assess the security level of publishing information on the Internet, for example on Facebook,	⇑	⇑	PL	SK
entering information using a template on the web	⇑	⇓	PL	-
Read and/or comment on the blog	⇑	⇓	PL	-

The self - esteem of Slovak students in the studied years increased in 42 cases, in six it did not change significantly, but only in the scope of 'Finding the Specific Information You Need on the Websites of Government Agencies' decreased significantly (Table 2). In the case of Polish students, whose self-esteem in 2018 was in almost all aspects higher than their Slovak counterparts, now it increased only in two cases and decreased in 29. This is probably the result of a collision with pandemic reality. In the era of lockdown, students had to use their IT competences to a broader extent and it probably allowed them to make self-esteem to make it easily, probably a bit inflated. The changes that we observe in both communities have also led to changes in perception in comparison of their competence in 2025. Currently, Polish students point to significantly higher competences in 11 areas (only) and Slovak students towards Polish in six. Therefore, we can talk about a diametrical change or competence, or their perception within seven years.

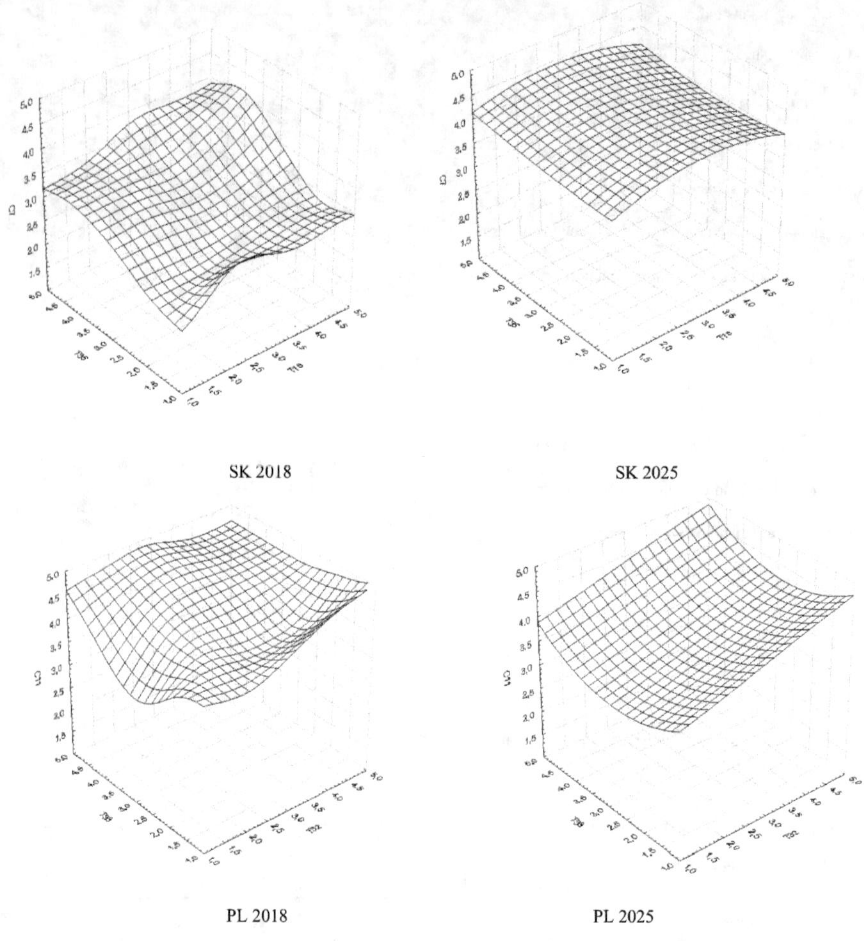

Fig. 1. Simulations smoothed by the smallest squares smoothed by distances for selected information and IT competences for Poland and Slovakia in 2018 and 2025.

The fact that information and IT competences intertwine, complement and strengthen each other does not require wider justification. However, in this study, the selected (most varied in assessments) competences for both communities studied were subject to analysis. In the case of Slovakia, these were: Programming in a Specialist Language (T18), Making Commercial Transactions Using Languages Other than Native (T35) and Organizer The Information Found by E.G. Placing them in List and Tables A for Poland Connect to the Internet Using a Mobile Phone (T32), Order and Buy Tickets

Online (T33) and Read and/Or Comment on the Blog (C11). In both cases, we are dealing with a higher level of smoothing the studied relationships in 2025 compared to the 2018 study. This allows for a slightly more accurate interpretation. In the case of C11 competences in Poland, it depends almost linearly on the increase in competences T32 and exponentially on T33. Compared to 2018, we observe a clear normalization of these relationships and their high stability. In Slovakia, for C7 (Organize the information found by e.g. placing them in lists and tables) competences, we are observing a clear weakening of the impact of competences T18 and T35 in 2025 compared to 2018. We are currently almost independence, with excessive self-esteem of the competence T18 causes a slight decrease in the self-assessment of C7 competence.

Table 3. Use of desktop computers and laptop computers/laptops among students

	Poland		Slovakia	
Question under investigation	t	p	t	p
1. How often do you use a computer while studying at university?	-1.393	0.082	-5.647	0.000
2. How often do you use a computer outside of university?	7.900	0.000	-7.050	0.000
3. How often do you use mobile devices?	7.091	0.000	-2.147	0.016
4. Do you have a portable computer/laptop?	-2.906	0.002	-3.492	0.000
5. Do you use a laptop or laptop while studying at university?	-5.886	0.000	-7.140	0.000

It turns out that the period of the last five years has changed the perception of electronic instruments in both Poland and Slovakia (Table 3). Interestingly, these changes often went in other directions. In Slovakia, students are now significantly more likely to use a computer (several times a day) at university ($t = -5.647$; $p < 0.001$) in Poland, this difference is significant no more than at the level of a statistical trend ($t = -1.393$; $p = 0.082$). Outside of university, students in Poland use computers (several times a day) significantly less often than before the pandemic ($t = 7.900$; $p < 0.001$), and in Slovakia significantly more often ($t = -7.050$; < 0.001). Students in Poland are significantly less likely to use mobile devices ($t = 7.091$; $p < 0.001$) in Slovakia, this change is observed in the opposite direction ($t = -2.147$; $p = 0.016$). The availability of a laptop computer and available at college every day in both groups was subject to changes-increasing the share of ownership and frequency of use of laptops in college.

It turns out that in terms of activity in commenting on what is happening on the web, there was a significant decrease among Polish students (in all aspects studied), while significant changes from the Slovak side were mainly associated with an increase in this activity, with the exception of a decrease in the frequency of commenting on books ($t = 1.699$; $p = 0.045$). Interestingly, the level of activity in this regard (despite the decline) is significantly higher in all cases on the part of foreign students (Table 4).

Table 4. Online activity among students

Question under investigation	Poland		Slovakia	
	t	p	t	p
Products purchased online	4.533	0.000	-10.098	0.000
Facebook messages from your friends	10.973	0.000	-12.086	0.000
Facebook posts of public figures or companies	6.518	0.000	1.200	0.115
Films	4.101	0.000	-20.277	0.000
Books	4.404	0.000	1.699	0.045
Interesting places	5.440	0.000	-14.894	0.000
I share online reports about places/events I participate in	2.937	0.002	-8.523	0.000

It turns out that the students studied probably do not equate digital competencies in themselves with the use of an electronic device. Correlations between the use of these devices and self-assessment of digital competence were investigated.

Table 5. The presence of positive and negative correlations between information and information competence and the use of electronic devices

Use:	Slovakia			Poland		
	positive correlation	non-significant	negative correlation	positive correlation	non-significant	negative correlation
computer as part of classes at university	1	45	3	7	42	0
computer outside the university	0	46	3	4	44	1
mobile devices such as smartphones	0	49	0	0	48	1
portable computer/laptop during stay at university	1	46	2	0	49	0

It turns out that there are very few statistically significant correlations between the use of individual electronic devices and self-assessment of digital competence (Table 5). Interestingly, for Slovak students, statistically significant correlations were found only in 10 cases out of 196, and 8 of them were negative correlations. None of them related to the connection of competencies with smartphone support. For Polish students, there were 13 cases, including two negative ones. These results hardly indicate the absence of correlations (they do not confirm their existence), but they are the result of very different responses from respondents with different competencies. It is worth noting that the number of correlations that are significant at the level of the statistical trend ($\alpha = 0.1$) is almost twice as high (29 for Slovakia and 25 for Poland, respectively). Perhaps expanding the study to a representative group of Polish (about 1,000 people) and Slovakia (about 600), which is our goal in the future, will provide more answers to our questions.

Table 6. The presence of positive and negative correlations between information and informational competence and comments on the studied issues in the network.

Question	Slovakia			Poland		
	positive correlation	non-significant	negative correlation	positive correlation	non-significant	negative correlation
Interesting places	4	44	1	0	14	35
Films	1	46	2	1	11	37
Books	3	46	0	0	16	33
Products purchased online	3	46	0	0	48	1
I share online reports about places/events I participate in	1	47	1	0	47	2
Facebook posts of public figures or companies	4	45	0	1	15	33

In the case of university students, we observe a noticeably higher number of statistically significant-negative-correlations between information and information competence and commenting on the issues studied on the web. In addition to commenting on products purchased online and sharing relationships, these correlations affect almost 2/3 of cases (Table 6). It follows that students in Poland who comment on individual questions online, at the same time, rate their information and information skills lower. For students from Slovakia, statistically significant dependencies often have an added sign, but in general they are very small and far from each other. The lack of significant correlations for comments on online purchases may be due to the fact that responses confirming activity in this regard were given by a relatively small number of respondents (less than 20%).

It is worth noting that in almost all activities (except for public posts on Facebook and books) in the field of commenting on various issues on the network, Slovak students showed much more activity compared to students from Poland (Fig. 2). More than half of Polish students generally commented on products ordered online and shared their relationships. In other cases, Polish students are unlikely to participate in comments on these issues, while their Slovak colleagues participate much more often, and in the case of film comments, half of them participate in discussions very often. For Facebook posts and comments on interesting places, a quarter of students in Slovakia are very often involved in the discussion.

The overall structure of competency assessments in Poland and Slovakia is very similar. The level of similarity of structures for almost all competencies exceeds 74%. For "find websites that contain the necessary information", it is 98.1%, and for "use drawing /graphics applications, such as Power Point" - 98.4%. The lowest-isolated in comparison with others-57.6% refers to web development, here Slovaks assess their competence diametrically differently-higher in relation to their Polish colleagues.

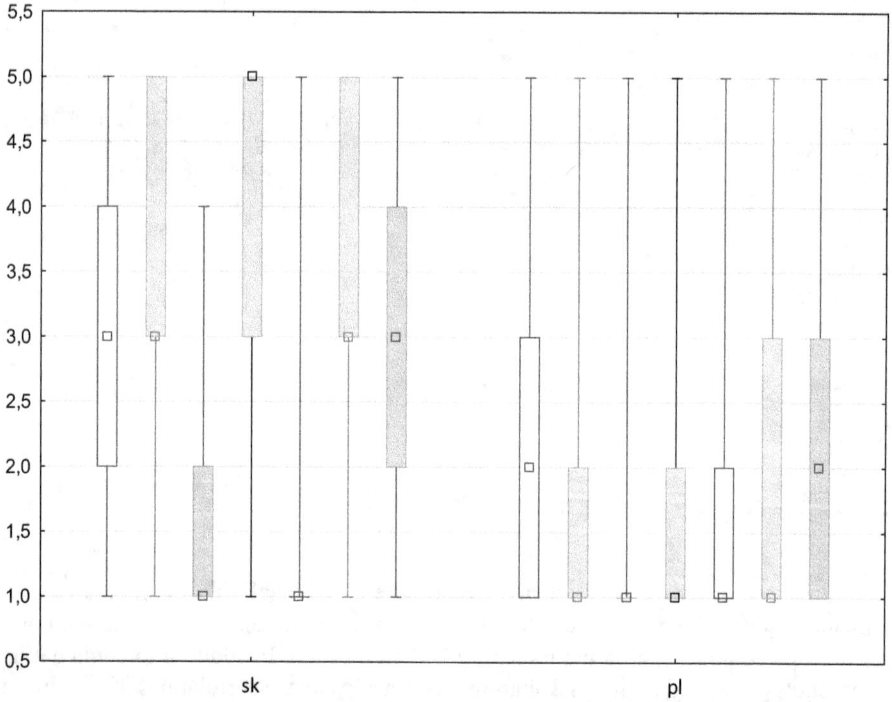

Fig. 2. Distribution of responses in the field of network activity of students from Poland and Slovakia. ▣ Products purchased online; ▣ Posts on Facebook of your friends; ▣ Posts on Facebook of public figures or companies; ▣ Films; ▣ Books; ▣ Interesting places; ▣ I share on-line reports from places/events in which I take part.

In addition, the method of changing the self-assessment of these competencies is diametrically different. While the average scores of Poland and Slovakia in the first edition of the study were correlated to the degree of 0.77, the current scores are linked at the level of 0.55. This is the result of changes in the Slovak group. The correlation of results for the Polish first and second editions is high-0.92, and for Slovakia-0.31. Looking at the' ranking ' of competence scores in Poland, the largest increase was associated with a jump of 22 positions (from 24 to 2) and a drop of 18 (from 11 to 29). In Slovakia, the largest increase was associated with a jump of 40 positions (from 49 to 92) and a fall of 42 (from 1 to 43). Moreover, in Slovakia we have sixteen changes above twenty positions, and in Poland-one. Thus, it is necessary to assume that the method of implementing information policy, digitization, etc. in both countries is different, at least in many areas.

5 Conclusions

This work focused on examining the level of self-assessment of digital competences among Polish and Slovak students and their use of electronic tools and online activities. The obtained research results indicate that a change has been observed in self-esteem in two groups of students - Polish and Slovak. Polish students in the second study do not assess their competences in many cases higher (see Table 1). It is worth emphasizing here that the results of the first study showed that Polish students assess their digital competences higher and at a less differential level than students from Slovakia [16]. In turn, students from Slovakia will assess a larger group of competences much higher in the second study. In addition, in the second study, both Polish and Slovak students, assessed their competences at a fairly balanced level, which is shown by flattened curves obtained from the results of strongly diverse assessments in the first study (Fig. 1). Thus, the presented results show that students' self-esteem (in our opinion) is currently more mature and indicates clear connections between individual competences and their self-esteem and a more objective approach to this assessment. This could have been influenced by Pandemia Covid-19 and the actual need to use their digital competences in conditions in which many educational, professional and private life activities were limited only to digital services.

The results of our research allow us to state that the self-esteem of digital competences of the studied groups of students is an increasingly conscious process, which can translate into their learning [27], development and provide young people with a better base to undertake professional work and broadly understood functioning in the digital world.

As for the use of electronic tools among students, the results showed a change in the growth direction of those who own and use laptops more frequently in college. This also applies to computers available in college. It is worth noting that students from Slovakia use a computer at the university much more often (several times during the day). The results of the study showed that the students studied do not equate digital competencies with the use of a particular electronic device, as evidenced by statistical correlations between the use of individual electronic devices and self-assessment of digital competencies. In turn, online student activity has significantly decreased among Polish students, but is still significantly higher in the analyzed cases than Slovak students.

This research has some restrictions. First of all, diverse groups of students, as to the size, fields of study, origin or work during study, may affect the shape of the results obtained and hinder the term relationship between them, which, among others show research Zhao et al. [45]. Secondly, the list of digital competences has not been updated at an angle, e.g. skills related to the use of AI, which can be a direction of further research. Thirdly, our study, both in the first and second stage, does not measure the actual level of digital competences but is only based on self-esteem, which also affects the assessment of the actual level of digital competences of students.

The indicated limitation, despite the fact that they do not allow generalization of obtained results for other groups of students, they constitute an important basis for developing new research related to the development and shaping of digital competences, especially in terms of IT solutions using AI and are important for educational institutions, indicating the direction of changes and development of the improvement of curricula for real market conditions for future employees. This issue is also emphasized in their work by Barboutidis and Stiakakis [4], who believe that achieving a sufficient level of digital competences is also in terms of educational institutions, which should adapt their curricula so as to enable their students a proper level of digital skills.

To sum up, our research did not take into account the dynamics of the development of digital competences related to the prosperity of new technologies, such as AI, augmented reality or data analysis, but clearly indicates the level of competences that still requires appropriate shaping and thus constitutes a starting point for identifying areas of deficit digital competences. It is also worth paying attention to the skills of using ICT technologies in a way that supports sustainable development (e.g. digital solutions for environmental protection, remote collaboration, minimizing the carbon footprint) and the development of new skills, which includes such competences as: the ability to use AI, a critical approach to content generated by AI, understanding the impact of ICT technologies on society, the environment or the labor market, as well as the ethical and safe use of these solutions. Additionally, the issue of digital education policy should also take into account aspects related to the level of advancement of ICT development in a given region, country and supporting equality and eliminating digital exclusion, which is particularly true for older people or people living in smaller towns. All the issues indicated above should be consistent with the challenges of digital transformation and the goals of the 2030 Agenda.

References

1. Ananiadou, K., Claro, M.: 21st century skills and competences for new millennium learners in OECD countries (OECD Education Working Papers No. 41). Paris, France: OECD Publishing (2009). https://doi.org/10.1787/218525261154
2. Calvani, A., Fini, A., Ranieri, M., Picci, P.: Are young generations in secondary school digitally competent? A study on Italian teenagers. Comput. Educ. **58**, 797–807 (2012)
3. Ayale-Perez, T., Joo-Nagata, J.: The digital culture of students of pedagogy specialising in the humanities in Santiago de Chile. Comput. Educ. **133**, 1–12 (2019)
4. Barboutidis, G., Stiakakis, E.: Identifying the factors to enhance digital competence of students at vocational training institutes. Technol. Knowl. Learn. **28**(2), 613–650 (2023)

5. Boyd, D.: It's complicated. London: Yale University Press. Edvard Hatlevik, O., Bj€ork Guðmundsdottir, G., Loi, M.: Digital diversity among upper secondary students: a multilevel analysis of the relationship between cultural capital, self-efficacy, strategic use of information and digital competence. Comput. Educ. 81, 345–353 (2015)
6. Catts, R., Lau, J.: Towards information literacy indicators. UNESCO Publishing, Paris, France (2008)
7. Cheng, J., Lei, J.: A description of students' commenting behaviours in an online blogging activity. London Rev. Educ. **19**(1) (2021). https://doi.org/10.14324/LRE.19.1
8. Chomiak-Orsa, I., Szajt, M., Smolag, K., Hajduová, Z.: Digital competences and the use of social media by Polish students – comparative analysis in 2018–2024. In: Hernes, M., Wątróbski, J., Rot, A. (eds.) Emerging Challenges in Intelligent Management Information Systems Proceedings of 27th European Conference on Artificial Intelligence ECAI 2024 – IMIS Workshop. Volume 2, Conference proceedings, pp. 69–83 (2024)
9. Chomiak-Orsa, I., Smolag, K.: Remote learning technologies in achieving the fourth sustainable development goal w: artificial intelligence. In: Nowaczyk, S., et al. (eds.) ECAI 2023 International Workshops. Part 2, pp. 140–147. Springer, Cham (2024)
10. Erstad, O.: Educating the digital generation. Nordic J. Digit. Lit. **1**, 56–70 (2010)
11. Eshet-Alkalai, Y.: Digital literacy. A conceptual framework for survival skills in the digital era. J. Educ. Multimedia Hypermedia **13**(1), 93–106 (2004)
12. Fenwick, T.: Learning as grounding and flying: knowledge, skill and transformation in changing work contexts. Paper presented at the conference: From grounded skills to sky qualifications, 17–19 August 2005, Kiruna, Sweden (2005)
13. Ferrari, A.: Digital Competence in Practice: An Analysis of Framework. Publications Office of the European Union, Luxembourg (2012)
14. Gulzar, M.A., Ahmad, M., Hassan, M., Rasheed, M.I.: How social media use is related to student engagement and creativity: investigating through the lens of intrinsic motivation. Behav. Inf. Technol. **41**(13), 2957–2970 (2022). https://doi.org/10.1080/0144929X.2021.1917660
15. Guzman-Simon, F., García-Jimenez, E., Lopez-Cobo, I.: Undergraduate students' perspectives on digital competence and academic literacy in a Spanish University. Comput. Hum. Behav. **74**, 196–204 (2017)
16. Hajduová, Z., Smoląg, K., Szajt, M., Bednárová, L.: Digital competences of Polish and Slovak students-comparative analysis in the light of empirical research. Sustainability **12**(18), 7739 (2020). https://doi.org/10.3390/su12187739
17. Hargittai, E., Hinnant, A.: Digital inequality: differences in young adults' use of the internet. Commun. Res. **35**(5), 602–621 (2008)
18. Hargittai, E.: Digital na(t)ives? Variation in internet skills and uses among members of the 'net generation.' Socio. Inq. **80**(1), 92–113 (2010)
19. Hatlevik, O.E., Bjork Guðmundsdottir, B., Loi, M.: Digital diversity among upper secondary students: a multilevel analysis of the relationship between cultural capital, self-efficacy, strategic use of information and digital competence. Comput. Educ. **81**, 345–353 (2015)
20. Hazar, E.: A comparison between European digital competence framework and the Turkish ICT curriculum. Univ. J. Educ. Res. **7**(4), 954–962 (2019). https://doi.org/10.13189/ujer.2019.070406
21. Johansson, J., Abrahamsson, L., Bergvall Kåreborn, B., Fältholm, Y., Grane, C., Wykowska, A.: Work and organization in a digital industrial context. Manag. Revue **28**(3) 281–297 (2017)
22. Lai, C.Y., Yang, H.L.: The reasons why people continue editing Wikipedia content-task value confirmation perspective. Behav. Inf. Technol. **33**(12), 1371–1382 (2014). https://doi.org/10.1080/0144929X.2014.929744
23. Lessig, L.: Remix: Making art and commerce thrive in the hybrid economy. Penguin Press, New York, NY (2008)

24. Levin, S.M.: Review on the influence of student engagement and interaction in online learning communities on academic achievement. Vectors of well-being: economy and society (2(53)), 72–85 (2024)
25. Liszczyk, K., Chomiak-Orsa, I.: The importance of eco-commerce in the con-text of sustainable development: a case study analysis. In: Nowaczyk, S., et al. (eds.) Artificial Intelligence. ECAI 2023 International Workshops. ECAI 2023. Communications in Computer and Information Science, vol. 1948. Springer, Cham (2024). https://doi.org/10.1007/978-3-031-50485-3_23
26. Loveless, A.: Creativity, new technologies and learning: A review of recent literature [An update]. Bristol, UK: Futurelab (2007)
27. Mehrvarz, M., Heidari, E., Farrokhnia, M., Noroozi, O.: The mediating role of digital informal learning in the relationship between students' digital competence and their academic performance. Comput. Educ. **167**, 104184 (2021)
28. Murawski, M., Bick, M.: Digital competences of the workforce – a research topic? Bus. Process. Manag. J. **23**(3), 721–734 (2017)
29. Ogonowska, A., Walecka-Rynduch, A.: Kompetencje cyfrowe młodych dorosłych: modele ramowe a rzeczywiste profile kompetencji. Analiza przyczynkowa; Annales Universitatis Paedegogicae Cracoviensis, folia 360 (2020) https://doi.org/10.24917/20811861.20.39
30. Pérez-Mateo, M., Romero, M., Romeu, T.: Collaborative Construction of a Project as a Methodology for Acquiring Digital Competences, Comunicar, n. 42, v. XXI, 2014, Media Education Research Journal 15–23 (2014). ISSN: 1134-3478
31. Prinza, Ch., Morlocka, F., Freitha, S., Kreggenfelda, N., Kreimeiera, D., Kuhlenkötter, B.: Learning Factory modules for smart factories in Industrie 4.0. Procedia CIRP **54**, 113–118 (2016)
32. Rambouseka, V., Štípeka, J., Vaňková, P.: Contents of digital literacy from the perspective of teachers and pupils. Procedia Soc. Behav. Sci. **217**, 354–362 (2016)
33. Rogacka, M.: Kompetencje cyfrowe w miejscu pracy – perspektywa pracowników i pracodawców. e-mentor **5**(97), 53–64 (2022) https://doi.org/10.15219/em97.1591
34. Schmid, R., Petko, D.: Does the use of educational technology in personalized learning environments correlate with self-reported digital skills and beliefs of secondary-school students? Comput. Educ. **136**, 75–86 (2019)
35. Shariman, T.P.N.T., Razak, N.A., Noor, N.F.M.: Digital literacy competence for academic needs: an analysis of Malaysian students in three universities. Procedia Soc. Behav. Sci. **69**, 1489–1496 (2012)
36. Somyürek, S., Karabulut Coskun, B.: Digital competence: Is it an innate talent of the new generation or an ability that must be developed? Br. J. Edu. Technol. **44**(5), 163-E166 (2013)
37. Song, G., Ling, C.: Users' attitude and strategies in infor mation management with multiple computers. Int. J. Hum.-Comput. Interact. **27**(8), 762–792 (2011). https://doi.org/10.1080/10447318.2011.555307
38. Społeczeństwo Informacyjne w liczbach 2014, wyd. Ministerstwo Administracji i Cyfryzacji, Warszawa 2014, s. 17
39. Szajt, M., Smolag, K., Hajduová, Z.: The use of social media by Polish and Slovak students and their digital competences, Procedia Computer Science, International Conference on Knowledge-Based and Intelligent Information and Engineering Systems (KES 2023), Ateny, Greece (06 do 08 IX 2023 r.), vol. 225, pp. 1211–1220 (2023)
40. Tarkowski, A., et al.: Taksonomia funkcjonalnych kompetencji cyfrowych oraz metodologia pomiaru poziomu funkcjonalnych kompetencji Kompetencje cyfrowe w miejscu pracy... trendy w zarządzaniu 64 e-mentor nr 5 (97) cyfrowych osób z pokolenia 50+, (2015). 10.2139/ssrn.2708226

41. Van Deursen, A.J.A.M., van Dijk, J.A.G.M., Peters, O.: Rethinking Internet skills. The contribution of gender, age, education, internet experience, and hours online to medium- and content-related internet skills. Poetics **39**, 125–144 (2011)
42. Wang, Q.: Using online shared workspaces to support group collaborative learning. Comput. Educ. **55**(3), 1270–1276 (2010). https://doi.org/10.1016/j.compedu.2010.05.023
43. Yu, A.Y., Tian, S.W., Vogel, D., Kwok, R.C.W.: Can learning be virtually boosted? An investigation of online social networking impacts. Comput. Educ. **55**(4), 1494–1503 (2010). https://doi.org/10.1016/j.compedu.2010.06.015
44. Zhao, Y., Sánchez Gómez, M.C., Pinto Llorente, A.M., Zhao, L.: Digital competence in higher education: students' perception and personal factors. Sustainability **13**(21), 12184 (2021)
45. Zhong, Z.-J.: From access to usage: the divide of self-reported digital skills among adolescents. Comput. Educ. **56**(3), 736–746 (2011)

The Current Intensity of AI Use and the Perception of AI Applications in Strategic Decision-Making

Katarzyna Grzesik(✉), Renata Brajer-Marczak, Michał Nadolny, Katarzyna Liwarska-Fulczyk, Iwona Chomiak-Orsa, and Anna Marciszewska

Wroclaw University of Economics and Business, Komandorska 118/120, 53-345 Wroclaw, Poland

katarzyna.grzesik@ue.wroc.pl

Abstract. The aim of the article was to determine the impact that the frequency of AI use has on managers' perceptions of AI applications in strategic decision-making. A measurement construct with four latent variables (AI Literacy, Trust, Perceived Benefit and Perceived Risk) was used to assess the perception of AI applications in strategic decision-making. Empirical research was conducted on a purposefully selected group of 218 enterprises in Poland, the respondents were people making strategic decisions in these enterprises. Statistical analysis indicated significant relationships between the frequency of current AI use and AI Literacy and Trust. Respondents who often use AI in many tasks or almost always work using AI, assess their AI Literacy competences higher and have greater trust in AI. The obtained results indicate the need for further education in the use of AI and building trust in AI. This is particularly important in the group of people with low intensity of using AI solutions. At the same time, the research results confirmed the irreversibility of the process of adapting AI solutions in the approach to strategic decision-making. Thus, it enriches scholarly discourse and expands knowledge in this area.

Keywords: AI · strategic decision-making (SDM) · AI applications · AI literacy · AI trust

1 Introduction

Organizations are increasingly using artificial intelligence (AI) tools in decision-making processes to optimize resource allocation and improve operational efficiency (Yesufu and Alajlani, 2025). Over time, the use of AI in making strategic decisions is also increasing (Chaturvedi et al., 2025). The challenge facing organizations today appears to be moving AI solutions from the operational level, where the main goal is to perform similar activities better than competitors, to the strategic level, where the goal is to deliver a unique set of values to customers (Biloslavo et al., 2024).

Published by Springer Nature Switzerland AG 2026
E. Mercier-Laurent et al. (Eds.): AI4S 2025, IFIP AICT 780, pp. 218–230, 2026.
https://doi.org/10.1007/978-3-032-21184-2_17

Strategic decision-making (SDM) is by nature uncertain, ambiguous, risky and complex. Strategic decisions have far-reaching effects and are difficult to reverse in the short term. Integrating AI into SDM offers organizations a range of opportunities, including increased innovation and improved organizational agility and adaptability (Davenport et al., 2020; Stone et al., 2020). SDM has traditionally relied on human cognitive abilities, such as judgement, intuition, reflective thinking, visionary insights, as well as on possessed experience. In recent years, however, the integration of AI into strategic decisions has begun to transform the way these decisions are made, introducing new capabilities and changing the role of humans in the process (Johnson, 2022; Wu et al., 2023; Chaturvedi et al., 2025). Integrating AI into an organization's strategic decision-making process goes beyond simply improving decision-making tools. The use of AI results in a transformation of the way strategic decisions are conceptualized, made and executed (Mahmud et al., 2022; Wu et al., 2023). It is also important to note that the application of AI in SDM is not only about it being a tool to assist the decision-maker, but it also becomes an active participant in the decision-making process, able to offer informed recommendations (Trunk et al., 2020), which creates an area for collaboration between the decision-maker and AI. The decision-maker, in addition to acting as an overseer ensuring that AI systems operate within ethical and organizational boundaries, also becomes a partner for AI. This creates a growing need for human-AI collaboration, in which human expertise and AI computing power work together to make more effective strategic decisions (Haesevoets et al., 2021; Inkpen et al., 2023; Chaturvedi et al., 2025).

It can be noted that many challenges related to the use of AI in decision-making processes are oriented towards cooperation between humans and AI. This cooperation is conditioned by a number of factors (Cao et al., 2021), among which knowledge, skills, attitudes, personality traits, perception of available AI solutions, as well as the experience the decision-maker has in using AI solutions are important. Understanding the factors influencing the human perception of the use of AI in decision-making processes plays a key role in the evaluation and acceptance of these solutions (Lee, 2018).

The aim of the article was to determine the impact that the frequency of artificial intelligence (AI) use has on managers' perceptions of AI applications in strategic decision-making (SDM). Empirical research was carried out to achieve the article's aim. The study had a pilot character and was conducted on a intentionally selected group of 218 companies in Poland. In the study, data were collected using a measurement construct with four latent variables (AI literacy, trust, perceived benefit and perceived risk) to assess the perception of AI applications in strategic decisions.

2 Background

Research findings indicate that the use of AI solutions allows enterprises to increase productivity, innovation, and effectiveness (Sahoh and Choksuriwong, 2023; Kaggwa et al., 2024). This is due to the fact that AI can analyze vast amounts of data, spot patterns and trends, and provide actionable insights that facilitate decision-making (Duan et al., 2019). This capability is particularly crucial in a VUCA environment, where the speed and accuracy of decision-making can significantly impact the success of an organization

(Biloslavo et al., 2024). The rapid advancement of AI technologies poses a critical challenge for businesses looking to integrate these tools into their SDM. Despite AI's potential benefits in improving analytical capabilities and generating actionable insights from vast data sets, many organizations face challenges in successfully implementing AI-based solutions that align with their current business models (Marimira and Gumel, 2025).

The role of AI in decision-making can be summarized in three categories: decision automation, decision augmentation and decision support. Decision automation provides scalability, speed, and consistency through the use of prescriptive or predictive analytics. Decision augmentation uses predictive analytics to suggest several decision options. Decision support uses predictive and diagnostic tools to support human decision-making, combining human intelligence with data-driven insights (Langer and Landers, 2021). The latter plays a major role in the strategic decision-making process. Research shows (Ekellem, 2024) that managers' perception of AI plays a key role in SDM, emphasizing the need for transparency, explainability, and collaboration. As organizations increasingly adopt AI tools to enhance their strategic decision-making capabilities, it is essential to cultivate a culture of familiarity and trust in AI to enable decision-makers to properly exploit the opportunities and challenges of AI technologies (Klingbeil et al., 2024).

Perceptions of AI in SDM have evolved over the decades, reflecting technological advances and societal attitudes. Initially, AI systems were received with skepticism, but as AI technologies began to demonstrate tangible benefits, managerial attitudes and perceptions shifted towards recognizing their potential in improving decision-making processes. Users' perception of AI in decision-making is shaped by several key factors, among which AI literacy, trust and perceived benefits and risks of AI are important. AI literacy refers to the knowledge and skills necessary to understand, evaluate and effectively use AI systems and tools in a variety of contexts. It also includes the ability to critically evaluate their outputs, understand their mechanisms and consider their ethical and social implications (Ng et al., 2021; Pinski and Benlian 2024).

Trust in AI is a key factor influencing how individuals and organizations interact with and use AI systems. Trust in AI largely depends on the user's perception of the AI capabilities, which is determined by the quality of the input data, the representation of the mathematical problem, and the algorithms used to make decisions (Kessler et al., 2017). To establish trust in AI, people need to be confident that their desires, needs, and rights are met by AI. Understanding the factors that influence trust is important, but there is also a need to understand the relationships between these factors. In addition to well-known parameters such as performance, explainability, transparency, compliance with certain regulations and standards, and ethical considerations, challenges such as bias, discrimination and privacy must also be taken into account to further enhance trust and technology adoption. A basic prerequisite is to identify the trust determinants, their relationship and how they interact to make an AI system trustworthy (Afroogh et al., 2024).

The use of AI in business decision-making offers a number of benefits, but at the same time comes with the possibility of threats (risks) associated with its use. There are indications that AI has the potential to increase the speed, quality and scale of strategic analysis, as well as enabling new approaches such as virtual strategy simulations.

Traditionally, strategic decision-making has been a purely human endeavor, relying on the expertise, intuition and cognitive abilities of experienced strategists. However, rapid advances in AI may begin to challenge this paradigm (Csaszar et al., 2024). The benefits expected by decision-makers are also not insignificant. AI can streamline operations, reduce human error and facilitate data-driven decision-making, ultimately leading to more accurate predictions and improved organizational performance (Chaturvedi and Dasgupta, 2024). On the other hand, AI and algorithmic decision-making processes are increasingly being criticized for their "black-box nature." People are naturally likely to wonder why and how AI systems and algorithms make given decisions (Shin and Park, 2019). As the complexity of AI systems and algorithms increases, people increasingly view them as 'black-box' systems that make it difficult to understand the reasons behind decisions. More and more specialized knowledge and experience is required to apply AI, and this increasing complexity results in a lack of transparency that makes it difficult to understand and negatively impacts trust. The issues associated with AI decision-making as "black-box" are exacerbated for everyday users who do not have technical knowledge and still need to interact with AI systems (Shin, 2021). The effectiveness of AI systems is limited by the inability of AI to explain its decisions and operations to users, which translates into key issues such as fairness, accountability, transparency, and explainability (Shin et al., 2020).

The synergy between human judgement and AI has the potential to significantly improve decision-making in organizations, but its full use depends on the acceptance of AI by decision-makers (Edwards et al., 2000). Previous research has not sufficiently explored managers' attitudes and intentions related to the implementation of AI-based solutions. In particular, there is a lack of empirical analyses attempting to explain managers' behavioral intentions regarding the use of AI (Duan et al., 2019; Cao et al., 2021; Dwivedi et al., 2021). Given the increasing use of AI solutions in decision-making processes, it seems important to understand the relationship between the frequency of AI use and perceptions of its usefulness and intentions to use AI in strategic decision-making. This viewpoint contributed to the formulation of the following hypothesis:

H: The perception of AI applications in strategic decision-making depends on the current intensity of AI use at work.

3 Research Method

The aim of this pilot study was provide an initial assessment of the perceptions of the role of AI in strategic decision-making. It was conducted on a group of 218 companies in Poland in the period March-November 2024. The sampling was purposive, ensuring a sufficiently high proportion of medium and large enterprises (the criterion of number of employees) so that the results of the study could relate to decision-making processes in organizations with a developed management system. The main premise of the sample selection was to ensure the appropriate size of the layers of the analyzed population. This was to ensure a minimum sample size for the designed statistical procedures. The method of observing phenomena was based on classical test theory (Furr, 2021). Four latent variables were defined in the model - AI Literacy, Perceived Benefit, Perceived Risk, Trust (see Fig. 1), along with one observable variable – Current Intensity of AI use. The latent and observable variables were assessed with, in total, 11 items rated on a 5-point Likert scale:

- AL.1: I understand how AI works.
- AL.2: I know how to apply AI for strategic decision-making.
- AL.3: I understand the capabilities and limitations of AI for strategic decision-making.
- AL.4: In general, how would you rate your own knowledge and experience about AI?
- PB.1: I see many advantages in using AI in strategic decision-making.
- PB.2: AI use in strategic decision-making increases productivity of the decision-maker.
- PR.1: I see many risks in using AI in strategic decision-making.
- PR.2: Using AI for strategic decisions in-creases the risk of these decisions.
- Tr.1: I believe AI-based systems are trust-worthy.
- Tr.2: I trust in the ability of AI to generate valuable and reliable insights and recommendations.
- Tr.3: I trust/will trust AI-generated insights and recommendations.

The constructs and survey questions were inspired by the work of Gerlach and Lutz, 2021; Davis et al., 1989; Mariani and Nambisan, 2021 Chatterjee et al., 2021. The study also assessed the frequency of AI use in managerial decision-making.

Latent construct verification was based on Confirmatory Factor Analysis (CFA). Cronbach's alpha coefficient was calculated to establish the internal consistency of the multi-item latent variables. Factor loading values above 0.7 were considered statistically significant. Scores for the latent variables were derived by averaging the ratings of the items associated with each construct. In the verification of the research hypothesis, the W-Shapiro-Wilk (W-S-W) test of small samples was used to assess the normality of distributions, followed by the nonparametric Kruskal-Wallis ANOVA.

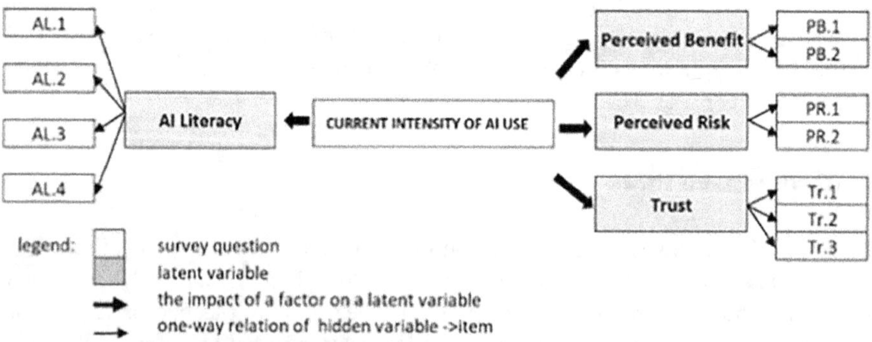

Fig. 1. Visualization of relationships between variables for hypothesis testing.

The survey was conducted using a mixed methods survey form. Approximately 19% of the respondents were representatives of medium-sized companies (10–49 employees). About 42% were micro or small companies (1–9 employees), 38.5% were companies with more than 50 employees. In the case of high-level managers (18.4%), a direct interview was conducted, after prior appointment. This applies to CEOs and presidents of companies, whose representation in the sample is high. Department directors, management board and managers represent approximately 24.7% of the sample. These respondents tended to prefer the CAWI survey. The remaining respondents, especially from companies with up to 10 employees, were surveyed using a random sample telephonic interview method. These respondents declared that they take an active part in the strategic decision-making process.

4 Results

The intensity of AI use in decision-making is described by a variable expressed on a four-point ordinal scale (see Table 1). Almost 47% of respondents did not use any AI tools in their company's decision-making at the time of the survey. Over 19% of respondents used AI in almost every aspect of the company's decision-making.

Table 1. Current intensity of AI use.

Are you currently using AI at work?	Number of observations	[%] of total
no	102	46.8
for some tasks	40	18.3
for multiple tasks	34	15.6
at almost every stage	42	19.3

Table 2 presents eleven measurable variables included in the measurement model (see Fig. 1). These questions define individual latent variables: AI Literacy describes the respondent's belief that they understand how AI works. This variable consists of 4 items of the survey form. The AI Literacy measurement does not determine the respondent's objective knowledge. It was designed to assess the respondent's self-awareness regarding how AI works. The remaining latent variables should be interpreted in the same way.

Examining the distributions of the survey items reveals that respondents generally held a favorable view of AI as a decision-making tool in the enterprise. The ratings are well above 3, which means that the respondents agree with the individual statements (see mean and median in Table 2). Participants also demonstrated awareness of the risks involved in using AI, showing some level of criticism toward its application. The individual items do not exhibit a normal distribution (see skewness and kurtosis in Table 2) and are also characterized by a low degree of left-sided skewness.

The positive verification of the measurement model was evidenced by high factor loadings of the latent constructs and, given the shortness of the measurement instrument, good values of the Cronbach's alpha coefficients. The distributions of latent variables cannot be considered normal (see W-S-W), therefore the verification of the research hypothesis was based on ANOVA Kruskal-Wallis, the calculations were performed in the Statistica 13 package.

Table 2. Basic statistics for Likert scale measurable variables (3 = neutral).

Construct name	Cronbach alpha	W-S-W* ($p < 0.05$)	CFA-ML** loadings	Model ID	Mean	Median	Skewness	Kurtosis
AI Literacy	0.87	0.96	0.86	AL.1	3.90	4.0	-0.76	0.31
			0.89	AL.2	3.72	4.0	-0.38	-0.26
			0.90	AL.3	3.72	4.0	-0.59	0.13
			0.72	AL.4	3.24	3.0	-0.09	-0.46
Perceived Benefit	0.70	0.94	0.88	PB.1	3.66	4.0	-0.38	-0.17
			0.88	PB.2	3.76	4.0	-0.48	0.08
Perceived Risk	0.66	0.92	0.86	PR.1	3.41	3.0	-0.01	-0.66
			0.86	PR.2	3.25	3.0	-0.14	-0.24
Trust	0.87	0.95	0.92	Tr.1	3.67	4.0	-0.04	-0.65
			0.90	Tr.2	3.74	4.0	-0.27	-0,72
			0.86	Tr.3	3.62	4.0	-0.15	-0.40

* W-Shapiro-Wilk (W-S-W) test; ** Confirmatory Factor Analysis (CFA) using the Maximum Likelihood (ML)

The intensity of the use of AI in decision-making processes was a subjective assessment by the respondent expressed on a four-point ordinal scale (see Fig. 2). The results of the analysis show that there is a significant relationship between the use of AI at work in decision-making and AI Literacy and Trust (see Table 3). This relationship does not apply to Perceived Benefits and Risks, both variables, although slightly differentiated in the sample, are the same for all individuals with respect to the study population, regardless of whether they use AI or ignore it altogether. The distinction for AI Literacy and Perceived Benefits relates to two groups of respondents: those who do not use AI at work at all (answer 'no') and those who declare that they use in many tasks or almost always. Those who do not use AI declare significantly lower AI Literacy and Trust scores. The most likely factor for this is a lack of experience in using the tool for business decision support. This is usually accompanied by low levels of knowledge and trust in unfamiliar procedures and methods.

The adaptation of AI for strategic decision-making purposes seems to be an inevitable and irreversible process. This is indicated by the results of this study (see Table 4). In the group of people who currently do not use AI at all, as many as 58% declare that they will probably use this technology in the future. If only the sample from this study was considered, the group of people who use AI in the decision-making processes of companies would increase to over 75%. It can be seen that the process of abandoning AI is unlikely. Only about 5% of respondents who currently use AI declare that they will probably not use this technology in the future. Under current technological landscape, this may mean that the process of adapting AI in strategic decisions in companies is practically irreversible.

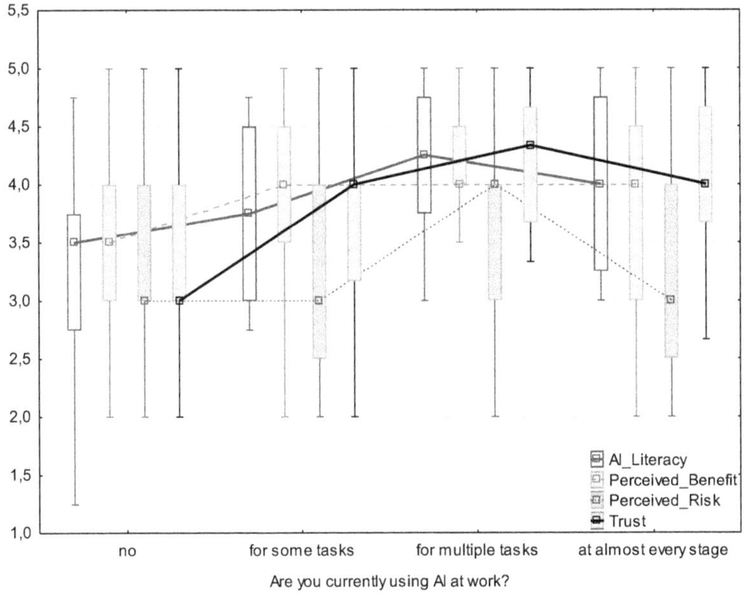

Fig. 2. Box plot of the distributions of latent variables by the intensity of AI use in decision-making processes. The continuous line corresponds to a statistically significant relationship between variables.

Table 3. The influence of the current intensity of AI use at work on the respondent's assessment of AI.

Current intensity of AI use at work	Kruskal-Wallis ANOVA p-value	Minimum p-value for multiple comparisons	Decision
AI Literacy	0.001	0.019	People who do not use AI at work give lower ratings than those who use AI in many or almost all tasks
Trust	0.004	0.001	
Perceived Benefit	0,069	0.117	No differences
Perceived Risk	0.261	0.511	

Table 4. Use of AI in strategic decisions and adaptation plans [% of total]

What are the chances that you will accept AI as an equal decision-making tool in the company?	Are you currently using AI at work?				Total in a row
	at almost every stage	for multiple tasks	for some tasks	no	
very un-likely	0.00	0.92	0.00	1.83	2.75
rather unlikely	2.75	0.00	1.83	3,67	8.26
I don't know	7.34	0.92	5.50	16.51	30.28
rather likely	5.50	12.84	9.17	22.02	49.54
very likely	3.67	0.92	1.83	2.75	9.17
Total in a column	19.27	15.60	18.35	46.79	

5 Discussion

The aim of the study was to determine the impact of AI usage frequency on the perception of AI applications in SDM by managers. Statistical analysis of the obtained data indicated significant relationships between the frequency of current AI use and AI Literacy and Trust in AI. This shows that respondents who often use AI in many tasks or almost always work with AI, rate their AI Literacy competences higher and have greater trust in AI.

These results correspond with the research results of other authors, which indicate that in the case of strategic decisions, factors such as perceived trustworthiness, algorithmic fairness, and explainability of AI systems have a strong impact on users' willingness to rely on these technologies when making strategic decisions (Akmaeva et al., 2020). Trust, transparency of obtained results, ethical considerations, and hybridization of decision-making processes have a key impact on the success of AI initiatives in strategic contexts and on users' perceptions (Delbufalo et al., 2022). Other studies show that people who felt that they had more control over their own online information (online autonomy) are more likely to consider AI as fair and useful (Kaggwa et al., 2024).

The study did not confirm a statistical relationship between the frequency of current AI use and Perceived Benefits and Risks. In relation to the studied population, both variables are the same for all people, regardless of whether they use AI or completely ignore this solution. This is surprising because research by various authors indicates significant benefits that organizations can receive by supporting the strategic decision-making process with AI solutions. It is claimed that, for example, large language models (LLM) match or even exceed human skills in tasks requiring reasoning skills similar to the competences of strategists (Bubeck et al., 2023; Katz et al., 2023). Among the benefits, the democratization of SDM and access to the so-called "McKinsey in a box" are also indicated. This signals the need to understand the integration of AI with strategic decision-making (Csaszar et al., 2024).

Researchers also strongly emphasize the risk of using AI in SDM. It concerns the danger of duplicating the same concepts, but also ethical issues. AI models are based on data from the past, so there is a risk of duplicating conventional and widely accepted strategies, potentially reducing the attribute of novelty, diversity, and uniqueness of strategic ideas (Felin and Holweg, 2024). It may also result in a failure to adequately reflect the unique needs and contexts of individual companies. Concerns are also expressed about the bias of AI algorithms and the possibility of AI making decisions that may be inconsistent with human values or ethical standards.

It should be emphasized that the relationship between AI and strategic decision-making is complex and evolving, with AI acting as both a catalyst and a potential change agent in the decision-making system (Chaturvedi et al., 2025). It can be seen as a game changer that affects the speed of strategic decision-making. A study conducted by Stipić (2021) indicates that a methodological approach to strategic planning that includes AI can significantly impact the profitability and success of a company. Integrating AI with strategic planning is crucial for companies to effectively navigate complex business environments and make high-quality, effective decisions that lead to successful business outcomes. The current socio-economic and political processes, accelerated by the pandemic, require a re-evaluation of corporate management strategies in many cases. AI can help in the transition from traditional hierarchical models to more agile approaches that quickly respond to the changing business landscape (Akmaeva et al., 2020).

6 Conclusion

The use of AI in SDM is a relatively unexplored topic. This article contributes to the development of research in this area by examining the role of AI user perception. The authors hope that the results of the study will provide a better understanding of the importance of the impact of AI use frequency on the perception of its use in SDM.

In terms of practical implications, the study provides managers with insight into the possibilities offered by the use of AI in the process of making strategic decisions. The obtained results indicate the need for further education in the use of AI and building trust in AI. This is especially important in the group of people with low intensity of using these solutions. Equipping these people with the appropriate knowledge and skills can contribute to increasing the use of AI in SDM.

Taking into account the high dynamics of AI development, the obtained research results confirmed the irreversibility of using AI in SDM. This requires a new perspective on strategic decision-making processes in enterprises, which corresponds to the results of other studies indicating that AI systems influence human actions by changing the role and scope of human participation in the strategic decision-making process and the way this process is implemented (Duan et al., 2019; Johnson, 2022; Mahmud et al., 2022; Wu et al., 2023; Kaggwa et al., 2024). This issue, however, requires a deeper exploration of the pragmatic applications of AI and the development of an effective framework for SDM (Chaturvedi et al., 2025).

It should also be noted that the presented research process had its limitations. One of them was the use of a broad concept, which is AI. The research focused on the use of

AI in decision-making without distinguishing between specific types of AI and without taking into account the role and scope of AI use in making strategic decisions. Future research could refer more closely to the types of AI, indicating, for example, the use of generative artificial intelligence, large language models (LLM), etc. Another limitation was the reliance on subjective declarations of respondents when assessing the variables studied. Taking into account the obtained results, it is advisable to further explore the undertaken topic by conducting broader quantitative research, taking into account the specificity of various industries, as well as qualitative research, in particular using the case study method or in-depth interviews with managers making strategic decisions.

Disclosure of Interests. The authors have no competing interests to declare that are relevant to the content of this article.

References

Afroogh, S., Akbari, A., Malone, E., Kargar, M., Alambeigi, H.: Trust in AI: progress, challenges, and future directions. Humanit. Soc. Sci. Commun. **11**(1), 1568 (2024)

Akmaeva, R., Arykbaev, R., Epifanova, N., Glinchevskiy, E.: Influence of digitalization upon formation of corporate strategy and new business models of modern organizations. In: International Scientific-Practical Conference "Transformation of Corporate Governance Models under the New Economic Reality", SHS Web of Conferences, vol. 89, p. 03003 (2020)

Biloslavo, R., Edgar, D., Aydin, E., Bulut, C.: Artificial intelligence (AI) and strategic planning process within VUCA environments: a research agenda and guidelines. Manag. Decis. (2024). https://doi.org/10.1108/md-10-2023-1944

Bubeck, S., et al.: Sparks of artificial general intelligence: early experiments with GPT-4. Preprint, submitted March 22 (2023). https://arxiv.org/abs/2303.12712

Cao, G., Duan, Y., Edwards, J.S., Dwivedi, Y.K.: Understanding managers' attitudes and behavioral intentions towards using artificial intelligence for organizational decision-making. Technovation **106**, 102312 (2021)

Chatterjee, S., Rana, N.P., Dwivedi, Y.K., Baabdullah, A.M.: Understanding AI adoption in manufacturing and production firms using an integrated TAM-TOE model. Technol. Forecast. Soc. Chang. **170**, 120880 (2021)

Chaturvedi, A., Yadav, N., Dasgupta, M.: Tech-driven transformation: unravelling the role of artificial intelligence in shaping strategic decision-making. Int. J. Hum.-Comput. Interact. 1–20 (2025)

Csaszar, F.A., Ketkar, H., Kim, H.: Artificial intelligence and strategic decision-making: evidence from entrepreneurs and investors. Strategy Sci. **9**(4), 322–345 (2024)

Davenport, T.H., Guha, A., Grew, D., Bressgott, T.: How artificial intelligence will change the future of marketing. J. Acad. Mark. Sci. **48**(1), 24–42 (2020)

Davis, F.D.: Perceived usefulness, perceived ease of use, and user acceptance of information technology. MIS Q. 319–340 (1989)

Delbufalo, E., Di Bernardo, M., Risso, M.: Human-machine interaction and AI for competitive business in the digital era. Symphonya **2**, 134–143 (2022)

Duan, Y., Edwards, J.S., Dwivedi, Y.K.: Artificial intelligence for decision making in the era of big data – evolution, challenges and research agenda. Int. J. Inf. Manage. **48**, 63–71 (2019)

Dwivedi, Y.K., et al.: Artificial Intelligence (AI): multidisciplinary perspectives on emerging challenges, opportunities, and agenda for research, practice and policy. Int. J. Inf. Manage. **57**, 101994 (2021)

Edwards, J.S., Duan, Y., Robins, P.C.: An analysis of expert systems for business decision making at different levels and in different roles. Eur. J. Inf. Syst. **9**(1), 36–46 (2000)
Ekellem, E.A.F.: Strategic Alchemy: The Role of AI in Transforming Business Decision-Making. Authorea Preprints (2024). https://doi.org/10.36227/techrxiv.24707151.v2
Felin, T., Holweg, M.: Theory is all you need: AI, human cognition, and causal reasoning. Preprint, submitted April 4 (2024). https://doi.org/10.2139/ssrn.4737265
Furr, R.M.: Psychometrics: an introduction. SAGE Publications (2021)
Gerlach, J.M., Lutz, J.K.: Digital financial advice solutions–evidence on factors affecting the future usage intention and the moderating effect of experience. J. Econ. Bus. **117**, 106009 (2021)
Haesevoets, T., De Cremer, D., Dierckx, K., Van Hiel, A.: Human-machine collaboration in managerial decision making. Comput. Hum. Behav. **119**(Feb), 106730 (2021)
Inkpen, K., et al.: Advancing human-AI complementarity: the impact of user expertise and algorithmic tuning on joint decision making. ACM Trans. Comput.-Hum. Interact. **30**(5), 1–29 (2023)
Johnson, J.: Delegating strategic decision-making to machines: Dr. Strangelove Redux? J. Strategic Stud. **45**(3), 439–477 (2022)
Kaggwa, S., Eleogu, T.F., Okonkwo, F., Farayola, O.A., Uwaoma, P.U., Akinoso, A.: AI in decision making: transforming business strategies. Int. J. Res. Sci. Innov. **10**(12), 423–444 (2024)
Katz, D.M., Bommarito, M.J., Gao, S., Arredondo, P.: GPT-4 passes the bar exam. Preprint, submitted March 15, (2023). https://doi.org/10.2139/ssrn.4389233
Kessler, T., Stowers, K., Brill, J.C., Hancock, P.A.: Comparisons of human-human trust with other forms of human-technology trust. Proc. Hum. Factors Ergon. Soc. **61**(1), 1303–1307 (2017)
Klingbeil, A., Grützner, C., Schreck, P.: Trust and reliance on AI—an experimental study on the extent and costs of overreliance on AI. Comput. Hum. Behav. **160**, 108352 (2024)
Langer, M., Landers, R.N.: The future of artificial intelligence at work: a review on effects of decision automation and augmentation on workers targeted by algorithms and third-party observers. Comput. Hum. Behav. **123**, 106878 (2021). https://doi.org/10.1016/j.chb.2021.106878
Lee, M.K.: Understanding perception of algorithmic decisions: fairness, trust, and emotion in response to algorithmic management. Big Data Soc. **5**(1), 2053951718756684 (2018)
Mahmud, H., Islam, A.K.M.N., Ishtiaque, S., Smolander, K.: What influences algorithmic decision-making? A systematic literature review on algorithm aversion. Technol. Forecast. Soc. Chang. **175**, 121390 (2022)
Mariani, M.M., Nambisan, S.: Innovation analytics and digital innovation experimentation: the rise of research-driven online review platforms. Technol. Forecast. Soc. Chang. **172**, 121009 (2021)
Marimira, N., Gumel, B.I.: The role of artificial intelligence in strategic decision-making. Asian J. Econ. Bus. Account. **25**(3), 316–327 (2025)
Ng, D.T.K., Leung, J.K.L., Chu, S.K.W., Qiao, M.S.: Conceptualizing AI literacy: an exploratory review. Comput. Educ. Artif. Intell. **2**, 100041 (2021)
Pinski, M., Benlian, A.: AI literacy for users–a comprehensive review and future research directions of learning methods, components, and effects. Comput. Hum. Behav. Artif. Hum. 100062 (2024)
Sahoh, B., Choksuriwong, A.: The role of explainable artificial intelligence in high-stakes decision-making systems: a systematic review. J. Ambient. Intell. Humaniz. Comput. **14**(6), 7827–7843 (2023)
Shin, D.: The effects of explainability and causability on perception, trust, and acceptance: implications for explainable AI. Int. J. Hum. Comput. Stud. **146**, 102551 (2021)
Shin, D., Park, Y.J.: Role of fairness, accountability, and transparency in algorithmic affordance. Comput. Hum. Behav. **98**, 277–284 (2019)
Shin, D., Zhong, B., Biocca, F.A.: Beyond user experience: what constitutes algorithmic experiences? Int. J. Inf. Manage. **52**, 102061 (2020)

Stipić, V.V.: Interaction of strategic management processes and achieved corporate profitability: Evidence from Croatia. BH Ekonomski Forum **13**, 133–149 (2021)

Stone, M., et al.: Artificial intelligence (AI) in strategic marketing decision-making: a research agenda. Bottom Line **33**(2), 183–200 (2020)

Trunk, A., Birkel, H., Hartmann, E.: On the current state of combining human and artificial intelligence for strategic organizational decision making. Bus. Res. **13**(3), 875–919 (2020)

Wu, C., Zhang, R., Kotagiri, R., Bouvry, P.: Strategic decisions: survey, taxonomy, and future directions from artificial intelligence perspective. ACM Comput. Surv. **55**(12), 1–30 (2023)

Yesufu, L., Alajlani, S.: Enhancing strategic decision-making with AI: unveiling the untapped potential. In: Economic and Political Consequences of AI: Managing Creative Destruction, IGI Global Scientific Publishing, pp. 415–436 (2025)

A Simulation-Based Approach to Processes in the Municipal Waste Management Sector

Aleksander Binsztok(✉), Beata Butryn, Krzysztof Hauke, and Maciej Pondel

Wroclaw University of Economics and Business, Komandorska 118/120, 53-345 Wrocław, Poland

{aleksander.binsztok,beata.butryn,krzysztof.hauke, maciej.pondel}@ue.wroc.pl

Abstract. In the era of rapid urbanization, a growing population, and increasingly stringent environmental regulations, municipal waste management is becoming one of the key components of sustainable urban development. Waste management systems must operate not only in an economically efficient manner but also in accordance with the principles of environmental protection and the circular economy. In this context, simulation tools are gaining increasing importance, enabling the analysis, planning, and optimization of processes related to the generation, collection, and transportation of municipal waste.

Waste management poses one of the major challenges faced by contemporary cities, requiring the optimization of collection, transport, and processing operations while taking into account environmental, economic, and social aspects.

The aim of this article is to present the potential of using computer simulation as a tool to support efficient municipal waste management at the level of a residential housing estate.

The paper presents a simulation model that reflects residents' behaviors, waste collection schedules, and the spatial distribution of waste containers. The model allows for the analysis of various scenarios, such as changes in collection frequency, reorganization of waste collection routes, or ecological education initiatives targeted at residents. The simulation experiments made it possible to assess the impact of selected actions on container fill levels.

The results indicate that computer simulation can serve as a valuable decision-making tool for local governments and estate managers, supporting the planning of sustainable municipal waste management strategies.

Keywords: Simulation · Municipal waste · Municipal waste management

1 Introduction

Efficient waste management constitutes one of the key challenges at both global and local levels. According to the World Bank report *What a Waste 2.0* (2018), global municipal solid waste (MSW) generation currently exceeds 2 billion tons annually, and it is projected to increase by approximately 70% by 2050, reaching 3.4 billion tons per year

Published by Springer Nature Switzerland AG 2026
E. Mercier-Laurent et al. (Eds.): AI4S 2025, IFIP AICT 780, pp. 231–244, 2026.
https://doi.org/10.1007/978-3-032-21184-2_18

compared to the 2016 baseline [1]. In the context of accelerating urbanization, population growth, and evolving consumption patterns, the effective planning and optimization of waste collection and treatment processes are becoming essential for maintaining the quality of life in urban environments and protecting the natural ecosystem.

In practice, waste collection enterprises face a range of significant operational challenges. These include:

- overfilled waste containers, which lead to illegal dumping,
- misalignment of collection schedules with the variable needs of residents,
- non-optimized collection routes, resulting in increased fuel consumption and greenhouse gas emissions [2].

As research in this field has demonstrated, inefficient route planning can raise operational costs by up to 20% [3] and may also contribute to greater vehicle wear and higher emissions levels.

However, the aforementioned challenges do not exhaust the list of operational issues faced by waste collection enterprises. From a logistics and management perspective—areas of primary interest in this study—several additional difficulties can be identified:

- **temporal variability in waste volumes** (waste generation is highly irregular; it tends to increase significantly during holidays, mass events, or tourist seasons. Standard collection schedules often fail to accommodate such fluctuations, as noted by Wilson, Velis, and Cheeseman [4]),
- **lack of real-time data on container fill levels** (many waste management systems still rely on fixed-route collection models, which do not incorporate dynamic monitoring of container capacity [5),
- **labor shortages and lack of skilled drivers** (the waste management sector faces increasing difficulty in recruiting qualified personnel, which impacts the execution of transportation plans [6]),
- **aging and failure-prone equipment** (specialized waste collection vehicles are expensive and challenging to maintain; mechanical breakdowns lead to service delays and operational downtime [7]),
- **ineffective waste sorting at source** (mixed waste instead of properly separated streams hinders both the collection process and subsequent recycling or disposal efforts [8]),
- **variable road and weather conditions** (rain, snow, or urban traffic congestion can drastically affect collection times, complicating precise route planning [9]),
- **high fuel costs and energy price volatility** (rising fuel prices have a direct impact on the profitability of waste collection operations),
- **increasing societal expectations and regulatory demands** (both residents and municipal authorities demand higher service quality—e.g., more frequent selective waste collection—while simultaneously expecting reduced environmental impacts [10]).

In light of the arguments presented above, it becomes evident that traditional methods of waste logistics management—relying on fixed schedules and manual route planning—are frequently inadequate in the dynamic context of urban environments. These approaches often lack the flexibility required to respond to real-time changes, resulting in reduced operational efficiency and increased costs [11].

In response to these challenges, modern decision-support tools, such as computer simulations, are receiving growing attention. The simulation of logistical processes—taking into account variables such as fluctuating container fill levels, differences in waste types, and traffic variability—allows for the modeling of alternative operational scenarios, testing of proposed solutions, and optimization of resource allocation without the need for real-world deployment [12].

The aim of this study is to assess the usefulness of employing computer simulation for planning and optimizing the operations of a company engaged in waste and refuse collection. By comparing the outcomes of a traditional operational scenario with those of an optimized one, the authors seek to demonstrate the potential benefits of implementing such technology in the context of municipal waste management practice.

2 Literature Review

Environmental management has leveraged various simulation and decision support tools to optimize waste management practices. These tools facilitate effective decision-making by providing insights through modeling, assessment, and analysis of waste management systems and processes.

A prominent approach in waste management systems is the use of Life Cycle Assessment (LCA) tools, which scrutinize the environmental impacts associated with all stages of a product's life from raw material extraction to disposal. The Easewaste model is noteworthy in this realm as it assesses waste flow, resource consumption, and environmental emissions [13].

Moreover, Izdebski et al. underscore the utility of LCA in managing municipal waste collection logistics, noting that integrating logistics into waste management enhances its effectiveness throughout the entire product life cycle [14].

Decision support tools tailored for specific waste streams, such as for example construction and demolition (C&D) waste, have been extensively researched. Oyenuga and Bhamidiarri present various modelling tools like Cost-Benefit Analysis (CBA), Multi-Criteria Decision Analysis (MCDA), and Material Flow Analysis (MFA) that can effectively support C&D waste management decisions [15]. These models allow stakeholders to evaluate management strategies based on technical feasibility, economic viability, and environmental repercussions, rendering comprehensive decision-making frameworks applicable to demolition waste scenarios [16].

A noteworthy application of simulation tools in waste management is in modelling organic waste management systems. For instance, the study by Setyaningrum and Sari explores the use of the Arena simulation software for analysing the flow of organic waste from collection through disposal. The capacity of simulation to predict outcomes based on various scenarios enables managers to identify improvements and enhance the efficiency of waste supply chains, particularly crucial for organic waste management [17]. Additionally, efforts such as those presented by Buffière et al. highlight the utilization of simulation models to effectively assess waste management strategies, specifically in the domain of anaerobic digestion [18].

The intersection of technology and waste management is increasingly evident through the rise of integrated systems that employ decision support tools. For example, systems using Building Information Modelling (BIM) enhance decision-making by facilitating the efficient management of waste generated during construction activities [19]. Moreover, studies indicate that Artificial Intelligence (AI) technologies are being successfully employed to optimize waste sorting and recycling processes, enhancing overall waste management efficiency [20].

Research by Vea et al. underscores the necessity of various decision support tools for assessing circular biowaste management systems. These tools highlight the potential of innovating waste management practices that are aligned with circular economy principles, illustrating a vital shift in how waste systems are perceived and managed [21].

Simulations and decision support tools are not only valuable for technical decision-making but also for fostering community engagement in waste management practices. Tools like the Zero Waste Index (ZWI) provide metrics for assessing waste management systems' performance regarding virgin material substitution, which is significant for achieving community participation and awareness in sustainable practices [22]. This interconnectedness aligns with findings from Fiksel and Bakshi, which illustrate how strategic planning facilitated by simulation can integrate stakeholder interests in sustainable development [23].

As electric vehicle adoption surges globally, the associated increase in lithium-ion batteries a significant component of these vehicles poses distinct challenges due to the impending waste they generate at the end of their life cycle. Research in this area has evolved to address both the environmental and economic dimensions of battery waste management, underlining the importance of efficient recycling systems and the exploration of secondary uses for depleted batteries.

The rapid proliferation of electric vehicles has led researchers to analyze various waste management systems. The bibliometric study by Nurdini et al. provides insights into the growth of literature concerning waste management within the realm of electric vehicle batteries spanning from 1995 to 2023. This period highlights the increasing recognition of the environmental challenges posed by battery waste, as governmental policies and consumer interests shift toward more sustainable solutions [24].

The game-theoretic approach presented by Yao et al. advocates for the strategic classification of used batteries, facilitating their optimization for secondary uses based on quality and expected economic returns [25]. This cross section of research emphasizes that the effective management of battery waste not only mitigates environmental damage but can also yield significant economic benefits through the reuse of valuable materials.

The establishment of closed-loop supply chains for battery management, as discussed by Guo et al., is essential to coordinate the responsibilities of manufacturers and consumers in promoting battery recycling and reducing waste generation [26]. These systemic approaches are critical for developing a comprehensive understanding of waste management that operates on multiple levels—from consumer behaviour to overarching legislative frameworks.

The application of simulation and decision support tools in environmental management illustrates significant advancements in optimizing waste management strategies.

By offering integrated assessments, continually evolving technologies, and fostering community involvement, these tools enhance the efficacy, sustainability, and resilience of waste management systems. The ongoing development and application of these models across various sectors highlight the importance of data-driven insights in shaping future environmental management policies.

In the context of waste management systems, the existing literature suggests significant advancements and varying methodologies. Nonetheless, there are substantial gaps that need to be addressed in order to enhance sustainability and effectiveness within this field. The following synthesis of potential references outlines these gaps, identifying areas where further research and exploration would be warranted.

Warlina and Listyarini utilized a dynamic systems approach to estimate landfill capacity [27]. While such methodologies promise precision, there remains an identified gap in exploring the long-term implications of dynamic modeling on landfill management, particularly in shifting socio-political landscapes. Further studies could objectively assess how changes in waste generation patterns affect landfill management strategies over time, enhancing risk assessment and decision-making in waste disposal.

Hao et al. introduced a simulation model utilizing system dynamics for waste management [28]. However, many studies on waste focus primarily on the quantitative aspects of waste generation rather than integrating qualitative factors, such as stakeholder perceptions and behaviors influencing waste generation and management. This gap indicates a need for integrated approaches that combine quantitative waste measurements with qualitative stakeholder input to foster comprehensive waste management solutions.

Areas such as material flow analysis [29], community engagement [30], and the integration of emerging technologies offer substantial opportunities for advancing waste management practices in increasingly complex and diverse environments.

3 The Importance of Computer Simulation in Managing Municipal Waste Generated by Residential Communities

Contemporary municipal waste management systems face growing challenges associated with the intensifying process of urbanization, demographic diversity, and increasingly strict environmental and legislative requirements. The rising number of residents in settlement units, along with changes in consumption patterns, contributes to a growing volume and variety of waste generated at the household level. In response to these challenges, the application of analytical tools that enable forecasting and optimization of waste management system performance has become essential. One of the most effective approaches is the use of computer simulation, which allows for the representation of complex processes occurring in real systems and for examining the impact of individual variables on system efficiency.

Description of the Simulation Model Structure

The proposed simulation model aims to reflect the functioning of a waste management system within a residential housing estate, taking into account numerous spatial-demographic, operational, and technical determinants. The most significant input factors include:

1. Housing Estate Structure - The model assumes a precise representation of the urban structure of the housing estate, defined by:
   - the number of residential buildings, divided into single-family homes and multi-family apartment buildings (Table 1).

Table 1. Area parameters

Name of the estate/street	Number of houses	Number of garbage shelters
Housing estate/street 1	10	3
Housing estate/street 2	5	2
Housing estate/street 3	7	2
Housing estate/street 4	8	3
Housing estate/street 5	10	5
Housing estate/street 6	5	5
Housing estate/street 7	7	4
Housing estate/street 8	10	2
Housing estate/street 9	10	5
Housing estate/street 10	3	2

This structure determines the spatial distribution of waste collection points and provides the basis for aggregating waste production data.

2. Number of Residents per Household - For each type of building, the model assumes an average number of people per household. This parameter allows for the dynamic calculation of the quantity of waste generated over time and supports realistic scaling of waste generation phenomena within the estate (Table 2).

Table 2. Resident parameters

Number of residents in a single property/house	4 – 40 (random)

3. Daily Waste Generation per Capita - Waste production is modeled according to established quantitative indicators, expressed in kilograms per person per day. The following waste fractions are included:
   - biodegradable waste (e.g., organic residues, kitchen waste),
   - metals and plastics (e.g., packaging, cans, PET bottles),
   - paper and cardboard (e.g., newspapers, boxes),
   - glass (e.g., bottles, jars),
   - mixed municipal waste (residual fraction not suitable for segregation) (Table 3).

Table 3. Daily generated waste divided into categories [kg/person]

Category	from	to
Bio	0,30	1,00
Metals and plastics	0,10	0,50
Paper and cardboard	0,10	0,30
Glass	0,20	0,60
Other (mixed)	0,50	1,10

The model allows for the individual adjustment of generation rates for each fraction depending on the season, household type, or local policy frameworks.

4. Waste Container Capacity - Each waste fraction is associated with a designated type of container with a specific capacity (in liters or kilograms). This parameter is used to calculate the time required to fill a container and to predict overflow events. It enables assessment of whether the container placement and number are appropriate to meet actual residential needs (Table 4).

Table 4. Container capacity divided into categories [kg]

Category	Load capacity
Bio	100
Metals and plastics	200
Paper and cardboard	100
Glass	500
Other (mixed)	250

5. Waste Collection Schedule - The model incorporates the waste collection schedule, specifying the number of days per week allocated for collecting each waste fraction. Simulation makes it possible to examine the consequences of changing collection frequencies and their impact on container overflows and logistical efficiency (Table 5).
6. Container Fill-Level Indicators - Each container has an individual fill-level indicator that reflects, in real time, the degree of its utilization. This feature enables early identification of problematic areas and the implementation of response mechanisms, such as emergency collections or container relocation (Table 6).
7. Waste Collection Vehicle Capacity - The model integrates technical parameters of waste collection vehicles, including:

- maximum load capacity (in kilograms and/or cubic meters),
- number of waste fractions that can be transported simultaneously,
- average service time per collection point (Table 7).

Table 5. Waste collection days divided into categories

Category	Mon	Tue	Wed	Thu	Fri	Sat	Sun
Bio	Y	Y	Y	Y	Y		
Metals and plastics		Y	Y		Y		
Paper and cardboard	Y			T			
Glass			Y				
Other (mixed)	Y	Y		Y	Y		

Table 6. Container filling factors generating a garbage truck call

Category	Fill indicator
Bio	0,50
Metals and plastics	0,80
Paper and cardboard	0,95
Glass	0,80
Other (mixed)	0,75

Table 7. Garbage truck capacity divided into categories [kg]

Category	Capacity
Bio	1000
Metals and plastics	3000
Paper and cardboard	1000
Glass	3000
Other (mixed)	3000

This allows for analysis of route efficiency, operating costs, and potential delays due to exceeding the permissible load capacity.

Applications and Analytical Potential of the Model

The developed simulation model is applicable in both academic research and in engineering and urban planning practice. It enables:

- scenario analysis of the impact of demographic and spatial changes on the functioning of the waste collection system,
- testing of alternative collection schedules in terms of cost, emissions, and operational efficiency,
- support in planning investments in waste infrastructure,
- prediction of critical moments related to container overflows,

- assessment of the effectiveness of waste segregation policies.

The model can be integrated with Geographic Information Systems (GIS) and real telemetry data from container fill-level sensors, further enhancing its predictive and operational value.

4 Research Results in the Field of Computer Simulation for Municipal Waste Management for Specific Parameters

Analysis of Municipal Waste Management Efficiency Based on Container Fill Levels (January–April 2025)

Based on the presented data, a number of conclusions can be drawn regarding the efficiency of municipal waste management in the analyzed housing estate, with particular attention given to the fill levels of containers across various waste fractions during the period from January to April 2025 (Table 8).

Table 8. Percentage filling of municipal waste containers in 2025 from January to April

Average Completion	Category					
Year / Month	Bio	Glass	Metals and plastics	Other (mixed)	Paper and cardboard	Average
2025	**109,05%**	**43,14%**	**47,49%**	**60,68%**	**58,95%**	**63,86%**
I - January	115,24%	45,40%	45,59%	58,70%	57,07%	64,40%
II - February	82,27%	36,50%	46,02%	53,45%	56,54%	54,95%
III - March	119,14%	42,70%	49,44%	68,56%	58,95%	67,76%
IV - April	120,40%	48,83%	49,19%	61,81%	64,27%	68,90%
Average	**109,05%**	**43,14%**	**47,49%**	**60,68%**	**58,95%**	**63,86%**

1. Overfilled Biodegradable Waste Containers – A Warning Sign

The most pronounced phenomenon observed in the dataset is the systematic overfilling of biodegradable waste containers. The average fill level exceeds 100% in every month except February (with an overall average of 109.05%, and as high as 120.40% in April). These findings indicate a significant underestimation of both container capacity and the frequency of waste collection for this fraction. Excess bio-waste may lead to serious sanitary and epidemiological consequences, especially during spring and summer months when decomposition processes accelerate.

Recommendation: A revision of the biodegradable waste collection schedule and/or an increase in the number or capacity of containers for this fraction is necessary.

2. Underutilization of Glass Containers

Glass waste containers demonstrate the lowest fill levels in the entire dataset, averaging 43.14%, with a minimum of 36.50% in February. This suggests that the current infrastructure (in terms of container capacity and/or collection frequency) is oversized relative to the actual generation rate of glass waste. It may also reflect insufficient public engagement in glass waste segregation.

Recommendation: Consider reducing the number of glass containers or extending the interval between collections to achieve operational savings. Simultaneously, assess whether public education on proper glass segregation requires reinforcement.

3. Moderate but Stable Load for Metals and Plastics Fraction
Containers designated for metals and plastics show a relatively stable fill level, averaging 47.49%, with minor monthly variability. Although overflows are not observed, it is important to note that this fraction is highly variable in terms of volume and may be subject to seasonal fluctuations (e.g., increased packaging waste during holiday periods).

Recommendation: Maintaining the current collection system appears justified, provided seasonal increases continue to be monitored.

4. Mixed Waste and Paper Fractions – Approaching Critical Thresholds
Mixed waste and paper/cardboard fractions reach average fill levels of 60.68% and 58.95%, respectively. While these are not yet critical values, the March increase in mixed waste to 68.56% and the April rise in paper waste to 64.27% may signal an upward trend, possibly linked to heightened resident activity in spring (e.g., cleaning, shopping, renovations).

Recommendation: Ongoing monitoring of these fractions is recommended, taking into account seasonality and potential adjustments to the collection schedule during peak usage months.

5. Average System-Wide Efficiency
The average fill level across all waste fractions during the analyzed period was 63.86%. From a logistical perspective, this reflects relatively efficient system operation (i.e., minimal unnecessary empty trips). However, the data simultaneously highlights a marked imbalance among different waste categories.

Final Conclusions
Both simulated and empirical data on container fill levels serve as a valuable tool to inform managerial decision-making. The findings indicate the need for:

- Expanded access to biodegradable waste containers,
- Infrastructure optimization for the glass fraction,
- Continued monitoring and operational flexibility for the remaining categories.

Data-driven waste management enables dynamic adaptation of the system to residents' actual needs while minimizing operational costs and environmental risks.

Quantitative Analysis of Municipal Waste Mass Generated by Residents (January–April 2025): Strategic Implications for Local Waste Management
Based on the presented data concerning the mass (in kilograms) of municipal waste generated by residents of the housing estate from January to April 2025, a multifaceted quantitative analysis can be performed. This enables the identification of dominant waste

streams, assessment of seasonality, and formulation of strategic recommendations for waste management at the local level (Table 9).

Table 9. Municipal waste in kilograms in 2025 from January to April

Total - Mass of generated waste [kg.] Year/Month	Category Bio	Glass	Metals and plastics	Other (mixed)	Paper and cardboard	Total
2025	229413,82	140784,57	105280,32	283028,53	70609,38	829116,62
I - January	64706,44	39776,25	29624,32	78534,76	19726,61	232368,38
II - February	43925,55	27237,27	20226,42	54499,75	13656,1	159545,09
III - March	67194,61	40838,03	30628,72	82699,96	20650,24	242011,56
IV - April	53587,22	32933,02	24800,86	67294,06	16576,43	195191,59
Total	229413,82	140784,57	105280,32	283028,53	70609,38	829116,62

1. Dominance of Mixed Waste – A Challenge for Selective Collection Systems
During the analyzed period, the largest waste mass was recorded in the mixed waste category (Other/mixed), totaling 283,028.5 kg, which accounts for over 34% of the total mass of all fractions. This significant share may be indicative of:

- Insufficient waste segregation at the source,
- Inadequate infrastructure for selective collection,
- Low levels of environmental awareness among residents.

Conclusion: There is a need to intensify educational efforts and potentially review the spatial distribution of selective collection points to reduce the share of residual waste in the overall stream.

2. High Mass of Biodegradable Waste – Considerable Potential for Composting
Biodegradable waste (Bio) ranks second in terms of mass, amounting to 229,413.82 kg (approx. 28% of the total). Particularly high volumes in March and April may reflect seasonal increases in gardening activity, green area maintenance, and higher consumption of fresh food.

Conclusion: This fraction holds substantial potential for local processing, e.g., through neighborhood composting facilities or expansion of the bio-waste collection system. Monitoring container overflows, especially during the spring, is essential.

3. Glass Waste – Relatively High Mass with Low Growth Dynamics
Glass constitutes the third-largest fraction by weight (140,784.57 kg). Monthly volumes are relatively stable, ranging between approximately 32,000 and 40,000 kg. The combination of high total mass and low fill-level indicators (as shown in earlier analyses) suggests large container volumes and low collection density.

Conclusion: Optimizing the frequency of glass collection is feasible without risking overflows, leading to potential reductions in logistical costs.

4. Metals and Plastics – A Variable but Significant Fraction

A total of 105,280.32 kg of metal and plastic waste was generated, representing around 13% of the total mass. This fraction showed a steady increase from February (20,226 kg) to March (30,628 kg), likely due to seasonal purchasing trends and increased consumption of packaged products.

Conclusion: Ongoing monitoring of this fraction is warranted, as it represents a key material stream in the circular economy. When properly collected and processed, it can yield significant environmental and economic benefits.

5. Paper and Cardboard – A Consistently Growing Stream

Paper and cardboard accounted for 70,609.38 kg (just over 8.5% of the total). While not dominant, this stream has shown a month-to-month increase—from 13,656 kg in February to 16,576 kg in April—potentially driven by rising online purchases and increased use of office and school materials.

Conclusion: Consider expanding the capacity of paper containers in the spring months and promoting best practices for the separate disposal of cardboard and paper waste among residents.

6. Total Waste Mass – Nearly 830 Tons in Four Months

The total mass of waste generated during the analysis period reached 829,116.62 kg, averaging over 207 tons per month. This substantial volume provides a foundation for:

- Calibrating waste collection and transport systems,
- Planning vehicle and container usage,
- Defining alert and intervention thresholds.

General Summary

The quantitative data highlights several key challenges and opportunities:

Challenges:

- Excessive volumes of mixed and biodegradable fractions – necessitating improved segregation and collection mechanisms.
- Balancing efficiency and cost for lower-intensity fractions (glass, paper).

Opportunities:

- Development of local bio-waste processing infrastructure.
- Optimization of logistics and enhanced resident education as tools for long-term system management.

5 Conclusion

The conducted research clearly demonstrates the high utility of computer simulation as a decision-support tool in managing municipal waste within residential housing estate environments. The implemented model enabled a detailed representation of real-world interdependencies between demographic structure, the temporal dynamics of household waste generation, and the operational parameters of collection and transport systems. By incorporating variables such as the number of residents, waste fractions, container and vehicle capacities, as well as collection schedules, the simulation facilitated the identification of critical areas—most notably, container overflows in the biodegradable waste category and underutilization in the case of glass.

The use of a simulation environment made it possible to test various organizational scenarios, including adjustments to collection frequency, redistribution of containers, and alternative waste segregation strategies. The analyses revealed that computer-based modeling not only supports data-driven decision-making but also allows for the prediction of the consequences of proposed changes prior to their real-world implementation. This approach significantly enhances the operational efficiency of the waste management system by reducing logistical costs and environmental risks while simultaneously improving the quality of services delivered to residents.

In the context of dynamic urban development and increasing environmental requirements, computer simulation emerges as a critical component of modern, adaptive waste management systems. The results obtained confirm the rationale for implementing modeling tools not only as instruments for strategic planning but also for real-time monitoring and ongoing optimization of the system at the local scale.

References

1. Kaza, S., Yao, L., Bhada-Tata, P., Van Woerden, F.: What a Waste 2.0: A Global Snapshot of Solid Waste Management to 2050. Urban Development. World Bank Group (2018)
2. Evangelista, P., Santoro, L., Thomas, A.: Environmental sustainability in logistics and supply chains. Springer (2018)
3. Nuortio, T., Kytöjoki, J., Niska, H., Bräysy, O.: Improved route planning and scheduling of waste collection and transport. Expert Syst. Appl. **30**(2), 223–232 (2006)
4. Wilson, D.C., Velis, C., Cheeseman, C.: Role of informal sector recycling in waste management in developing countries. Habitat Int. **30**(4), 797–808 (2006)
5. Faccio, M., Persona, A., Zanin, G.: Waste collection multi objective model with real time traceability data. Waste Manag. **31**(12), 2391–2405 (2011)
6. Waste360 Staff: Driver Shortages Are Hitting Waste Management Sector Hard, Waste360. https://www.waste360.com/collections/driver-shortages-are-hitting-waste-management-sector-hard. Accessed 27 Apr 2025
7. Chung, S.S., Lo, C.W.H.: Waste management practices in Hong Kong: reduction and recycling of municipal solid waste. Waste Manag. **24**(6), 563–574 (2004)
8. European Environment Agency: Municipal waste management across European countries. EEA Report No 2/2020. https://www.eea.europa.eu/publications/municipal-waste-management-across-european-countries/copy_of_municipal-waste-management-across-european-countries. Accessed 03 May 2025
9. Baptista, S., Oliveira, M., Zúquete, E.: Impact of weather conditions on urban waste collection operations. Environ. Sci. Pollut. Res. **27**, 12439–12452 (2020)
10. United Nations: Transforming our world: the 2030 Agenda for Sustainable Development. https://sdgs.un.org/2030agenda. Accessed 02 May 2025
11. Gallo, M., Guerriero, F., Guizzi, G.: Waste collection modeling for real-world applications: a review and new perspectives. Transp. Res. Part E Logist. Transp. Rev. **111**, 1–20 (2018)

12. Guerrero, L.A., Maas, G., Hogland, W.: Solid waste management challenges for cities in developing countries. Waste Manag. **33**(1), 220–232 (2013)
13. Christensen, T.H., et al.: Experience with the use of LCA-modelling (EASEWASTE) in waste management. Waste Manag. Res. **25**(3), 257–262 (2007)
14. Izdebski, W., Serafin, S., Tarasevych, V.: The application of the life cycle assessment method in managing the logistics of municipal waste collection. Evergreen **10**(03), 1242–1252 (2023)
15. Oyenuga, A., Rao Bhamidiarri, R.: Considering appropriate decision support models for construction and demolition waste management optimization: possibilities and limitation. SSRG Int. J. Econ. Manag. Stud. **2**(10), 1–5 (2015)
16. Berg, M., Hulsbeek, L., Voordijk, H.: Decision-support for selecting demolition waste management strategies. Build. Cities **4**(1) (2023)
17. Setyaningrum, R., Nopita, S.: Organic waste management based on supply chain management using arena simulation and macro ergonomics approach. OPSI **16**(2), 217–224 (2023)
18. Buffière, P., Loisel, D., Bernet, N., Delgenes, J.P.: Towards new indicators for the prediction of solid waste anaerobic digestion properties. Water Sci. Technol. **53**(8), 233–241 (2006)
19. Liphadzi, N.M., Musonda, I., Onosen, A.: The use of building information modelling tools for effective waste management: a systematic review. In: JOP Conference Series: Earth and Environmental Science, vol. 1101, Sustainable Lifecycle (2022)
20. Maniatis, P.: Artificial intelligence in waste sorting: advancing recycling processes in Greece through AI-driven solutions. Int. J. Clin. Case Rep. Rev. **23**(1), 1–10 (2025)
21. Vea, E.B., Sanchez, V.M., Thomsen, M.: A review of waste management decision support tools and their ability to assess circular biowaste management systems. Sustainability **10**(10), 3720 (2018)
22. Sujai, S., Juwana, I.: Waste management planning toward zero waste in Hotel XYZ Bandung with circular economy principles (case study: room service facility's solid waste). IOP Conf. Ser. Earth Environ. Sci. **940**(1), 012052 (2021)
23. Fiksel, J., Bakshi, B.: Industrial ecology network optimization with life cycle metrics. In: Proceedings of the 2010 IEEE International Symposium on Sustainable Systems and Technology (2010). https://ar.thai-books.sk/book/64774930/b688f4/ieee-2010-ieee-international-symposium-on-sustainable-systems-and-technology-issst-arlington-v.html. https://doi.org/10.1109/issst.2010.5507698. Accessed 03 May 2025
24. Nurdini, A., Nurcahyo, R., Prabuwono, A.S.: Waste from electric vehicle: a bibliometric analysis from 1995 to 2023. World Electric Veh. J. (WEVJ) **14**(11), 300 (2023)
25. Yao, C., Ma, X., Hu, T., Gu, X.: A game theory approach in quality classification and secondary use of electric vehicle used batteries. In: Journal of Physics: Conference Series, Volume 2224, 2021 2nd International Symposium on Automation, Information and Computing. https://iopscience.iop.org/article/https://doi.org/10.1088/1742-6596/2224/1/012138. Accessed 25 Oct 2023
26. Guo, R., He, Y., Tian, X., Li, Y.: New energy vehicle battery recycling strategy considering carbon emotion from a closed-loop supply chain perspective. Sci. Rep. **14**, Article no 688 (2024)
27. Warlina, R., Listyarini, S.: The study of estimation of landfill capacity through dynamic system approach. Scientifica **2022**(2), 1–8 (2022)
28. Hao, J., Hills, M.J., Huang, T.: A simulation model using system dynamic method for construction and demolition waste management in Hong Kong. Constr. Innov. **7**(1), 7–21 (2007)
29. Bureecam, C., Chaisomphob, T., Sungsomboon, P.Y.: Material flows analysis of plastic in Tthailand. Therm. Sci. **22**(6), 2379–2388 (2018)
30. Orfanou, A., Vakalis, S., Karelis, C.H., Alexopoulos, C.H., Charalabidis, Y., Lekkas, D.F.: Assessing the correlation between citizen science and carbon footprint: introduction of COMPAIR's CO2 calculator. Energies **16**(17), 6228 (2023)

AI and Sustainability – Canada Focus

Margaret Havey(✉)

International Federation for Information Processing (IFIP), 2 Guntramsdorfer Str. 103/14 2340, Mödling, Austria
margaret.havey2950@gmail.com

Abstract. The accelerating deployment of artificial intelligence (AI) is unfolding in parallel with the world's triple planetary crisis of climate change, biodiversity loss, and pollution. This paper examines how AI intersects with sustainable development, with a specific focus on Canada's national context, governance model, and emerging AI ecosystem. It traces the evolution of global sustainability efforts from the UN's early environmental initiatives to the adoption of the 2030 Agenda and its 17 Sustainable Development Goals (SDGs), and outlines how Canada has translated these commitments into a whole-of-government approach through its Federal Sustainable Development Strategy (FSDS). Against this backdrop, the paper analyzes how Canada is leveraging AI to accelerate clean energy, critical minerals development, industrial decarbonization, and ecosystem protection, while simultaneously navigating the trade-offs that accompany sustainable development—particularly those related to mining, AI data center infrastructure, water consumption, and energy demand. The societal implications of AI adoption, including labour market disruption, risks associated with generative AI, and impacts on creative industries and education, are explored as key considerations for responsible technology governance.

Ultimately, this paper argues that AI can be a powerful enabler of sustainability if deployed within strong ethical, environmental, and institutional frameworks, but warns that without careful oversight, AI's resource intensity and social impacts could undermine the very goals it seeks to advance.

Keywords: Artificial Intelligence · Sustainability · Canada

1 Introduction – A Triple Planetary Crisis

We are in the midst of a triple planetary crisis – a crisis characterized by nature and biodiversity loss, climate change, and pollution and waste [1].

Over time, we have witnessed the loss of many species, beginning with the Dodo bird in 1662 [2]; yet we have done little to stop the plundering and hunting practices that led to its demise. We have become aware of climate change through hearing about the annual UN's Climate conference (COP), through a near constant barrage of news and literature on the subject, or by simply noticing the no longer deniable changing weather patterns. However, when it comes to pollution and waste, many of us remain blissfully

Published by Springer Nature Switzerland AG 2026
E. Mercier-Laurent et al. (Eds.): AI4S 2025, IFIP AICT 780, pp. 245–262, 2026.
https://doi.org/10.1007/978-3-032-21184-2_19

unaware of the dangers. We are used to the air we breathe, mines are located somewhere else, and waste is shipped somewhere else.

This crisis has been caused by human activities, and without major changes, the ability to sustain life on Earth will end. Although moving to a new planet is an appealing thought, it is not realistic. Even if we do discover a planet that is capable of sustaining life, we cannot actually get there, because the existing supplies of energy on earth are only enough to fuel one ship per year [3].

There is no Planet B [4].

1.1 Sustainable Development as a Solution

It has taken a long time to get the world's attention turned towards sustainability. Protests in the late 1960s were followed by the rise of environmentally focused organizations in the1970's, including Greenpeace, the United Nations Environment Programme (UNEP), and the Cousteau Society for the Protection of Ocean Life. In 1987 the Montreal Protocol [5] on ozone depletion came into force, and the UN World Commission on Environment and Development delivered its report defining sustainable development as development that "seeks to meet the needs and aspirations of the present without compromising the ability to meet those of the future" [6].

By the time the 1992 United Nations Conference on Environment and Development took place in Rio de Janeiro, Brazil, interest in acting on environmental issues was strong, and it was recognized that individual countries could not solve these issues on their own. Often referred to as the Earth Summit, its post-Cold War timing was good, and it brought together delegates from 179 countries to focus on the impact of human activities on the environment [7].

The 1992 Summit is remembered for the launch of the United Nations Framework Convention on Climate Change (UNCCC) usually referred to by its annual Conference of the Parties or COPs; Agenda 21 which lists the dimensions of protections needed for sustainable life; and for an inspiring speech given by the legendary French explorer and environmentalist Jacques-Yves Cousteau.

Cousteau's message: ***The continuance of the human species relies on a set of rights for future generations*** [8].

Cousteau, who had already begun collecting signatures in support of the rights of children to inherit an uncontaminated planet [9], became the force behind the UNESCO Declaration on the Safeguarding of Future Generations adopted in November 1997. As the UNESCO Director-General put it in his report on the draft Declaration on the Safeguarding of Future Generations:

"The idea of the preparation of a declaration concerning the rights of future generations was first launched by the late Commandant Cousteau. He led a major global campaign in support of a declaration prepared by the Equipe Cousteau. He co-operated with UNESCO in the elaboration of the present Declaration" [10].

In the years since then, there have been other conferences and other commitments, including the Year 2000 UN Millennium Development Goals (MDGs) [11] focusing on assistance to developing countries. As that program was coming to a close in 2015, and after much discussion and many consultations, the UN's 2030 Agenda for Sustainable Development was developed.

2 Agenda 2030 [12]

Adopted at the UN General Assembly in September 2015 by the 193 UN Member States, Agenda 2030 is a framework for sustainable development. Through its 17 Sustainable Development Goals (SDGs), it encompasses the three interconnected aspects of sustainable development (economic, social, and environmental) while maintaining focus on the current generation's responsibility to safeguard Earth for future generations (UNESCO Declaration 1987). Agenda 2030 also incorporates a means of implementation of the Goals achieved through building a global partnership and establishes a global review and reporting cycle. It represents a call to the world to work together to protect the earth for those yet to be born, by ending the practices that are leading to its demise, while still supporting prosperity in the present.

It recognizes that "social and economic development depends on the sustainable management of our planet's natural resources" (Fig. 1).

Fig. 1. Source: https://sdgs.un.org/goals

By agreeing to Agenda 2030, member states are expected to progress targets related to meeting the Goals which are applicable to their country context and challenges, inclusive of their regional interests and Indigenous Peoples, and in accordance with their governmental framework. Member states are encouraged to conduct Voluntary National Reviews (VNRs) [13], with most having conducted at least one since 2017 [14], and 36 countries listed as VNR-presenting countries for the 2026 UN High-level Political Forum (HLPF) on Sustainable Development [13].

HLPFs [15] are held annually in New York under the auspices of the Economic and Social Council (ECOSOC), and every fourth year at the SDG Summit held during the UN General Assembly at the level of Heads of State and Government. The next SDG Summit will be held in September 2027 and will be the last one before the 2030 deadline.

2.1 Alignments to the UN Agenda 2030 and Its Sustainable Development Goals

Led by the United Nations, alignments to the SDGs are growing every year. Governments, academia, citizens, corporations, and industry are all becoming involved. For example:

- 193 of the 195 (or a few more) countries in the world [16] are member states of the UN and have made commitments to the SDGs.
- The UN ITU's (International Telecommunications Union) AI for Good Program is a growing global Initiative focused on identifying practical applications of AI to advance the UN SDGs, and through its program industry is becoming involved. Its 2025 Global Summit attracted 11,000+ in-person participants from 169 countries [17].
- Action lines of the World Summit on the Information Society (WSIS) co-organized by the ITU, UNESCO, UNDP, and UNCTAD support the implementation of SDGs to advance sustainable development through information and communications technologies [18].
- Technical Committees in UN-connected organizations are aligned to the SDGs, and a growing number of research projects are focused on AI and sustainability.
- As countries make progress on the Agenda, supporting laws and regulations come into effect, bringing corporations and industries into alignment.

3 Canada's Commitment to Sustainability – How Canada Makes It Work

3.1 The Canadian Context

Canada is a member of the Commonwealth of Nations headed by the King of England. It is a parliamentary democracy led by an elected Prime Minister and governed as a decentralized federation with powers and obligations split between the national (federal) government and the provinces and territories. The Prime Minister governs the country through Cabinet to set policies and priorities for Canada, while the day-to-day operations of the Government of Canada (GC) are performed by the 100+ federal departments and agencies.

Until very recently, Canada's main trading partner was the United States. An abrupt change in their former close relationship, along with current geopolitical tensions, has led Canada to take strong steps to both diversify its trading relationships and guard its sovereignty – from coast to coast to coast – and to fully meet its UN commitments. Canada's bounty of the natural resources required to advance sustainable development will be a significant contributor to achieving all.

3.2 Canada's Sustainable Development Program

Canada is one of the 193 member states that agreed to the UN's 2030 Agenda for Sustainable Development and meets its commitments through its Environment and Climate Change Canada (ECCC) department. ECCC approaches this through the development of a Federal Sustainable Development Strategy (FSDS) applicable to federal departments and agencies, while each of the departments develop a Departmental Sustainability Development Strategy (DSDS).

The 2022–2026 FSDS [19] is Canada's most recent Strategy, and it is the first FSDS to include all three perspectives of the Agenda (social, economic, and environmental) and all 17 SDGs. It is also the first to require participation from all 100+ federal organizations, and it more fully incorporates the perspectives of Indigenous Peoples, who have rights to 40% of Canada's land mass [20].

Additional inputs were collected from other Canadian levels of government, businesses, not-for-profit and volunteer organizations, and citizens; resulting in an FSDS that is a comprehensive version of the UN Sustainable Development Strategy tailored to Canada's reality (Fig. 2).

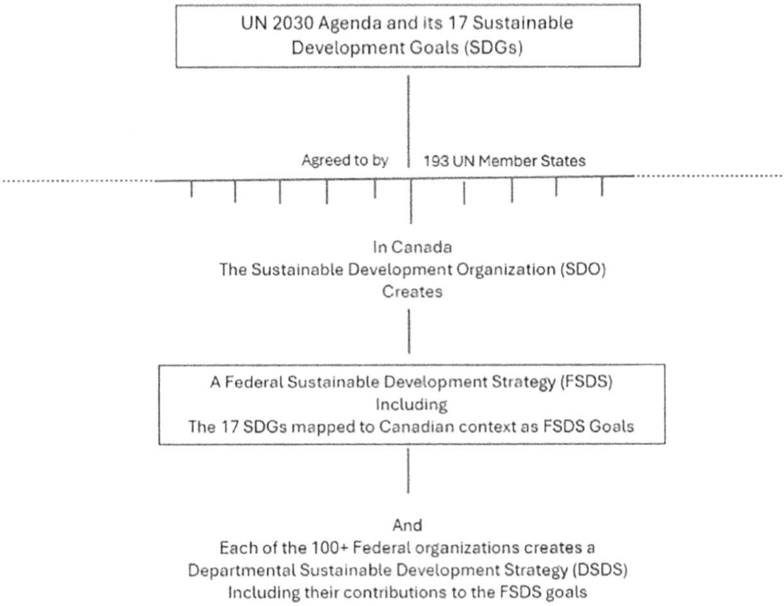

Fig. 2. The United Nations Agenda 2030 Sustainable Development Goals

At the end of the period covered by the FSDS Canada will participate in a VNR, with their next one expected in 2027.

By itself, this is simply a process. It is the people who run the departments that bring it to life and meet the goals and targets, and there are a few departments that deserve a special mention, starting with Environment and Climate Change Canada, which has a leading role (Table 1).

As noted in Agenda 2030, the SDGs are interdependent, as are the FSDS goals. For example, to achieve its climate goals, Canada is targeting a reduction of 40 to 45% Greenhouse Gas (GHG) emissions by 2030, and net-zero emissions by 2050. To do that Canada will need to increase its supply of clean energy – its hydroelectricity qualifies, increase the number of electric vehicles (EVs) on its roads, and provide adequate EV

Table 1. GC Departments key to Canada's Sustainable Development Goals [21]

Environment and Climate Change Canada (ECCC)	Sustainable Development Office Climate change program Weather and environmental conditions monitoring Pollution and waste management Monitor nature loss, species at risk
Natural Resources Canada (NRCan)	Minerals and mining, Canada's Critical Minerals Strategy Forests and forestry Energy efficiency, Energy sources and distribution
Department of Fisheries and Oceans (DFO)	Oceans conservation, protected areas, management Habitat protection for sustainable aquatic ecosystems *Role may grow with 2025 UN treaty on high seas marine biodiversity
Innovation, Science and Economic Development (ISED)	Science and Innovation; R&D and Innovation Global Innovation Clusters including Advanced Manufacturing, Scale AI, Ocean
National Research Council (NRC)	Challenge programs – industry partnerships to advance research priorities 12 Research Centers across Canada
Public Services and Procurement Canada (PSPC)	Ensures contracts include environmental and human rights clauses Real Property GHG reduction life-cycle cost analysis Linguistic services to Indigenous Peoples
Shared Services Canada (SSC)	Networks, Data Centers, Hosting Services GC AI Hub, LLMs and High-Performance Computing Co-creates digital and data solutions for science and research communities

charging infrastructure. Why focus on vehicles? Because burning fossil fuels is the leading source of GHGs (Table 2) [22].

As GC departments and agencies work to meet their obligations under the FSDS they will invariably involve research institutes and industry through existing relationships and contracts as well as with grants and challenge programs. When combined with the collaboration with many citizen, business, and Indigenous groups on the development of the FSDS, the priorities and viewpoints of virtually all sectors of Canadian society have been solicited.

Table 2. Agenda 2030 SDGs mapped to sample FSDS Goals and Targets SDGs 7, 9, 12, and 13 illustrate above interdependencies

Agenda 2030 SDG [12]	Canada FSDS Goal [19]	Sample FSDS Targets, language simplified [19]
SDG 6 CLEAN WATER AND SANITATION	FSDS Goal 6 Ensure clean and safe water for all Canadians	By 2027, action plans are in place to advance restoration and protection of major lakes and rivers in Canada
SDG 7 AFFORDABLE AND CLEAN ENERGY	FSDS Goal 7 Increase Canadian's access to clean energy	By 2030, 90%, and in the long term 100% of Canada's electricity is generated from renewable and non-emitting sources
SDG 9 INDUSTRY, INNOVATION AND INFRASTRUCTURE	FSDS Goal 9 Foster Innovation and Green Infrastructure in Canada	By March 31, 2026, 34,500 new electric vehicle chargers and 25 hydrogen refueling stations are completed
SDG 12 RESPONSIBLE CONSUMPTION AND PRODUCTION	FSDS Goal 12 Reduce Waste and transition to zero-emission vehicles	For the 2030 model year, at least 60% of new light-duty vehicle sales are zero-emission vehicles
SDG 13 CLIMATE ACTION	FSDS Goal13 Take action on climate change and its impacts	Achieve 40 to 45% greenhouse gas emission reductions by 2030, and net-zero greenhouse gas emissions by 2050
SDG 14 LIFE BELOW WATER	FSDS Goal 14 Conserve and protect Canada's Oceans	By 2026, at least 55% of Canada's key fish stocks are in the Cautious and Healthy zone
SDG 15 LIFE ON LAND	FSDS Goal 15 Protect and Recover Species, Conserve Canadian Biodiversity	Between 2023 and 2026, Canada's sustainable wood supply level exceeds the annual timber harvests

3.3 Canada's Use of AI in Sustainable Development

Canada has significant supplies of the resources needed for sustainable development and intends to become a clean energy superpower that will ensure its allies have access to its clean energy for theirs. To achieve this, it is ramping up the development of its critical minerals industry, supercharging its clean energy program, and funding research and development in supporting technologies. New critical minerals mines are being built and others are expanding, always with low emissions in mind; and new supply chains and port systems for future transport both inside and outside of Canada are being created [23].

An AI ecosystem is both an essential element of research and development and a fundamental ongoing component of operations for many of the industries developed.

Canada's AI Ecosystem

A core objective of Canada's Pan-Canadian Artificial Intelligence Strategy's [24] is to develop an AI ecosystem that advances AI for positive social, economic, and environmental impact, the three dimensions of sustainable development.

Building the Canadian AI ecosystem is well underway, is continuously evolving, and is moving into high gear under the new Carney government. It consists of a network of organizations collaborating to develop and commercialize AI research findings, the creation of a large-scale sovereign AI infrastructure (AI data centers, cloud platforms, foundation models, compute power), a pool of AI talent (AI developers, data scientists, research scientists), and the development of a governance structure. Core components of the AI Ecosystem Network are shown below (Table 3).

Table 3. Core Components of the Canadian AI Ecosystem Network

CIFAR [25] Canadian Institute for Advanced Research	**NRC** [26] National Research Council of Canada	**ISED** [27] Industry, Science, and Economic Development Canada
• Pan-Canadian AI Strategy • MOU with Amii, Mila, Vector National AI Institutes • AI Chairs Program • Funding for AI Safety Solution Networks	12 Specialized Research Centers including; Clean Energy Innovation Funded collaborative AI R&D and AI Solutions: • Materials for the Clean Fuels Challenge Program • Research Security Challenge Program for open-source intelligence (OSINT) due diligence	• Innovation Clusters: • Digital Technology Protein Industries Advanced Manufacturing Scale AI Ocean • Industrial Decarbonization • Canadian Sovereign AI Compute Strategy • Canadian AI Safety Institute

And the many Canadian AI start-ups and established AI innovators.

CIFAR, NRC, and ISED work together with industry in many ways. A Memorandum of Understanding (MOU) exists between CIFAR, the 3 National AI Institutes, and NRC to collaborate on R&D projects, training opportunities, and cross-sector initiatives; and the Canadian AI Safety Institute is supported by both the GC NRC research capabilities and the wider Canadian research community via CIFAR.

The Rest of the Ecosystem

A Sovereign Canadian AI Infrastructure

There is of course, already some AI infrastructure in Canada: "sovereign" and "Canadian" are the key words here. Canadian companies need to be able to use AI on large-language models (LLMs) holding Canadian information, and Canadian data must not be subject to another country's laws.

Projects in progress include an AI data center being built in Canada by Cohere Inc. that will enable businesses across Canada to access domestic compute capacity, a Canadian-owned and located supercomputer system for researchers and industry, and a smaller secure computing facility for government-related research and development [28].

Canada's AI Talent

- Canada has a top-tier AI academic and research community including notables like Geoffrey Hinton, Yoshua Bengio and more.
- Canada ranks first in the G7 for year-over-year growth of AI talent, and since 2019, first in the G7 for the number of AI-related papers published per capita [29].

Governance

Canada contributes to international discussions on AI and is a signatory to the first international AI Treaty, The Council of Europe Framework Convention on AI, Human Rights, Democracy, and the Rule of Law [30]; and to the UNESCO Recommendation on the Ethics of Artificial Intelligence [31]. Some of Canada's more impactful work in this area to date includes:

- Its joint proposal with France to establish the Global Partnership on Artificial Intelligence (GPAI). OECD member countries working on AI have recently joined GPAI, bringing its member country total to 44. All member countries must agree to advance the appropriate use of AI in agreement with human rights and shared democratic values, putting international cooperation to work creating trustworthy AI [32].
- The creation of a Canadian AI Safety Institute (CAISI) to investigate and act on the risks of advanced AI. Current research, coordinated by CIFAR, involves the themes "Safeguarding Courts from Synthetic AI Content" [33] and "Mitigating Dialect Bias" [34] in diverse linguistic communities.

Canada's Commitment to Climate Change

When it comes to sustainable development, Canada is focused on reducing its existing GHG emissions and at the same time advancing its clean energy industry, which will further reduce its emissions. Both will also have a broader global impact, as current and future buyers of Canada's clean/cleaner energy will also reduce their GHG emissions. Before moving away from the topic of Canada, it's worth looking at its top emitters - the oil and gas sector and the transportation sector (Fig. 3).

Oil and Gas

Much of Canada's prosperity comes from its natural resources, especially the oil and gas industry, both a significant source of jobs in Canada and its top GHG emitter, at 30% of Canada's total emissions in 2023 [35]. Canada has made some headway in decreasing its emissions since then (by 8% overall and by 40% in the oil sands (tailings ponds) [23], and plans to make further improvements through decarbonization using CCUS (carbon

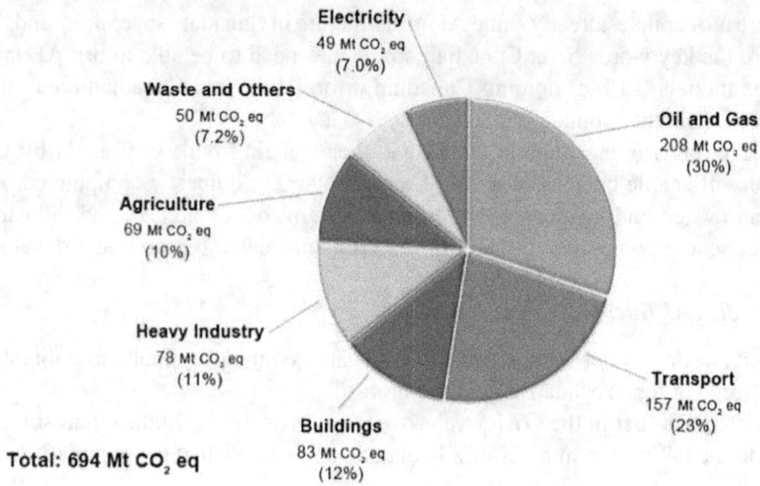

Fig. 3. Canada's GHG Emissions by Economic Sector (2023), **Source:** Greenhouse gas sources and sinks in Canada [35]

capture, utilization, and storage) technologies and by ensuring the supporting pieces of the industry are as low-carbon as possible [37].

When it comes to new oil and gas-related mines and industries, it's clear from the first ten projects being fast-tracked by the new Major Projects Office (MPO) that low emissions and clean energy projects are strongly supported. For example, the LNG Canada expansion is projected to both significantly improve the amount of LNG available to European customers, and result in emissions that are "35% lower than the world's best-performing LNG facilities and 60% lower than the global average" [36].

Transportation

To reduce emissions in the transportation sector, Canada's second-largest source of emissions at 23% in 2023 [35], the focus is on moving the fuel source for vehicles from gasoline to electricity, which requires lithium-ion batteries. As previously noted, Canada has the critical minerals needed to build the batteries, and another look at the projects being fast-tracked under the MPO includes a number of new mines being created to increase the supply of the critical minerals in the first 10 projects [36].

Equally important, there is also an NRC Critical Battery Materials Initiative that is establishing AI-enabled platforms to, for example, identify deep minerals deposits where near-surface deposits are declining [37].

A Note on Forestry

Forests act as carbon sinks by removing emissions from the air and storing them. Logging and forest fires release the stored emissions, becoming significant carbon sources.

In the 2024 Logging Emissions report produced by Nature Canada, the Natural Resources Defense Council, and Nature Québec, GHG emissions from industrial logging in 2022 represented at 21% of Canada's total emissions for that year, which would make logging the third-highest GHG emitter in the country [38]. Yet in Canada's official GHG reporting for both 2022 and 2023, forestry is included in the "Waste and Others" category contributing 7.2% to the total emissions [35].

Although emissions from wildfires are not included in Canada's GHG count for economic sectors, their over-sized impact on climate change, air quality, and habitat cannot be ignored. Identifying remote wildfires sooner using AI in conjunction with satellite data is in the planning stages and shows great promise.

4 The Trade-Offs that Come with Sustainable Development

The growing excitement around AI is reaching a fever pitch that is reminiscent of the California Gold Rush in the mid-1800s, when would-be miners "rushed in" from many areas of the United States and other countries in order to stake a claim and become rich. In the chaos of the time, when there were no laws protecting property rights, miners staked claims wherever they found nuggets. In the end, a few became very wealthy, and the rapid population increase led to an economic boom for California. Conversely, many miners were wiped out and left with empty pockets, and there were disastrous impacts on Native Americans [39].

We are currently in the chaotic phase of AI, and we need to take care to look beyond the current economic excitement to consider the long-term consequences. This requires understanding the difference between weak and strong sustainability, where weak sustainability is the concept that digging mines or creating parks qualifies as sustainability when their economic or social value equals the simultaneous value of biodiversity and environmental loss (nature capital). With strong sustainability the value of biodiversity and environmental loss considers the broader impact on the natural ecosystem. Planting trees in a new location to replace those cut down for a park in another location is not considered an equivalent trade, as the value of biodiversity loss is much greater than the number of trees removed [40].

The question to be answered then, is "Can sustainable development do both?" According to François Mancebo at the University of Reims does not think so [41].

Let's examine this more closely, starting with mining and AI data centers.

4.1 Mining

Mines create emissions throughout their lifecycle and all along their value chain. Given the urgent need to radically reduce GHG emissions and slow global warming, solutions for reducing mines emissions are of paramount importance. In addition to the decarbonization discussed earlier, fuel switching from fossil fuels to electricity will help, and an easy place to start is by changing out gasoline-fueled vehicles used to carry and transport the ore for electric vehicles – where energy stored in the vehicle's lithium-ion batteries supplies power to the motor.

However, lithium-ion batteries are manufactured from critical minerals (lithium, graphite, nickel, cobalt, copper, and rare earths), most of which are still in the ground. To support sustainable development, AI can be used to determine where the best deposits are, AI researchers can work with industry on creating sustainable supply chains, and sustainable practices can be used during mine construction.

Bottom Line: To reduce emissions and save the climate, we must first dig more mines. In the short term, that means more emissions.

4.2 Data Centers

Identifying innovations in support of sustainable development noted above is accomplished through the use of AI operating on LLMs, and that requires access to AI data centers.

AI Data Centers are not your everyday data centers. They are designed to meet the high-performance computing requirements of deep learning used in research and science, which requires massive numbers of GPUs (graphics processing units). Instead of housing racks of computers working on separate tasks, the GPUs of the AI data center are interconnected and share data across the data center. According to Chase Lochmiller, CEO of Crusoe, when talking about the OpenAI data center his company is building in Abilene, Texas; one way to think of an AI data center is that it doesn't just house computers, it *is* the computer [42].

Running the complex computations these data centers are used for requires a tremendous amount of power, generates a lot of heat, and requires a constant supply of clean water to prevent the GPUs from overheating – clean drinking water. In the Abilene facility the power capacity of the racks will be 130 watts vice 2–4 watts in a standard data center, and the chips will be water-cooled [42].

The increased stress on power grids, the loss of drinking water in a warming climate that is drying up existing sources, the 24/7 noise produced by AI data centers, and lower property values are common concerns. Water is perhaps the most serious of the complaints, as it can only be replaced by nature, and in our global warming reality, that is unlikely. Without major changes, the AI data centers beginning to dot the landscape will eventually reduce the world's water supply to unrecoverable levels.

It is not unusual for data center announcements to be met with a great deal of resistance, with water a recurring theme:

- About 270 miles north of Abilene in Amarillo, Texas, residents spoke out against The Fermi data center complex being built nearby, chiefly because of its proposal to purchase 2.5 million gallons of water per day from the city to operate, in an area where the aquifer's water supply is already stressed, and its level is decreasing much faster than it is can be replenished [43]. In October 2025, the Amarillo council approved the water deal with Fermi [44].
- 2,000 km northwest of Amarillo in Rocky View County, Alberta the local farmers had a better result when their local Council rejected a proposal for a data center on 450 hectares of farmland. The biggest issue: the community knew AI data centers were coming and had already selected an appropriate area for semi-industrial use where they could be built without major impacts. However, in the requested location there would be less water for crops, stressed cattle from the noise, and ultimately, less food for the world [45].

On the positive side, many data center builders are incorporating sustainable development in their construction. They are moving towards renewable power supplies and are looking for alternatives to water cooling of chips. Constructing data centers underwater, underground, in cooler climates, and even in Space is being explored, and are considered improvements in cooling requirements and less strain on supplies of drinking water. The downstream impacts on the ocean, the underground, and Space can only be imagined at this point.

5 AI's Impacts on Society

Most of us don't work in mines or in data centers, and our experience with the increasingly automated world we live in is minimal and gradual. Yet it is difficult not to notice the chatbot assistants and job losses based on efficiency gains as generative AI, agentic AI, and humanoid robots start to replace humans.

The concerns that worry many of us include the impact of AI on our jobs and livelihoods, the impact of generative AI on critical thinking, the risk synthetic data and media pose to the creative professions, and not knowing who or what can be trusted.

5.1 Jobs – Losses and Gains, Disruption Rippling Throughout Society

A major workforce transition is upon us. Many workers are beginning to be displaced by the efficiency gains of AI, and many more will follow. It has happened many times before in history, but never to us. Just as the advent of electricity replaced the lamp lighters of long ago, refrigeration the ice cutters, and automated switching the switchboard operators; AI will replace some of us.

A Case in Point [46]

The invention of the telephone in 1876 replaced some of the telegraph functions and led to manual telephone exchanges run by telephone switchboard operators, many of whom came from the telegraph companies. Eventually the local switchboard operator became the engine that kept small towns running smoothly as they generally knew where the town Doctor/banker/milkman was and could quickly find any one of them in an emergency. However, the queues for the operators soon became very long, would-be callers became frustrated by the long waits and crossed lines, and the inventors of the day began working on automated switching. Interestingly, the winning design was created by funeral director Almon B Strowger who became convinced that an operator was rerouting calls meant for him to another funeral director.

In the UK, the replacement of the manually operated telephone switchboards by automated telephone switching happened gradually from 1912 to sometime in the 1960s. Although the automated exchanges were attractive because of the savings they represented by making operators redundant, they were expensive to set up.

Parallels with Today

There is an interesting parallel with that transition and the one underway now as call center operators and other entry level jobs seem to be the first to go. Recent job cuts that have made the news include Amazon slashing 30,000 jobs, many of them by replacing warehouse operators with humanoid robots – something most of us will not even notice; and job losses in the Australia banking industry due to "efficiencies."

As a former Product Manager at the Commonwealth Bank of Australia (CBA) put it, "AI was brought in, and we taught the machines, and due to productivity gains, headcount was taken out" [47].

While it doesn't appear that today's transition will be as gradual as it was for the switchboard operators, it may be, as the reports of problems caused by generative AI errors are starting to dampen down the enthusiasm. According to Gartner [48], generative AI is entering the "trough of disillusionment," which is when the rush to adopt a technology significantly slows down. In addition, the World Economic Forum's Future of Jobs Report 2025, predicts that by 2030 more jobs will have been created than lost, with an overall gain in employment of 7% [49].

As optimistic as that sounds, it is cold comfort for those who are losing their jobs now, as the outgoing cashiers, clerks, call center and warehouse operators are unlikely to be hired for AI and machine learning specialist positions.

Speaking of the "Trough of Disillusionment" [48].

The Australian government was recently caught unawares when a Deloitte report on the Australian welfare compliance system was found to have substantial errors that were only noticed by chance, when Chris Rudge, a University of Sydney Law Professor was asked for his opinion on the report by someone who had been negatively impacted by the compliance system. As he was looking through the report, he noted that some of the books referenced did not exist. He was able to spot that quickly, as they were attributed to colleagues of his, and he then began to check more thoroughly. In total, he identified approximately 20 errors and went public with his findings. In the end Deloitte had to repay the government for some of their costs and include a note on the errors in the final report [50].

In the legal world, there have been a number of reports of lawyers using Generative AI in preparation for court. In one instance that occurred In February 2024, a Canadian lawyer was reprimanded after she had found and cited two cases in support of her argument in a divorce case. When the opposing counsel asked for copies of the cases and none were provided, he investigated further and found the cases did not exist. The lawyer had used ChatGPT as a search tool but had not checked to confirm the cases existed. She was required to pay the costs incurred by the opposing counsel when they undertook the search [51].

In both above examples the errors were most likely inference errors caused by generative AI's designed-in desire to fulfill the user's request, and when faced without the information needed to do so, simply made something up. These are often referred to as AI hallucinations, and they may cause some companies to tread lightly while the technology matures.

Impact on Creative Professions

When the (Screen) Writers Guild of America went on strike in 2023 [52] one of the significant issues was the impact that technology, particularly synthetic data and media, would have on their profession. They were concerned about the loss of work if the studios were able to create scripts using AI and the low residuals being paid for streaming. They also wanted guardrails implemented that would protect them in various scenarios. Their basic argument was that AI is only able to recreate what writers have already created, which is their intellectual property. Although they won the strike and received numerous concessions, the gains made may not last for long.

While web scraping of information that falls under copyright can be somewhat controlled once identified, misuse of synthetic data and synthetic media (deepfakes for example) is more difficult to manage. Although these issues are under discussion internationally, laws are slow to be enacted, and there will be some losses in the interim. In Canada's Budget 2025, planned amendments to the Copyright Act to protect artists' and creators' copyrights have been announced - it's a start [23].

Education and Critical Thinking

In the present, concerns are focusing on things like how to determine if a student has used generative AI to write their university papers, and how to meet teaching goals if students use AI instead of doing the work themselves. If someone who is a poor writer uses AI, how will they ever learn to write? And if they let AI do the thinking for them, how will they develop critical thinking skills? How will those who do know how to write and how to think be able to distinguish themselves?

At this point in time there are more questions than answers, and multiple possibilities.

Future Schools.

In a 2025 forum held at the Harvard Graduate School of Education on the schools of the future, two different visions were presented. One was that schools will consist of a few years learning the basics followed by teachers as coaches who introduce their students to ideas that challenge their thinking. The other was that future generations must be trained to orchestrate a team of AIs [53]. It's possible both could co-exist.

In the Meantime.

Extremely high levels of skills will be required to manage the transition to AI and to maintain and operate all the new technologies that accompany it, and educational institutions at all levels will need to adapt their curriculums to suit. Critical thinking will be very much in demand, and in a best-case scenario, significant effort and brainpower will be applied to ensuring that it stays that way.

6 Final Thoughts

Agenda 2030 and its 17 SDGs go far beyond their MDG predecessors to encompass the three dimensions of sustainable development: economic, social, and environmental; vice focusing exclusively on developing countries. This change acknowledges that protecting earth's natural resources without allowing for human prosperity will not work, and that social and economic development relies on the availability of natural resources, which in turn requires their being managed in a sustainable way [12]. In other words, if we don't protect our planet's natural resources, they will eventually be depleted to a point where Earth can no longer sustain life.

To make headway on the SDGs by 2030, artificial intelligence supported by research talent and industry interest may make the difference between success and failure, and that requires building AI data centers and creating AI ecosystems to develop and commercialize new technologies and supporting supply chains. The challenge is to ensure they are built and operated in a way that is neither detrimental to the environment nor undermines the successes achieved by others with respect to their SDGs.

This challenge is not limited to data centers, and to overcome it, a more comprehensive measurement scheme to determine a country's progress towards the SDGs is needed. Ensuring that our descendants will be able to live, love, and prosper requires considering both the present and future sustainability of the natural resources that support life, and projecting that across multiple generations.

There will be other challenges as well, as not all countries/governments see the importance of the bigger picture, some aren't interested, and others lack the needed resources. If countries could work together to make the kind of progress the Montreal Protocol has had on closing the hole in the ozone layer (projected to fully recover by the middle of the century) [5] we could see significant improvements by 2050.

Riches will be of no use to our descendants if there is no clean drinking water left.

References

1. United Nations Environment Program. UNEP – Your partner for people and planet. UNEP – your partner for people and planet | UNEP - UN Environment Programme (2025). https://www.unep.org/resources/publication/unep-your-partner-people-and-planet Last accessed 09/12/2025
2. Wikipedia contributors. Dodo. Wikipedia, The Free Encyclopedia (2025). https://en.wikipedia.org/wiki/Dodo. Accessed 09 Dec 2025
3. Berners-Lee, M.: There is No Planet B. Cambridge University Press (2021)
4. Berners-Lee, M.: There is No Planet B. Cambridge University Press. Term coined by John Locke (2021). https://bullfeet.com/en-ca/blogs/news/there-is-no-planet-b?srsltid=AfmBOop3arJHh6PFSiym1uj-fqcaL-Fzq8PTa-NmKJbt22bsQUWL_tEw/. Accessed 09 Dec 2025
5. UNEP About Montreal Protocol. UNEP. https://www.unep.org/ozonaction/who-we-are/about-montreal-protocol. Accessed 09 Dec 2025
6. Bruntland, G.H.: Our Common Future: Report of the World Commission on Environment and Development (1987)
7. Wikipedia contributors. Earth Summit. Wikipedia, The Free Encyclopedia. Earth Summit - Wikipedia (2025). Accessed 09 Dec 2025
8. Jacques, C.: Jacques Cousteau on the demographic implications of man's divorce from nature. Popul. Dev. Rev. **18**(4), 780–781 (1992). https://doi.org/10.2307/1973776
9. The Cousteau Society Legacy. https://www.cousteau.org/know/legacy. Accessed 09 Dec 2025
10. UNESCO. Report by the Director-General on the draft Declaration on the Safeguarding of Future Generations. UNESDOC Digital Library (1997). https://unesdoc.unesco.org/ark:/48223/pf0000108750. Accessed 09 Dec 2025
11. United Nations. The Millennium Development Goals Report. United Nations (2015). https://milenniumgoals/2015_MDG.Report/pdf/MDG%202015%20rev%20(July%201).pdf. Accessed 09 Dec 2025
12. United Nations Department of Economic and Social Affairs. Transforming our world: the 2030 Agenda for Sustainable Development (2015). https://sdgs/un/org/2030agenda. Accessed 09 Dec 2025
13. United Nations High-Level Political Forum on Sustainable Development Voluntary National Reviews. https://hlpf.un.org/vnrs. Accessed 09 Dec 2025

14. United Nations High-Level Political Forum on Sustainable Development Voluntary National Reviews (2025). https://view.officeapps.live.com/op/view.aspx?src=https%3A%2F%2Fhlpf.un.org%2Fsites%2Fdefault%2Ffiles%2F2025-08%2FVNRs%2520as%2520of%252029%2520July%25202025.xlsx&wdOrigin=BROWSELINK. Accessed 09 Dec 2025
15. United Nations High-Level Political Forum on Sustainable Development High-Level Political Forum. https://hlpf.un.org. Accessed 09 Dec 2025
16. World Population Review. How Many Countries are there in the World 2025? World Population Review (2025). https://worldpopulationreview.com/country-rankings/how-many-countries-are-there. Accessed 09 Dec 2025
17. ITU. ITU's AI for Good Global Summit energizes collaboration on AI governance, skills, and standards (2025). https://www.itu.int/en/mediacentre/Pages/PR-2024-07-11-AI-for-Good-closing.aspx. Accessed 8 Dec 2025
18. World Summit for the Information Society WSIS – SDG Matrix. https://www.itu.int/net4/wsis/sdg/. Accessed 09 Dec 2025
19. Government of Canada: Environment and Climate Change Canada (2025). The 2022-2026 Federal Sustainable Development Strategy. Introducing the strategy - Canada.ca. Accessed Apr 2026
20. Government of Canada: Modern Treaties. Modern treaties (2025). Accessed 09 Dec 2025
21. Government of Canada: Departments and Agencies (2025). https://www.canada.ca/en/government/dept.html. Accessed 9 Dec 2025
22. The Energy Institute: Statistical Review of World Energy 2025 (2025). Home | Statistical Review of World Energy
23. Government of Canada: Canada Strong Budget 2025 (2025). https://budget.canada.ca/2025/report-rapport/pdf/budget-2025.pdf. Accessed 9 Dec 2025
24. CIFAR: The Pan-Canadian AI Strategy (2025). https://cifar.ca/ai/. Accessed 09 Dec 2025
25. CIFAR (2025). https://cifar.ca/. Accessed 09 Dec 2025
26. Government of Canada: National Research Council Canada (2025). https://nrc.canada.ca/en. Accessed 9 Dec 2015
27. Government of Canada: Innovation, Science and Economic Development Canada (2025). https://ised-isde.canada.ca/site/ised/en. Accessed 9 Dec 2025
28. Government of Canada: Canadian Sovereign AI Compute Strategy (2025). https://ised-isde.canada.ca/site/ised/en/canadian-sovereign-ai-compute-strategy. Accessed 9 Dec 2025
29. Government of Canada Department of Finance: Deputy Prime Minister announces $240 million for Cohere to scale-up AI compute capacity (2024). https://www.canada.ca/en/department-finance/news/2024/12/deputy-prime-minister-announces-240-million-for-cohere-to-scale-up-ai-compute-capacity.html. Accessed 9 Dec 2025
30. Council of Europe: The Framework Convention on Artificial Intelligence (2025). https://www.coe.int/en/web/artificial-intelligence/the-framework-convention-on-artificial-intelligence. Accessed 9 Dec 2025
31. UNESCO Artificial Intelligence & Emerging Technologies Recommendation on the Ethics of Artificial Intelligence. https://www.unesco.org/en/artificial-intelligence/recommendation-ethics
32. OEDC.AI About the Global Partnership on Artificial Intelligence (GPAI). https://oecd.ai/en/about/about-gpai
33. CIFAR: Safeguarding Courts from Synthetic AI Content. CIFAR-Solution-Networks (2025). https://cifar.ca/ai/cifar-solutions-networks/safeguarding-courts-from-synthetic-ai-content
34. CIFAR: Mitigating Dialect Bias. CIFAR-Solution-Networks (2025). https://cifar.ca/ai/cifar-solution-networks/mitigating-dialect-bias/
35. Government of Canada: Greenhouse gas sources and sinks in Canada: executive summary 2025 (2025). https://www.canada.ca/en/environment-climate-change/services/climate-change/greenhouse-gas-emissions/sources-sinks-executive-summary-2025.html

36. Government of Canada: Major Projects Office Projects and Transformative strategies map (2025). https://www.canada.ca/en/privy-council/major-projects-office/projecs/map.html
37. National Research Council Canada Critical Battery Materials Initiative. Government of Canada. https://nrc.canada.ca/en/research-development/research-collaboration/programs/critical-battery-materials-initiative
38. Nature Canada: Logging Emerges as Canada's Third Largest Climate Polluter (2024). https://naturecanada.ca/news/press-releases/logging-emerges-as-canadas-third-largest-climate-polluter/
39. Wikipedia Contributors California Gold Rush. Wikipedia, The Free Encyclopedia. https://en.wikipedia.org/w/index.php?title=California_gold_rush&oldid=1324228040
40. Wikipedia Contributors Weak and strong sustainability. Wikipedia, The Free Encyclopedia. https://en.wikipedia.org/w/index.php?title=Weak_and_strong_sustainability&oldid=1325483138
41. Mancebo, F., Sachs, I.: Transitions to Sustainability. Springer Science+Business Media (2015)
42. Bloomberg Technology Interviewer: Crusoe, Blue Owl Capital, and Primary Digital Enter Second Phase of $15B Joint Venture. Youtube Bloomberg Technology (2025). https://www.youtube.com/watch?v=R1S6_oaAank
43. Carver, J.L.: Texas' Thirst for Rural Data Centers Collides With Water Scarcity. Governing (2025). https://www.governing.com/infrastructure/texas-thirst-for-rural-data-centers-collides-with-water-scarcity
44. Cuviello, M.: Amarillo council finalizes Fermi America water deal amid debate over value, transparency Rocky View County council votes against date centre complex. Amarillo Globe-News (2025). https://www.amarillo.com/story/news/2025/10/29/amarillo-council-approves-fermi-america-water-deal-with-city/86964641007/. Accessed 30 Oct 2025
45. Wilson, J.: Rocky View County council votes against data centre complex. CTV news (2025). https://www.ctvnews.ca/calgary/article/public-hearing-set-for-tuesday-over-proposed-rocky-view-county-technology-park/
46. Science Museum. Goodbye to the Hello Girls: Automating the Telephone Exchange. Science Museum (2018). https://sciencemuseum.org.uk/objects-and-stories/goodbye-hello-girls-automating-telephone-exchange
47. Whitson, R.: Australia in a 'dangerous place' as AI adoption ramps up, prompting calls to protect jobs. ABC News (2025). https://www.abc.nat.au/news/2025-09-25/Australia-in-dangerous-place-as-ai-adoption-ramps-up/105807430. Accessed… (would be after 24 Sep buts is that 2025
48. Gartner: Gartner Hype Cycle Identifies Top AI Innovation in 2025. Gartner Newsroom (2025). https://www.gartner.com/en/newsroom/press-releases/2025-08-05-gartner-hype-cycle-identifies-top-ai-innovations-in-2025
49. World Economic Forum: Future of Jobs Report 2025. World Economic Forum (2025). https://reports.weforum.org/docs/WEF_Future_of_Jobs_Report_2025.pdf
50. Jayes, L.: Deloitte report caught with AI generated reference errors. Sky News Australia (2025). https://youtube.com/watch?v=0u4x7IoCwKw
51. Procter, J.: B.C. lawyer reprimanded for citing fake cases invented by ChatGPT. CBC News (2024). https://www.cbc.ca/news/canada/british-columbia/lawyer-chatgpt-fake-precedent-1.7126393
52. Lee, W., James, M.: WGA and the studios reach tentative deal to end writers' strike. Los Angeles Times (2023). https://www.latimes.com/entertainment-arts/business/story/2023-09-24/writers-strike-over-wga-studios-reach-deal-actors
53. Boyles, S.: How AI could radically change schools by 2050. The Harvard Gazette (2025). https://news.harvard.edu/gazette/story/2025/09/how-ai-could-radically-change-shools-by-2050/

Data-Driven Sustainability: The Role of Urban Platforms in Participatory Smart-City Governance

Lukasz Przysucha(✉) and Adam Sulich

Faculty of Business Management, Wroclaw University of Economics and Business, Komandorska 118/120, 53345 Wroclaw, Poland
lukasz.przysucha@ue.wroc.pl

Abstract. Accelerating urbanisation has led to wider adoption of the smart-city (SC) model in pursuit of sustainable development (SD). This study examines how urban digital platforms that integrate sensing, data governance, analytics, and participatory tools translate citizen inputs into sustainability outcomes and green-job (GJ) formation. A comparative multiple-case study is conducted across Amsterdam, Athens, Seville, and Osaka. Evidence is compiled from municipal strategies, platform portals and open-data catalogues, programme evaluations, technical and procurement documentation, and peer-reviewed sources. Cross-case synthesis and pattern matching are applied to assess how platform features (open application programming interfaces—APIs, standards, privacy and accountability arrangements) and participation modalities (reporting, co-design, participatory budgeting) influence measurable outcomes in energy, mobility, circularity, and air quality, as well as associated GJ pathways. The results indicate that (1) platforms with transparent pipelines from citizen input to implementation show stronger links to SD indicators and ecosystem activity; (2) robust data-governance mechanisms broaden participation and enhance the durability of results; and (3) close alignment between platform analytics and procurement/operations fosters local green-and-digital employment. Implications for platform design and governance are outlined, together with research needs on participation quality, impact measurement, and transferability across institutional contexts.

Keywords: Green Jobs · Crowdsourcing · Participatory Governance · Smart City · Sustainable Development · Urban Platforms

1 Introduction

Rapid urbanization [1] and climate-related pressures [2] have increased demand for governance models that deliver measurable sustainability outcomes in cities [3]. The Smart City (SC) paradigm positions data-enabled infrastructures and socio-technical innovation to improve accessibility [2], mobility, and quality of life while advancing Sustainable Development (SD) [4]. At the operational layer, urban digital platforms federate heterogeneous data (sensor, administrative, citizen-generated), provide access and

Published by Springer Nature Switzerland AG 2026
E. Mercier-Laurent et al. (Eds.): AI4S 2025, IFIP AICT 780, pp. 263–272, 2026.
https://doi.org/10.1007/978-3-032-21184-2_20

trust mechanisms (APIs, standards, privacy and accountability) [5], and host participatory tools that structure agenda-setting, co-creation, and feedback [6]. A persistent gap concerns the mechanisms by which platform-mediated participation is translated into concrete SD outcomes and Green Job (GJ) formation [7, 8].

The objective of this article is to explain how platform features (e.g., open APIs, standards, privacy and accountability arrangements) and participation modalities (e.g., reporting, co-design, participatory budgeting) relate to outcomes in energy, mobility, circularity, and air quality, together with associated GJ pathways [9].

A comparative multiple-case design is employed, focusing on Amsterdam, Athens, Seville, and Osaka. The cities were selected for variation in governance traditions and digital maturity [10], explicit sustainability ambitions [11], the presence of operational urban platforms with open-data provisions, and the availability of public documentation enabling triangulation. Evidence is drawn from municipal strategies, platform portals and open-data catalogues [12], programme and project evaluations, as well as technical and procurement documentation [13]. Cross-case synthesis and pattern matching are used to identify recurring mechanisms linking platform design and governance to SD indicators and GJ dynamics.

The contribution is fourfold: (1) a mechanism-oriented account of how platform governance and participation translate into measurable SD outcomes; (2) an analytic framework connecting platform features and participation modalities to GJ pathways; (3) comparative evidence across diverse institutional contexts; and (4) actionable implications for platform design, data governance, and alignment with procurement and operations. The article proceeds as follows: Sect. 2 presents materials and methods; Sect. 3 reports the cross-case results; Sect. 4 discusses theoretical and managerial implications; Sect. 5 concludes with limitations and directions for future research.

2 Materials and Methods

A comparative multiple-case design was applied to explain how urban digital platforms [14] translate citizen inputs into sustainability outcomes and GJ formation [15]. The design enables mechanism-focused analysis within distinct institutional and technological settings and subsequent cross-case synthesis of recurring patterns [16]. Four cities - Amsterdam, Athens, Seville, and Osaka - were purposively selected to maximise variation in governance traditions and digital maturity while ensuring explicit sustainability ambitions, operational urban platforms with open-data provisions and participatory tools, and sufficient public documentation to permit triangulation. Geographic diversity spanning the European Union and Japan increases external heterogeneity without compromising comparability in terms of smart-city orientation.

The evidentiary window covered 2015–2025. Data were drawn exclusively from publicly available documents [17], including municipal strategies and policy papers; platform portals and open-data catalogues with API documentation and usage dashboards [18]; programme and project evaluations [19]; as well as technical and procurement documentation that links analytics to operational processes [20]. Peer-reviewed publications were used only insofar as they reported city programme results [21] or platform features relevant to the cases [19, 21, 22]. No human-subjects data were collected.

Key constructs were defined and operationalised ex ante [23]. Platform features encompassed openness of APIs and data catalogues, adherence to standards, and privacy and accountability arrangements such as impact assessments, auditability, and role-based access controls [24]. Participation modalities covered reporting (issue and observation intake) [25], co-design (workshops and pilot sandboxes) [26], and participatory budgeting, with specific attention to the presence of feedback loops and public status tracking [27]. Sustainability domains comprised energy, mobility, circularity, and air quality. Outcome indicators were recorded as reported by official sources - e.g., energy savings; electric-vehicle (EV) charging capacity [28] or sessions; modal-share or fleet metrics; recycling and diversion rates; PM2.5 or NO_2 levels [29, 30]. Where precise quantitative values were unavailable, directional evidence was retained as descriptive. GJ pathways were inferred from documented initiatives, procurement texts, and programme descriptions, focusing on roles and skills plausibly mobilised by platform-enabled interventions, such as photovoltaics (PV) installation and operations and maintenance (O&M), EV-charging operations [31], data engineering and analytics [32], and cybersecurity [33].

Analysis proceeded in two stages. Within each case, a common rubric was applied to synthesise evidence on platform governance, participation instruments, sustainability outputs and outcomes, and associated GJ pathways. Cross-case synthesis then employed pattern matching and explanation building around three focal mechanisms: transparency of the pipeline from citizen input to implementation; robustness of data-governance arrangements; and alignment between platform analytics and procurement or operations. Each mechanism was coded on an ordinal three-level scale (low, medium, high) based on explicit documentary indicators - for example, public issue trackers, service-level agreements (SLAs), and implementation logs for transparency; formal privacy policies, impact assessments, and audit references for governance; and contractual key performance indicators (KPIs) and API clauses linking data insights to service delivery for alignment.

Reliability and validity were supported through a structured codebook that specified construct definitions, inclusion rules, and anchors for the ordinal scales; a subset of coded items was cross-checked for consistency, with discrepancies resolved by refining coding rules. Triangulation across document types mitigated single-source bias, and a chain of evidence was maintained to trace claims to documentary excerpts. Given the reliance on public materials and the observational nature of the design, findings are interpreted as analytically generalisable patterns consistent with plausible mechanisms rather than causal estimates. Differences in indicator definitions and reporting practices across cities constrain strict comparability [34], and measurement precision is bounded by the quality and granularity of municipal disclosures. Ethical approval was not required due to exclusive use of publicly accessible documents.

3 Results

The cross-case comparison revealed consistent mechanisms linking platform governance to sustainability outcomes and GJ pathways. Core patterns are summarised in Table 1, which reports three ordinal indicators coded from public documentation: M1—transparency of the pipeline from citizen input to implementation; M2—data-governance robustness (privacy, standards, auditability); and M3—alignment of analytics with procurement and day-to-day operations. Ratings (Low/Medium/Medium–High/High) were assigned on the basis of explicit evidence: API documentation and data requirements (for M2), public status trackers or implementation logs (for M1), and procurement/KPI references that tie data products to service delivery (for M3). For illustration, Amsterdam's DataPunt/DSO API documentation and API-key policy underpin strong M2–M3 scores; Athens's OGP action plan indicates maturing participation and open-data ambitions; Seville's IDE/Open Data portal establishes standardised, API-backed spatial data; and Osaka's CKAN-based city portal together with the national "Japan Mobility Dataspace" provide a data-sharing backbone for mobility services.

Table 1. Platform governance mechanisms and participation instruments across cases

City	M1: Transparency	M2: Data-governance robustness	M3: Analytics–operations alignment
Amsterdam	High	High	High
Athens	Medium	Medium	Medium
Seville	Medium	Medium	Medium–High
Osaka	Medium–High	High	High

Source: Authors elaboration based on city reports

To streamline Table 1, the dominant participation tools are described here. Amsterdam combines reporting applications with co-design pilots and neighbourhood-level participatory budgeting supported by public status tracking. Athens emphasises incident/reporting portals and co-design within resilience and building-retrofit programmes. Seville foregrounds co-design around mobility and public space alongside targeted participatory budgeting. Osaka operates pilot sandboxes and transport reporting within innovation districts [18]; the CKAN registry and national mobility-dataspace documentation illustrate the underlying data-sharing architecture.

Indicators M1–M3 isolate minimum conditions repeatedly cited in municipal documentation as gatekeepers for translating crowdsourced inputs into implemented solutions: public traceability and follow-through (M1), trustworthy data handling and standardisation (M2), and operational hooks—contracts, KPIs, workflows—that ensure data products alter services on the ground (M3). The indicators privilege mechanisms rather than intentions and remain comparable across institutional contexts using documentary evidence.

Amsterdam [35] has a mature API estate (DataPunt/DSO) with an enforced API-key policy, discoverable datasets, and implementation-visible tooling supports high M1–M3: inputs can be traced to fixes or deployments; governance is standardised and auditable; analytics are referenced in operations and procurement. Directional signals align with the expansion of smart charging and the scaling of energy-efficiency and circular pilots, which stimulate demand for PV, O&M, charging-network operations, and data services.

In Athens participation is anchored in reporting portals and co-design within resilience and energy retrofits; the OGP 2025–2028 plan targets participatory budgeting and accessible open-data portals. Evidence supports medium M1–M3: ambitions and pilots are clear, while traceability and procurement hooks appear programme-selective. Anticipated job pathways cluster around retrofit skills, energy audits, and municipal data roles.

The IDE/Open Data portal in Seville exposes spatial and mobility datasets with API access, reinforcing medium M2 and medium–high M3 where procurement specifies data-driven criteria (notably PV rollout and mobility management). Participation mixes co-design, reporting, and targeted participatory budgeting. Resulting job pathways emphasise PV deployment/O&M, mobility operations, and analytics-enabled services.

The Oska city's CKAN portal and the national mobility-dataspace initiative yield high M2–M3 and medium–high M1 in transport and green-infrastructure domains. AI-supported mobility deployments and a sustained privacy/cybersecurity posture translate into demand for green-AI engineering, smart-mobility systems, and cybersecurity roles [12].

To make outcome signals and employment implications explicit, Table 2 consolidates directional sustainability results and indicative green-job (GJ) pathways drawn from the same documentary record.

Table 2. Directional SD outcomes and green-job creation pathways across cases

City	Reported SD outcomes (directional)	Indicative GJ pathways
Amsterdam	Expansion of smart charging; scaling of energy-efficiency and circular pilots; improved issue-resolution traceability	PV installation & O&M; EV charging operations; data engineering/analytics; platform services
Athens	Gains in building efficiency; improved service responsiveness; incremental mobility improvements	Building retrofits; energy auditing; municipal data roles; operations support
Seville	Acceleration of PV deployment; enhancements in sustainable mobility management; selective air-quality initiatives	PV deployment & O&M; mobility operations; data-informed service roles
Osaka	AI-supported mobility deployments; green-infrastructure integration; sustained cybersecurity posture	Green-AI engineering; smart-mobility systems; cybersecurity & privacy engineering

Source: Authors elaboration based on city reports

Cities with high M3 (Amsterdam, Osaka) exhibit the clearest scaling signals—growth in smart charging, AI-supported mobility, and consistent data curation/security—because analytics are wired into procurement and operating KPIs. Where M1 is medium (Athens, Seville), participation volumes are visible but the last mile from input to implementation is less systematic unless programme-level tracking and contractual clauses are explicit; this constrains the breadth of GJ pathways, skewing them towards operations and maintenance and retrofit roles rather than higher-value analytics and engineering.

Programmes lacking public issue trackers or implementation logs (low/medium M1) show weaker conversion of engagement into service change, undermining responsiveness and legitimacy over time. Fragmented data governance (lower M2)—characterised by diffuse privacy policies, limited PIA/DPIA references, and inconsistent standards—discourages participation among privacy-sensitive stakeholders, narrowing the input signal and shortening the durability of improvements. Pilots without operationalisation (low/medium M3)—i.e., not tied to procurement frameworks, interoperability/API clauses, or service KPIs—tend to stall at demonstration; conversely, embedding data requirements in contracts and SLAs accelerates scaling beyond pilots. Finally, heterogeneity in indicator definitions (e.g., "charging session," "energy saved") limits strict comparability; hence, outcomes are reported directionally rather than as pooled effect sizes. Mitigation steps included triangulating multiple document types per claim and maintaining a chain of evidence to specific portal pages, API documentation, and programme briefs. Collectively, the patterns support analytic generalisation at the mechanism level and indicate that transparency (M1), governance robustness (M2), and operational alignment (M3) function as complementary conditions for translating platform-mediated participation into measurable sustainability outcomes and associated green-job formation.

4 Discussion

The findings indicate that three governance mechanisms: M1 (transparency of the pipeline from citizen input to implementation), M2 (data-governance robustness), and M3 (alignment of analytics with procurement and operations) - jointly shape whether platform-mediated participation yields measurable sustainability outcomes and green-job pathways. Across the four cases, M3 appears as the strongest proximate driver of scaling, while M1 and M2 condition the credibility, inclusiveness, and durability of effects. The pattern is consistent with a complementarity logic: M1 reduces informational frictions by making responsiveness auditable; M2 standardises and de-risks data flows, supporting reuse across programmes and vendors; and M3 couples data products to contractual and operational routines so that signals translate into service changes rather than remaining as pilots or proofs of concept.

Two recurrent pathways are suggested by the evidence. An operations-led pathway emerges where reporting with public status tracking under at least medium M1 combines with adequate M2, producing near-term gains in service responsiveness and demand for operations-and-maintenance roles (for example, EV-charging operations and PV O&M). An innovation-led pathway is observed where co-design and pilot sandboxes are accompanied by strong procurement hooks and outcome KPIs under mature M2, enabling the scaling of data-intensive interventions such as AI-supported mobility; this pathway is associated with higher-skill analytics and engineering roles. Cities displaying high M3 and at least medium M1–M2 move more consistently along both pathways, whereas weaknesses in any one mechanism tend to break the chain from participation to implementation.

Theoretically, the analysis advances a mechanism-oriented account of participatory smart-city governance. Rather than treating participation volume as the primary driver, the results foreground an impact chain in which visibility and traceability (M1) establish an accountability substrate; standardised and privacy-preserving governance (M2) creates reliable and reusable assets; and contractual and operational alignment (M3) institutionalises feedback by wiring analytics into everyday delivery. This extends platform-governance scholarship by identifying procurement clauses, interoperability requirements, and outcome KPIs as critical coupling devices between socio-technical platforms and urban service systems, thereby explaining why nominally similar participatory efforts produce divergent implementation trajectories.

Managerial and policy implications follow directly. Codifying transparency under M1—for instance, publishing machine-readable status feeds and implementation logs—enables residents and oversight bodies to audit responsiveness and reduces room for discretionary drift. Hardening M2 through standards catalogues, role-based access, and the publication of privacy-impact assessments strengthens trust and lowers integration costs, which is particularly important for engaging privacy-sensitive groups. Embedding analytics into contracts and SLAs under M3 - including interoperability and data-access clauses and explicit outcome KPIs - appears to be the most immediate lever for moving beyond pilots, as it forces the alignment of vendor incentives, budget cycles, and operational workflows with data-driven change. Planning for skills and jobs should be integrated into this architecture by mapping each intervention to the competencies it requires and using procurement to stimulate local training and hiring pipelines.

Measurement considerations arise from the same mechanisms. Practical indicators include the share of citizen inputs with publicly visible status and the median time from input to implementation for M1; the publication rate of standards and privacy assessments and the frequency of API audit events for M2; and the proportion of service contracts with explicit data/KPI clauses and the count of workflows consuming platform APIs for M3. Outcome metrics should remain domain-specific - charging sessions or kWh for electromobility, added PV capacity for energy, modal shifts for mobility, and pollutant concentrations for air quality - while green-job formation can be proxied by the volume and content of platform-related tenders, contracted roles, and aligned training enrolments. Such a measurement frame allows cities to monitor progress on mechanisms and outcomes without assuming strict cross-city comparability.

External validity depends on institutional readiness. Transferability is higher where baseline data quality and cataloguing support M2, where procurement frameworks can accommodate interoperability and KPI clauses required by M3, and where civic-tech capacity can sustain public status tracking and implementation logs under M1. Jurisdictions with fragmented data ownership or rigid procurement procedures may benefit from a staged approach - first standardising datasets and API gateways (M2), then deploying transparency tooling (M1), and finally embedding data clauses into contracts and SLAs (M3). Limitations of the present study - reliance on public documentation and non-uniform indicator definitions - recommend caution in causal claims; nevertheless, the consistency of patterns across heterogeneous contexts supports analytic generalisation at the level of mechanisms.

Taken together, the discussion supports an implementation-centric view of participatory smart-city governance. Transparency (M1) and robust data governance (M2) enable credible, inclusive participation, but consistent, scalable sustainability outcomes and broader green-job pathways materialise only when analytics are contractually and operationally coupled to service delivery under M3.

5 Conclusions

The analysis across Amsterdam, Athens, Seville, and Osaka indicates that three governance mechanisms - M1 (transparency of the pipeline from citizen input to implementation), M2 (data-governance robustness), and M3 (alignment of analytics with procurement and operations) - jointly determine whether platform-mediated participation translates into measurable sustainability outcomes and green-job pathways. Among these, M3 emerges as the strongest proximate driver of scaling, while M1 and M2 condition legitimacy, breadth of participation, and durability of effects. Cities that combine high M3 with at least medium levels of M1 and M2 display clearer expansion in domains such as smart charging, AI-supported mobility, energy efficiency, and circularity, alongside more diverse employment pathways that extend beyond operations and maintenance into analytics and engineering roles. Conversely, medium M1 without explicit implementation tracking or contracts, fragmented data governance (M2), and pilots not wired to procurement or KPIs (M3) constrain conversion from citizen input to service change and narrow job effects to low-leverage roles.

The study contributes a mechanism-oriented framework that privileges how platform features and governance choices couple participation to service delivery, offers comparative evidence across diverse institutional contexts, and proposes operational indicators that cities can monitor to assess progress. Practically, the conclusions point to a sequencing logic: codify public traceability to make responsiveness auditable (M1); standardise and secure data assets to enable trustworthy reuse (M2); and embed analytics into contracts, SLAs, and workflows so that data products systematically alter frontline operations (M3). While findings rest on publicly available documentation and are therefore subject to reporting heterogeneity and non-uniform indicator definitions, they support analytic generalisation at the level of mechanisms. Future work should assemble longitudinal contract- and telemetry-based datasets with harmonised outcome definitions to test the strength of the M3 and scaling relationship and to quantify green-job formation with greater precision. Overall, the evidence favours an implementation-centric view of participatory smart-city governance: transparency and robust data governance enable credible participation, but only contractual and operational coupling of analytics ensures consistent, scalable sustainability outcomes and broader green-job creation.

References

1. Toh, C.K.: Tokyo's city sustainability: strategy and plans for net zero emissions by 2050. IET Smart Cities **4**, 81–91 (2022). https://doi.org/10.1049/smc2.12033
2. Przysucha, L., Sulich, A.: Crowdsourcing as a tool for smart city within sustainable development. In: Mercier-Laurent, E., Kayakutlu, G., Owoc, M.L., Wahid, A., Mason, K. (eds.) Artificial Intelligence for Knowledge Management, Energy and Sustainability. pp. 204–219. Springer, Cham (2024)

3. Mercier-Laurent, E., Talens, G., Thivant, E.: Developing a knowledge base on climate change for metropolitan cities. IFIP Adv. Inf. Commun. Technol. **614**, 130–143 (2021). https://doi.org/10.1007/978-3-030-80847-1_9
4. Kasztelan, A., Sulich, A.: Green Transformation of the Polish Economy. Emerald Group Publishing Ltd. (2023). https://doi.org/10.1108/978-1-83753-654-220231004
5. Paasi, J., Wiman, H., Apilo, T., Valkokari, K.: Modeling the dynamics of innovation ecosystems. Int. J. Innov. Stud. **7**, 142–158 (2023). https://doi.org/10.1016/j.ijis.2022.12.002
6. Przysucha, Ł.: Implementation of crowdsourcing as an integral element of communication engineering in the Smart City concept. RISTI - Rev. Iber. Sist. e Tecnol. Inf. 97–108 (2020)
7. Kozar, ŁJ., Matusiak, R., Paduszyńska, M., Sulich, A.: Green jobs in the EU renewable energy sector: quantile regression approach. Energies **15**, 6578 (2022). https://doi.org/10.3390/en15186578
8. Sołoducho-Pelc, L., Sulich, A.: Role of business intelligent systems in sustainable strategic management for green jobs creation. In: Procedia Computer Science, pp. 4461–4469. Elsevier B.V. (2024). https://doi.org/10.1016/j.procs.2024.09.296
9. Sulich, A., Sołoducho-Pelc, L.: The circular economy and the Green Jobs creation. Environ. Sci. Pollut. Res. **29**, 14231–14247 (2022). https://doi.org/10.1007/s11356-021-16562-y
10. Sivarethinamohan, R., Samiksha Reddy, R.: Digital twin for smart city resilience and solutions, pp. 605–619 (2025). https://doi.org/10.1002/9781394303564.ch25
11. Mercier-Laurent, E.: What technology for efficient support of sustainable development? In: Paprzycki, M., Maciaszek, L., Ganzha, M., Maciaszek, L. (eds.) Proceedings of the 2015 Federated Conference on Computer Science and Information Systems, FedCSIS 2015, pp. 1651–1654. Institute of Electrical and Electronics Engineers Inc. (2015). https://doi.org/10.15439/2015F010
12. Weichbroth, P., Wereszko, K., Anacka, H., Kowal, J.: Security of cryptocurrencies: a view on the state-of-the-art research and current developments. Sensors **23** (2023). https://doi.org/10.3390/s23063155
13. Voorwinden, A.: Regulating the smart city in European municipalities: a case study of Amsterdam. Eur. Public Law. **28**, 155–180 (2022). https://doi.org/10.54648/euro2022008
14. Rastvortseva, S.N., Panina, E.V: Municipal digital platforms in the urban socio-economic development (a case study of Amsterdam and Berlin). Sovrem. Evr. 92–104 (2025). https://doi.org/10.31857/S0201708325020081
15. Sulich, A., Zema, T.: Green jobs, a new measure of public management and sustainable development. Eur. J. Environ. Sci. **8**, 69–75 (2018). https://doi.org/10.14712/23361964.2018.10
16. Kutty, A.A., Küçükvar, M., Cihat Onat, N.C., Ayvaz, B., Abdella, G.M.: Measuring sustainability, resilience and livability performance of European smart cities: a novel fuzzy expert-based multi-criteria decision support model. Cities **137** (2023). https://doi.org/10.1016/j.cities.2023.104293
17. Karagkouni, A., Dimitriou, D.: The overton window in smart city governance: the methodology and results for mediterranean cities. Smart Cities **8** (2025). https://doi.org/10.3390/smartcities8030098
18. Sethi, M., Mittal, S., Suwa, A.: Developing a smart tool for integrated climate action planning (ICLAP 2050) in Asia-Pacific cities and its application to Japanese cities, pp. 22–48 (2025). https://doi.org/10.4324/9781003471417-3
19. Irawan, B., Syafrudin, S., Budihardjo, M.A.: A review of smart city initiatives: assessing the integration and impact on sustainable environmental management in urban heritage sites. In: Proceedings - International Conference on Informatics and Computational Sciences, pp. 370–375. Institute of Electrical and Electronics Engineers Inc. (2024). https://doi.org/10.1109/ICICoS62600.2024.10636898

20. Pereira, B.A., Lohmann, G., Houghton, L.: Technology trajectory in aviation: innovations leading to value creation (2000–2019). Int. J. Innov. Stud. **6**, 128–141 (2022). https://doi.org/10.1016/j.ijis.2022.05.001
21. Anagnostopoulos, T.V., Ramson, S.R.J.: A vehicle ride-sharing algorithm assessing passenger satisfaction according to spatial, temporal, and social behavior context based on real data sources. Futur. Transp. **5** (2025). https://doi.org/10.3390/futuretransp5020056
22. Abu-Rayash, A., Dincer, I.: Development and application of an integrated smart city model. Heliyon. **9** (2023). https://doi.org/10.1016/j.heliyon.2023.e14347
23. Renn, O., Goble, R., Kastenholz, H.: How to apply the concept of sustainability to a region. Technol. Forecast. Soc. Change. **58**, 63–81 (1998). https://doi.org/10.1016/S0040-1625(97)00072-3
24. Mattijssen, T.J.M.M., Hennen, W.H.G.J., Buijs, A.E., de Dooij, P., van Lammeren, R.J.A., Walet, L.: Urban greening co-creation: Participatory spatial modelling to bridge data-driven and citizen-centred approaches. Urban For. Urban Green. **94** (2024). https://doi.org/10.1016/j.ufug.2024.128257
25. To, C.K.M.: A quad model for assessing innovation potential: toward a theory of innovation orchestration quality. Int. J. Innov. Stud. **1**, 37–56 (2017). https://doi.org/10.3724/SP.J.1440.101004
26. Turner, C., Oyekan, J.: Personalised production in the age of circular additive manufacturing. Appl. Sci. **13** (2023). https://doi.org/10.3390/app13084912
27. Karaoulis, M.C., Diamanti, N.G., Bremmer, C.N., Slob, E.C., Ngan-Tillard, D.J.M., Karamitopoulos, P.: Application of Ground Penetrating Radar and Electrical Resistivity Tomography for Recognizing Cavities in Critical Urban Areas: The Case Study of Muntplein (Amsterdam, the Netherlands) (2025). https://doi.org/10.1109/IWAGPR65621.2025.11109010
28. Zema, T., Sulich, A., Grzesiak, S.: Charging stations and electromobility development: a cross-country comparative analysis. Energies **16** (2023). https://doi.org/10.3390/en16010032
29. Algaba, M.H.P., Huyghe, W., van Leeuwen, K.J., Koop, S.H.A., Eisenreich, S.J.: Assessment and actions to support integrated water resources management of Seville (Spain). Environ. Dev. Sustain. **26**, 7347–7375 (2024). https://doi.org/10.1007/s10668-023-03011-8
30. Jeleński, T., Dendys, M., Tomaszewska, B., Pająk, L.: The potential of RES in the reduction of air pollution: the SWOT analysis of smart energy management solutions for Krakow functional area (KROF). Energies **13** (2020). https://doi.org/10.3390/en13071754
31. Grzesiak, S., Sulich, A.: Car engines comparative analysis: sustainable approach. Energies **15** (2022). https://doi.org/10.3390/en15145170
32. Łuszczyk, M., Sulich, A., Siuta-Tokarska, B., Zema, T., Thier, A.: The development of electromobility in the European union: evidence from Poland and cross-country comparisons. Energies **14** (2021). https://doi.org/10.3390/en14248247
33. Sulich, A., Zema, T., Kulhanek, L.: Towards a secure future: a bibliometric analysis of the relations between cybersecurity and sustainable development. Procedia Comput. Sci. **225**, 1448–1457 (2023). https://doi.org/10.1016/j.procs.2023.10.133
34. Zeng, Z., Lan, J., Hamidi, A.R., Zou, S.: Integrating Internet media into urban flooding susceptibility assessment: a case study in China. Cities **101** (2020). https://doi.org/10.1016/j.cities.2020.102697
35. Parlapalli, V., Ganeeb, K.K., Bidkar, D.M., Veerapaneni, P.K., Sankiti, S.R., Harve, B.M.: AI-powered smart cities: data-driven approaches for urban innovation and sustainability, pp. 232–237 (2025). https://doi.org/10.1109/CCNCPS66785.2025.11135701

Algorithmic Governance in ESG Reporting: A Comparative Study of AI Tools' Effectiveness in Mitigating Greenwashing

Michał Petri(✉), Iwona Chomiak-Orsa, and Łukasz Przysucha

Wroclaw University of Economics and Business, 53-345 Wroclaw, Poland
{michal.petri,iwona.chomiak-orsa,lukasz.przysucha}@ue.wroc.pl

Abstract. Environmental, Social, and Governance (ESG) reporting has become a pivotal mechanism for promoting transparency and accountability in the pursuit of sustainable development within the global corporate sector. Recent legislative reforms within the European Union, particularly the implementation of mandatory ESG disclosures under the Corporate Sustainability Reporting Directive (CSRD), have significantly transformed corporate reporting expectations and practices. These developments prompt a critical examination of the role of artificial intelligence (AI) in ESG disclosure systems. This article investigates the dual nature of AI as both a facilitator and potential disruptor of transparency in ESG reporting, a phenomenon conceptualized herein as the "transparency paradox." While AI has the potential to enhance data accuracy, automate reporting processes, and enable more comprehensive sustainability assessments, it also introduces challenges related to algorithmic bias, interpretability, and unintended opacity in decision-making. The study critically evaluates how AI systems may both mitigate and exacerbate issues such as greenwashing and unreliable disclosures.

Keywords: ESG reporting · Artificial Intelligence · Transparency Paradox · Algorithmic Bias · Sustainability Disclosure · Regulatory Compliance · Greenwashing

1 Introduction

1.1 Theoretical Background

Corporate reporting has evolved from purely financial disclosures toward comprehensive environmental, social, and governance (ESG) assessments. The European Union has implemented transformative regulatory changes through the Corporate Sustainability Reporting Directive (CSRD), converting ESG reporting from a voluntary initiative into a mandatory requirement.

By applying artificial intelligence (AI) in ESG reporting, we can potentially enhance the accuracy and efficiency of sustainability disclosures. This integration can revolutionize the field, making ESG data more transparent and reliable. However, it also introduces new challenges related to algorithmic bias, data integrity, and ethical considerations, creating the "transparency paradox [1]." This research addresses the following questions:

Published by Springer Nature Switzerland AG 2026
E. Mercier-Laurent et al. (Eds.): AI4S 2025, IFIP AICT 780, pp. 273–282, 2026.
https://doi.org/10.1007/978-3-032-21184-2_21

1. How do current AI techniques mitigate algorithmic bias in ESG reporting?
2. What is the effectiveness of AI technologies in reducing greenwashing compared to conventional methods?
3. What ethical implications arise from using AI for mandatory ESG reporting?
4. How does EU ESG legislation affect the implementation and effectiveness of AI tools in compulsory reporting?

H1: Implementing AI-powered tools in mandatory ESG reporting significantly enhances the effectiveness of detecting and mitigating greenwashing practices compared to traditional reporting and auditing methods, resulting in improved transparency and reliability of ESG disclosures.

The research methodology encompasses a systematic literature review, content analysis of regulatory frameworks, and evaluation of AI implementations. It aims to contribute to developing more effective and trustworthy ESG reporting frameworks in an increasingly digital regulatory environment.

1.2 The Evolution of ESG Reporting and Regulatory Frameworks

As cities aim to become more sustainable and data-driven, smart grids supported by AI technologies are gaining increasing relevance. Recent research highlights the importance of IoT-based monitoring and AI-enhanced analysis in enhancing energy efficiency and reducing emissions in urban environments [2]. ESG reporting has evolved significantly over the past decades, transitioning from voluntary disclosures to increasingly structured and mandatory frameworks. This evolution reflects growing stakeholder demands for transparency regarding organizations' environmental impacts, social responsibilities, and governance practices. Non-financial disclosure regulations, including ESG protocols, aim to bolster transparency and consistency in reported data, with mandatory ESG disclosures potentially stimulating improvements in corporate social responsibility and ESG performance [3]. The European Union's Corporate Sustainability Reporting Directive (CSRD) represents a pivotal development in this evolution, establishing comprehensive requirements for sustainability disclosures and extending the scope of companies obligated to report. This regulatory shift aims to address limitations in previous frameworks and enhance the comparability and reliability of ESG information available to stakeholders, investors, and the public. However, increased volume of ESG disclosures under mandatory regimes does not necessarily translate to improved quality of such disclosures [4]. Strengthening regulatory oversight on ESG disclosures is imperative to counter greenwashing practices. Regulatory initiatives, such as the European Union's Sustainable Finance Disclosure Regulation (SFDR), aim to enhance clarity in ESG reporting and impose penalties for false advertising [5]. Policy-makers must work toward establishing more precise definitions and guidelines surrounding greenwashing to hold companies accountable for misrepresentations.

1.3 AI Applications in ESG Reporting Experiences

Artificial intelligence offers numerous applications within ESG reporting, including data collection and analysis, pattern recognition, anomaly detection, and predictive modeling. These applications could enhance the accuracy, comprehensiveness, and authenticity of ESG disclosures. The digital transformation enabled by AI systems can inhibit greenwashing behaviors by providing robust frameworks for tracking corporate performance and environmental impacts [6, 7]. AI tools can process large volumes of structured and unstructured data, including corporate communications, social media sentiment, and operational metrics, to generate insights into ESG performance. This enables more comprehensive assessments and reduces reliance on superficial marketing typical of greenwashing [5]. Techniques based on large language models (LLMs) and retrieval-augmented generation (RAG) support such advanced data analysis and disclosure practices (Table 1).

Table 1. Comparison of NLP Tools for Greenwashing Detection

Tool/Model	Description	Accuracy
Greenwashing Severity Index (GSI) [8]	Combines NLP techniques with ESG Focus Scores to measure discrepancies between ESG communication and actual performance ("ESG talk" vs. "ESG walk")	(Not applicable (semantic index)
ESG-BER [9]	A BERT-based classifier trained on sustainability reports to detect greenwashing patterns	~82%
ClimateBERTi [10]	Fine-tuned transformer model designed to assess climate-related statements and greenwashing risk	~84%
ReportParse [11]	A parsing tool for extracting structured ESG indicators from corporate sustainability reports	~68%
ChatGPT Comparison [12]	Evaluation of ChatGPT's performance versus keyword-based approaches for ESG data extraction	~74% (with prompt optimization)

2 The Transparency Paradox and Algorithmic Bias

The "transparency paradox" emerges at the intersection of AI and ESG reporting. While AI enhances data processing capabilities and can uncover patterns invisible to human analysts, it introduces challenges related to algorithmic bias, interpretability, and accountability [13]. Note that while AI enables better data collection, reliance on algorithm-driven analyses could introduce biases that distort reported outcomes, leading stakeholders to question the authenticity of the information.

A critical challenge in deploying AI for ESG reporting is the interpretability of its outputs. For AI to effectively support sustainable development, its models must be interpretable and transparent [14]. This insight directly applies to ESG reporting, where opaque algorithms risk undermining stakeholder trust and may conceal rather than expose greenwashing. While AI models, especially those based on machine learning, can identify complex patterns and potential indicators of greenwashing, these models often function as opaque systems—so-called "black boxes". This lack of transparency can undermine trust in ESG disclosures, particularly when stakeholders cannot easily trace or understand how specific insights were derived. Researchers argue in the context of healthcare analytics [15]:

This observation extends beyond medicine. In ESG reporting, too, explainability remains crucial—not only for technical audits but also to satisfy ethical, legal, and public accountability requirements. Leveraging interpretable models or supplementing complex AI systems with explainable frameworks is, therefore, a strategic imperative.

Algorithmic bias in ESG reporting can manifest in various forms, including sample bias, measurement bias, and aggregation bias. These biases can significantly impact the fairness and accuracy of ESG assessments. Organizations must integrate fairness and transparency into the software development lifecycle of AI systems, highlighting that technical tools and engineering methods can support identifying and mitigating biases before their impact on ESG decision-making [16] (Table 2).

Table 2. The Transparency Paradox in AI-Enabled ESG Reporting Source: own elaboration

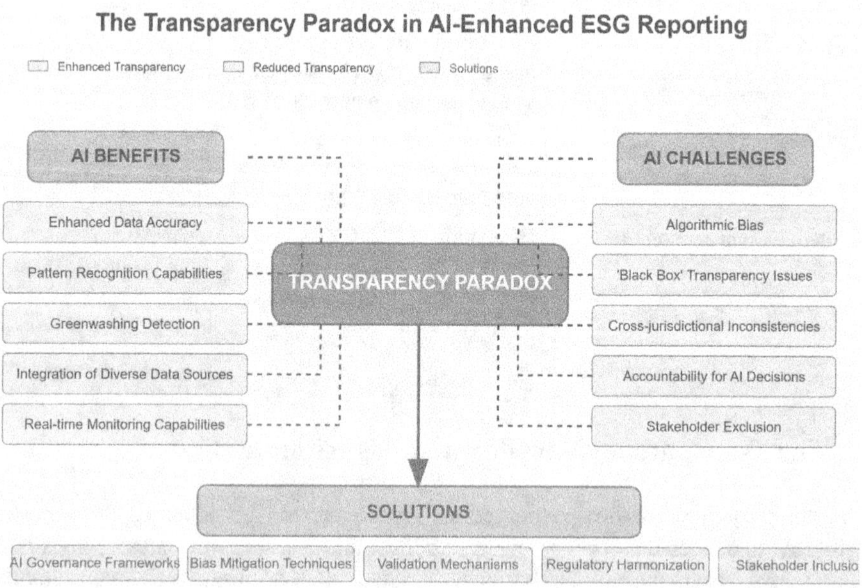

2.1 Ethical Frameworks for AI in ESG Reporting

Ethical considerations are paramount in the application of AI to ESG reporting. These considerations encompass transparency, accountability, fairness, and stakeholder engagement. It is crucial to ensure that AI models are trained on diverse datasets to prevent the embedding of biases into ESG assessments, which could lead to misleading conclusions about a firm's sustainability practices [13]. Various frameworks have been proposed to address these ethical considerations, including AI governance structures, fairness metrics, and transparency mechanisms. Research explores AI fairness measures and highlights the importance of regularly auditing AI systems to assess and correct potential biases [17]. Additionally, establishing transparent frameworks for AI operations that outline methodological approaches and decision logic fosters accountability and trust.

Effective governance in smart urban environments increasingly relies on participatory knowledge exchange mechanisms, such as crowdsourcing [18]. Crowdsourcing serves as a dynamic tool for bridging the information gap between local authorities and stakeholders, facilitating a bidirectional flow of knowledge that supports more informed decisions. This approach not only fosters community engagement but also generates what is termed "Collective Intelligence"—a synthesis of civic and administrative insights that improves public decision-making. These insights are highly transferable to the ESG context, where trust, transparency, and stakeholder alignment are equally critical in combating greenwashing and ensuring accountability.

2.2 European Union Regulatory Framework for AI-Enhanced ESG Reporting

The interpretation and implementation of ESG regulations vary significantly across different jurisdictions, creating challenges for the application of AI in global ESG reporting. Countries are often classified based on their development stages, which directly affects the robustness of their ESG regulations, with developed nations typically having more stringent and well-defined ESG standards [19]. These variations in regulatory frameworks impact data requirements, enforcement practices, and compliance expectations, thereby influencing the effectiveness of AI tools in ESG reporting. Challenges stemming from diverse ESG frameworks result in inconsistencies in reporting and difficulties for AI systems when deployed in varying contexts [20]. The theoretical background presented above provides the foundation for investigating the complex interplay between AI, ESG reporting, and regulatory frameworks. By understanding these theoretical underpinnings, this research aims to contribute to addressing the challenges and leveraging the opportunities presented by AI in mandatory ESG reporting.

3 Methodology

3.1 Research Design

This study employs a mixed-methods research design to comprehensively examine the role of artificial intelligence in mandatory ESG reporting. The research approach combines qualitative and quantitative methodologies to address the complex nature of the research questions. A systematic literature review forms the foundation of the research, supplemented by case studies and content analysis of relevant European Union regulatory frameworks and AI applications in ESG reporting (Table 3).

Table 3. Research procedure

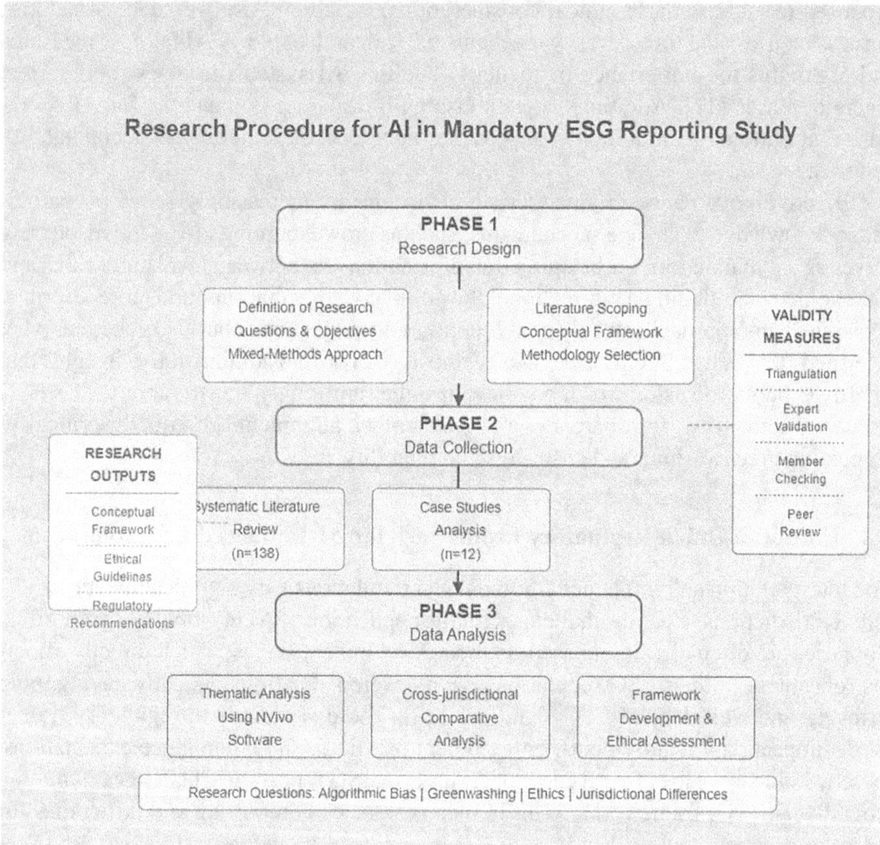

3.2 Systematic Literature Review

A comprehensive systematic literature review established the theoretical foundation of AI applications in ESG reporting. The search utilized the Web of Science, Scopus, IEEE Xplore, and Business Source Complete databases, combining keywords related to artificial intelligence, ESG reporting, and regulatory terms. From an initial 487 publications (2015–2024), 138 relevant studies were selected for in-depth analysis. Content analysis was conducted on EU ESG reporting regulations, specifically focusing on the CSRD, ESRS, and the European Green Deal framework, to identify requirements related to data quality, transparency, verification, and technological applications.

4 Data Analysis

The landscape of cross-jurisdictional ESG reporting presents several critical challenges:

- Regulatory Fragmentation: Our analysis revealed that the lack of a standardized approach to ESG reporting across jurisdictions creates significant compliance hurdles. This inconsistency complicates implementation, particularly for multinational corporations that must navigate diverse regulatory requirements across different jurisdictions. Regional variability increases the risk of greenwashing, where companies may selectively adhere to less stringent standards or present misleading information [21].
- Data Standardization Challenges: Perhaps the most frustrating obstacle we encountered was the inconsistency in metrics and reporting formats across regions. This variance makes meaningful comparative analysis nearly impossible in specific sectors. For instance, when examining climate-related disclosures, we found that carbon accounting methodologies varied significantly between the EU and the Asia-Pacific regions, creating data gaps that AI systems struggled to reconcile. These inconsistencies not only hinder algorithmic performance but also undermine stakeholder trust in the resulting assessments.
- Harmonization Efforts: While the challenges are substantial, I'm cautiously optimistic about several emerging initiatives. The work of the International Sustainability Standards Board (ISSB) shows particular promise, though its adoption remains incomplete. Industry-led interoperable data formats are also gaining traction, especially in the financial services sector. A framework for cross-border regulatory collaboration offers a practical pathway forward—but will regulators adopt it? The next few years will be crucial in determining whether these harmonization attempts can overcome entrenched regional differences [22].

5 The Duality of AI in ESG Transparency

Our research uncovered what I've come to call the "transparency paradox" in AI-driven ESG reporting. While AI systems significantly enhance analytical capabilities and data processing, they also introduce new forms of opacity that can erode stakeholder trust. This paradox manifested differently across our case studies. Large financial institutions, for instance, leveraged sophisticated AI tools that excelled at identifying subtle patterns in sustainability data, yet struggled to explain these insights to non-technical board members. In contrast, a mid-sized manufacturing firm we examined implemented a less advanced but more comprehensible system that, while processing less data, generated more actionable insights for stakeholders [23]. The framework for explainable AI offers valuable guidance here, although I found it needed adaptation for the specific contexts of ESG reporting [24]. Organizations that successfully navigate this tension typically implement transparent methodologies and maintain clear documentation, essentially demystifying the "black box" without sacrificing analytical power. But is this balance achievable at scale? That remains an open question worthy of further investigation.

This tension between predictive power and interpretability is not unique to ESG reporting. In the context of smart city systems, integrating artificial intelligence with

Digital Twin frameworks enables the creation of simulations that can forecast system behavior under varying scenarios [25]. Such models contribute to planning, diagnostics, and innovation by providing stakeholders with both predictive insight and operational clarity. Although applied to urban parking management, this framework has clear analogues in ESG contexts, where simulation-based AI could model and detect patterns of risk, such as greenwashing, while maintaining an interpretable and auditable process.

6 Key Findings

Our research yielded four principal findings that advance understanding of AI's role in mandatory ESG reporting:

First, effective bias mitigation in AI-driven ESG assessment requires an integrated approach. Organizations demonstrating best practices combine robust governance structures, technical solutions such as diverse training datasets, and external validation methods. Interestingly, we found that governance frameworks had the most significant impact; even relatively simple AI systems performed well when governed by thoughtful and inclusive processes.

Second, AI systems offer substantial advantages over traditional methods in detecting greenwashing, though significant limitations persist. The pattern recognition capabilities we documented (78% accuracy vs. 45% for manual reviews) represent a significant step forward in effectiveness. Still, challenges with qualitative assessment and strategic manipulation remain considerable barriers to complete reliability.

The effectiveness of AI in preventing greenwashing varies considerably by sector. The technology showed particular promise in environmentally intensive industries with quantifiable outputs, while struggling more with social dimensions of ESG that resist simple metrics.

Finally, regulatory fragmentation emerged as perhaps the most significant barrier to effective AI implementation across jurisdictions. The inconsistencies we documented not only complicate compliance but fundamentally undermine the comparative potential of AI-driven assessments.

6.1 Concluding Remarks

As I reflect on this research, I'm struck by both the promise and the challenges of integrating AI into mandatory ESG reporting frameworks. Incorporating sustainability into corporate governance requires a systematic navigation framework that aligns strategic, tactical, and operational levels of decision-making [20]. This model supports coherent, performance-driven sustainability implementation—an approach directly applicable to structuring ESG governance systems and AI-supported greenwashing detection. The technology offers unprecedented capabilities to enhance transparency and accountability, yet its implementation requires careful consideration of ethical dimensions and jurisdictional contexts.

What seems clear is that the path forward isn't purely technological. The most successful cases we examined balanced technological advancement with robust governance frameworks and stakeholder engagement. This balanced approach, rather than either uncritical AI adoption or excessive caution, offers the most promising route to genuinely sustainable corporate reporting.

The coming years will be formative for this field. As regulatory frameworks mature and AI capabilities evolve, the relationship between these technologies and ESG transparency will continue to develop. We hope is that this research contributes to shaping that relationship in ways that prioritize authentic sustainability and meaningful accountability rather than mere compliance or technological sophistication for its own sake.

Disclosure of Interests. The authors have no competing interests to declare that are relevant to the content of this article.

References

1. Hu, P., Zeng, Y., Wang, D., Teng, H.: Too much light blinds: the transparency-resistance paradox in algorithmic management. Comput. Hum. Behav. **161**, 108403 (2024). https://doi.org/10.1016/j.chb.2024.108403
2. Petri, M., Chomiak, I.: Towards a sustainable future: the use of electrical energy in smart cities, pp. 216–222 (2024)
3. Cicchiello, A.F., Marrazza, F., Perdichizzi, S.: Non-financial disclosure regulation and environmental, social, and governance (ESG) performance: the case of EU and US firms. Corp. Soc. Responsib. Environ. Manag. **30**, 1121–1128 (2022). https://doi.org/10.1002/csr.2408
4. Arvidsson, S., Dumay, J.: Corporate ESG reporting quantity, quality and performance: where to now for environmental policy and practice? Bus. Strategy Environ. **31**, 1091–1110 (2022)
5. Juthi, S., Kamrujjaman, M., Mistry, A., Alauddin, M.: Sustainable finance and data analytics: a systematic review of ESG data in investment decisions. Academic J. Bus. Adm. Innov. Sustain. **4**, 70–88 (2024). https://doi.org/10.69593/ajbais.v4i04.130
6. Weichbroth, P., Jandy, K., Zurada, J.: Toward sustainable development: exploring the value and benefits of digital twins. Telecom **5**, 774–791 (2024). https://doi.org/10.3390/telecom5030039
7. Maciej, P., et al.: AI tools for knowledge management – knowledge base creation via LLM and RAG for AI assistant. In: Marcin, H., et al. (eds.) Emerging Challenges in Intelligent Management Information Systems, pp. 3–15. Springer, Cham (2024)
8. Lagasio, V.: Measuring Greenwashing: The Greenwashing Severity Index (2023)
9. Lee, H., Lee, S.H., Park, H., et al.: ESG2PreEM: automated ESG grade assessment framework using pre-trained ensemble models. Heliyon **10**, e26404 (2024). https://doi.org/10.1016/j.heliyon.2024.e26404
10. Schimanski, T, Bingler, J., Hyslop, C., et al.: ClimateBERT-NetZero: detecting and assessing net zero and reduction targets. SSRN Electron. J. (2023). https://doi.org/10.2139/ssrn.4599483
11. Morio, G., In, S.Y., Yoon, J., et al.: ReportParse: a unified NLP tool for extracting document structure and semantics of corporate sustainability reporting, pp. 8749–8753 (2024)
12. Guo, B., Zhang, X., Wang, Z., et al.: How Close is ChatGPT to Human Experts? Comparison Corpus, Evaluation, and Detection (2023)
13. Sætra, H.S.: AI in context and the sustainable development goals: factoring in the unsustainability of the sociotechnical system. Sustainability **13**, 1738 (2021)
14. Mercier-Laurent, E.: Can artificial intelligence effectively support sustainable development? In: Artificial Intelligence for Knowledge Management: AI4KM 2021, pp. 15–28. Springer, Cham (2021)

15. Butryn, B., Chomiak-Orsa, I., Hauke, K., et al.: Application of machine learning in medical data analysis illustrated with an example of association rules. Procedia Comput. Sci. **192**, 3134–3143 (2021)
16. Batool, A., Zowghi, D., Bano, M.: AI governance: a systematic literature review. AI Ethics 1–15 (2025)
17. Ferrara, E.: Fairness and bias in artificial intelligence: a brief survey of sources, impacts, and mitigation strategies. Sci **6**, 3 (2023)
18. Melnykova, M., Gradoboieva, Y., Mirzodaieva, T., Ragulina, N.: Complex modernization of public infrastructure and hospitality as a factor in the sustainable development of the city in Ukraine. Eur. J. Sustain. Dev. **9**, 183–204 (2020). https://doi.org/10.14207/ejsd.2020.v9n1p183
19. Singhania, M., Saini, N., Shri, C., Bhatia, S.: Cross-country comparative trend analysis in ESG regulatory framework across developed and developing nations. Manage. Environ. Qual.: Int. J. **35**, 61–100 (2023). https://doi.org/10.1108/MEQ-02-2023-0056
20. Paridhi, A.A.: Sustainability reporting: current state and challenges. Bus. Strategy Dev. **6**, 362–381 (2023). https://doi.org/10.1002/bsd2.244
21. Kuok Ho, D.T.: A review of environmental, social and governance (ESG) regulatory frameworks: their implications on Malaysia. Trop. Aquatic Soil Pollut. **3**, 168–183 (2023). https://doi.org/10.53623/tasp.v3i2.282
22. Nazer, L.H., et al.: Bias in artificial intelligence algorithms and recommendations for mitigation. PLOS Digit Health **2**(6), e0000278 (2023). https://doi.org/10.1371/journal.pdig.0000278
23. Atf, Z., Lewis, P.: Is trust correlated with explainability in AI? A meta-analysis. IEEE Trans. Technol. Soc. 1–8 (2025). https://doi.org/10.1109/TTS.2025.3558448
24. van Wynsberghe, A.: Sustainable AI: AI for sustainability and the sustainability of AI. AI Ethics **1**, 213–218 (2021). https://doi.org/10.1007/s43681-021-00043-6
25. Chomiak-Orsa, I., Hauke, K., Perechuda, K., Pondel, M.: The use of Digital Twin in the sustainable development of the city on the example of managing parking resources. Procedia Comput. Sci. **225**, 2183–2193 (2023). https://doi.org/10.1016/j.procs.2023.10.209

Technological Challenges, Energy Poverty and Other Barriers to the Expansion of Energy Communities in Europe

Evaldo Costa[1](✉) and Eunika Mercier-Laurent[2]

[1] Iscte-Instituto Universitário de Lisboa, DINAMIA'CET, Av. das Forças Armadas, 1649-026 Lisbon, Portugal
Jose.Costa@iscte-iul.pt

[2] CReSTIC Centre de Recherche en Sciences et Technologies de l'Information et de la Communication - EA 3804, University of Reims Champagne Ardenne, Reims, France
eunika.mercier-laurent@univ-reims.fr

Abstract. This article investigates how renewable energy communities (RECs) are adopting cutting-edge technologies in Europe, what barriers they face to their expansion, and how these aspects are linked to energy poverty. Using a systematic literature review (SLR) approach, the research addresses three central questions related to the use of advanced technologies by RECs, examines the main barriers affecting the implementation of technologies by communities, and how the use of these technologies in the development of energy communities contributes to alleviating energy poverty. The results indicate that technologies such as smart grids and digital platforms are the most frequently adopted, while newer solutions such as blockchain, IoT, VPPs, and battery energy storage remain underutilized, requiring targeted policy support. Financial and technical barriers stand out as the main challenges, especially for low-income communities, limiting access to resources and delaying the implementation of RECs. Regulatory barriers also emerge as significant impediments to the progress of RECs in the European space. Despite the growing body of research, the limited number of publications and the uneven distribution of RECs across Europe—especially in southern countries with high renewable potential—reveal structural gaps. The study highlights the need for greater attention to supporting research and innovation, which can be achieved by harmonizing specific policies to promote inclusive and low-carbon energy transitions. Future studies are encouraged to investigate the technological maturity and scale of deployment to support the proliferation of RECs and reduce energy poverty.

Keywords: Renewable Energy Communities (RECs) · Advanced Energy Technologies · Energy Poverty · Systematic Literature Review (SLR) · Smart Grids and Digital Platforms · Policy and Regulatory Barriers · Inclusive Energy Transition

Published by Springer Nature Switzerland AG 2026
E. Mercier-Laurent et al. (Eds.): AI4S 2025, IFIP AICT 780, pp. 283–292, 2026.
https://doi.org/10.1007/978-3-032-21184-2_22

1 Introduction

Renewable Energy Communities (RECs) enable citizens to actively engage in the transition to clean energy sources. They drive the spread of renewable energy sources (RESs) while strengthening energy autonomy and resilience at the local level. By decentralizing energy systems, they help reduce dependence on fossil fuels and large utilities. Valued in European policies, they play a key role in meeting climate targets and promoting social inclusion [1]. The development of RECs focused on decentralization, decarbonization and democratization of the energy sector is crucial for energy transitions in Europe [1]. By fostering stakeholder collaboration, RECs accelerate the adoption of clean energy sources while increasing local energy resilience and autonomy [2].

Formed by citizens, local businesses and public authorities, RECs are using emerging technologies to increase energy efficiency, encourage energy sharing and address energy poverty [3]. Therefore, cutting-edge technologies are essential for the development of RECs, especially when the goal is to optimize local energy generation, sharing and consumption in a sustainable manner. Advanced technologies [4] serve as the backbone that makes modern energy communities technically viable, scalable and effective in Europe's transition to a clean energy mix [1, 5].

The adoption of technologies such as smart grids, blockchain, digital platforms, internet of things (IoT), energy storage, and artificial intelligence (AI) [6] has been fundamental to innovating business models [7] and creating new facilities for the efficient functioning of energy communities. These technologies enable, among other things, the intelligent management of production and consumption; the fair sharing of energy among members; the integration of intermittent RES; And the reduction of costs and increased access to energy [1, 8, 9].

Advanced technologies integrated into RECs have shown themselves to be valuable allies in tackling low-carbon energy poverty, while also fostering social justice, environmental sustainability, and grid resilience [10]. While renewable energy is essential for climate goals, recent studies show it does not automatically reduce energy poverty in Europe. Structural issues—such as housing quality, energy pricing, and income inequality—remain major barriers, calling for targeted policies to improve efficiency and affordability for vulnerable households. At the same time, a novel REC framework based on Italian case studies highlights the potential of energy communities as scalable tools for deep building renovations in urban areas. This structured model offers a clear pathway to support policymakers in expanding RECs while addressing environmental goals and socioeconomic disparities simultaneously [11, 12].

The relationship between communities and energy poverty involves several perspectives [3, 12]. For example, from a cost-saving standpoint, RECs represent feasible alternatives that can support households (particularly those with lower incomes) by lowering electricity expenses. By generating power locally through shared solar installations, transmission and distribution charges are reduced. Additionally, the cooperative business models of RECs enable members to participate in the resulting benefits of the system, such as revenue from energy production, which helps in making tariffs more accessible [10, 13].

In terms of energy access, RECs also play an important role, especially in areas not served by the electricity grid or where the existing infrastructure is precarious.

Through microgrids or off-grid systems, isolated rural areas or urban neighborhoods with fewer resources can gain access to RES. The benefits can also reach electricity grid operators, who often do not invest in expanding the traditional grid due to economic unfeasibility [1]. Inclusion in renewable energy systems is another point with a direct impact, especially for low-income families who do not have the financial resources to install solar systems in their homes. In this scenario, RECs act as instruments for democratizing access to energy and as allies in sustainability goals, by also contributing to the reduction of carbon emissions and the risks associated with climate change [3, 10].

A further equally important aspect is the fact that decentralized energy generation, promoted by RECs, contributes to reducing the overload of traditional networks, which favors the stability of the electrical grid, as well as increasing resilience in cases of failures or natural disasters [10]. In addition, the use of storage and demand management technologies is crucial to promoting the balance between energy supply and consumption, providing greater reliability and efficiency to the system. It is worth mentioning that RECs also promote social empowerment through the involvement of prosumers in governance and decisions on how to produce and consume energy [10, 11, 14].

However, the transition towards sustainable energy development faces significant obstacles, especially regarding the large-scale expansion of RECs in Europe. These challenges stem from difficulties in technological integration, high costs of equipment and components, outdated infrastructure and lack of supporting policies [13]. In addition to the technological barrier, there are regulatory, institutional, and socioeconomic aspects that make it difficult to combat energy poverty [14].

This study explores and synthesizes how RECs in European countries are adopting technologies, what are the main barriers to their expansion and how these issues relate to energy poverty, in addition to answering three questions: RQ#1: What are the most widely used technologies in energy communities in Europe? RQ#2: How are these technologies applied to reduce energy poverty in the context of energy communities? RQ#3: What are the main barriers that affect the implementation of technologies for the development of energy communities in European countries? In addition, the study aims to provide practical insights to optimize the use of technological solutions, promote energy inclusion and overcome technical and socioeconomic obstacles, contributing to the sustainable development of RECs in Europe space.

This work fills a gap in literature, considering the still limited number of studies that analyze, in an integrated manner, energy sharing from the perspectives of barriers to expansion, energy poverty and technological development. The findings can contribute to European energy planning and favor more inclusive routes towards a decentralized and low-carbon energy transition.

The next section presents the methodology used to describe the approach adopted in conducting the study. The results will offer answers to the research questions, with relevant insights into cutting-edge technology, energy poverty, and the barriers. Finally, the study's main conclusions will be presented.

2 Methodology

To understand how RECs in European countries are using technologies, identify the main barriers to their expansion and their connections with energy poverty, as well as answer the research questions, this study adopts a systematic approach [15] to identify, select and analyze scientific publications that address these topics.

The search was conducted in the Scopus and Web of Science (WoS) databases, which are widely used for scientific studies, covering the period from 2010 to 2025. Key terms related to renewable energy communities, emerging technologies (such as "smart grids", "blockchain", "IoT", "AI", "digital platforms", "energy storage", "virtual power plants"), energy poverty ("cost savings", "targeted access", "renewable energy access", "grid stability") and barriers ("economic barriers", "policy barriers", "technical barriers", "institutional barriers", "social barriers") were used.

Only peer-reviewed English-language documents (articles and reviews indexed in WoS and Scopus) were considered. Studies were included if they: Focused on European countries; Discussed the integration of technologies in RECs; Presented empirical, theoretical or conceptual approaches; Addressed at least one of the following topics: energy sharing, barriers to expansion or energy poverty.

Studies were excluded if they: Did not address energy communities; Did not mention technology adoption; Focused exclusively on non-European countries without comparative relevance; Were not in English or were not indexed (such as editorials, opinions or grey literature). The selection process in high-impact journals followed the PRISMA protocol, as illustrated in Fig. 1.

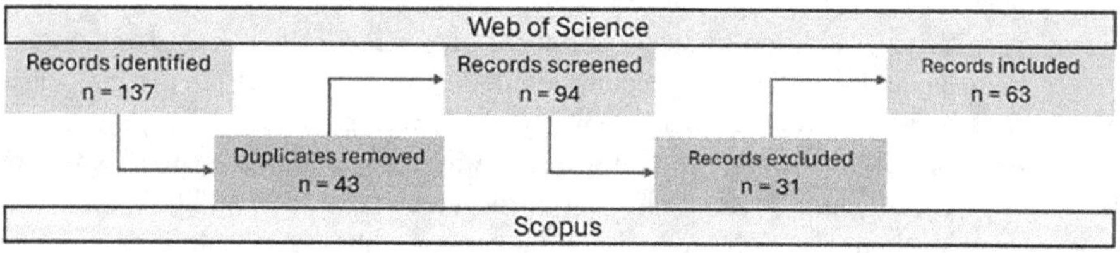

Fig. 1. Scheme of analysis

The database was categorized into four groups. The first group identifies the "year of publication", the "platform" from which the study was downloaded, and the "country" to which the study refers. The second group focuses on the technologies most adopted in REC studies, including "smart grids", "blockchain", "digital platforms", "AI", "virtual power plants" (VPPs), "energy storage", "IoT", and others. The third group relates to energy poverty targets, classified as "cost savings", "targeted access", "renewable energy access", "grid stability", and other aspects. The fourth group addresses the main barriers to the expansion of RECs, categorized as "financial & economic barriers", "regulatory & policy barriers", "technical & infrastructure barriers", "institutional & governance barriers", and "social & behavioral barriers".

3 Results

This section is presented in three main lines of analysis, starting with a general overview and narrowing down on issues related to technology, energy poverty and the main barriers to the expansion of energy communities.

3.1 Context of the Analysis Covering Temporal, Geographic and Bibliographic Scope

The "European Country" category, which groups together studies labeled as "General" and "European countries", has appeared with significant frequency over the years. This indicates a strong trend of research with a pan-European focus, addressing energy communities and energy poverty in an aggregated manner, without being restricted to a specific country. In addition, this predominance can be partly explained by the presence of conceptual studies, which explore the topic in a more theoretical or comprehensive way, without specific geographical delimitation. These studies tend to discuss models, policies and technologies applicable in multiple European contexts, contributing to the increase in the incidence of analyses classified as "European Country" (Fig. 2).

Temporal Distribution: There has been an increase in the number of countries studying energy communities since 2020, which may be related to the growth of collaborative energy policies in the EU, such as the European Green Deal, launched in 2019, and the post-COVID-19 recovery plans launched in 2020. In addition, the presence of studies in recent years (2023–2025) suggests that the topic remains high on the scientific agenda (Fig. 2).

Scientific studies on energy communities and energy poverty are well represented in high-impact platforms like WoS and Scopus, ensuring wide international dissemination. Countries such as Italy, the Netherlands, Spain, Germany, and Portugal appear frequently in the research, with southern European nations showing strong interest in solar and wind energy due to favorable climates. Notably, in 2024, Spain sourced over 50% and Portugal 71% of their electricity from renewables. Many studies adopt a comparative or collaborative approach across multiple countries.

3.2 Analysis of Technologies, Energy Poverty and Barriers

Here, the three research questions will be answered: technology, energy poverty and barriers (Fig. 2). Technology: The predominance of "smart grids", "energy storage", and "digital platforms" is consistent with the advancement of this technology in the European energy sector, especially following the launch of the European Green Deal and funding programmes such as Horizon Europe. Furthermore, the increasing use of "VPP" reflects the need to address the intermittency of renewable sources such as solar and wind, which is widely documented in EU technical studies and policies (Fig. 3).

It is worth noting that the lower presence of technologies such as "IoT" and "Blockchain" may not reflect their real adoption, but rather an underrepresentation in the studies analyzed, especially due to the substantial presence of conceptual or exploratory articles in the sample (Fig. 3).

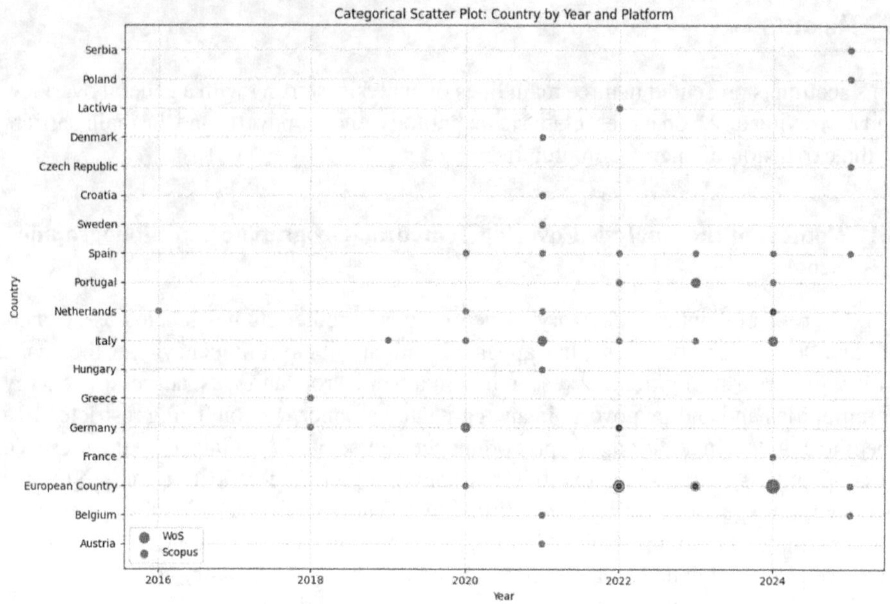

Fig. 2. Broad perspective on energy community engagement

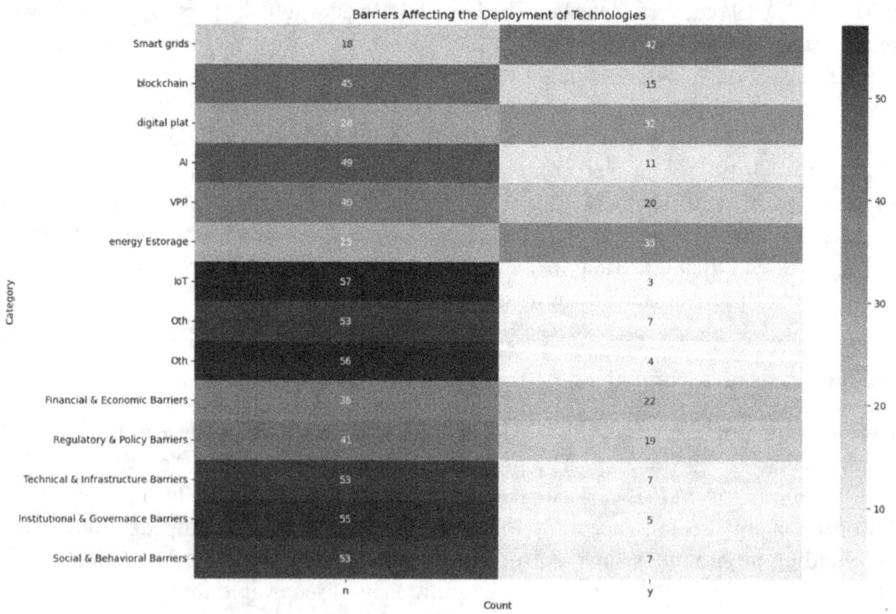

Fig. 3. Exploring technology and key barriers

Key barriers: "Financial" and "regulatory" barriers are widely recognized in literature as the main obstacles to the expansion of energy communities, especially in countries with lower investment capacity (Fig. 3). The presence of "technical" barriers also reinforces the need for harmonization of policies and incentives at European level.

Energy poverty: The heatmap (Fig. 4) analysis reveals key patterns in how different barriers affect aspects of energy poverty. Cost is the most frequently impacted aspect. RES faces a combination of financial, regulatory, and social barriers, indicating that high costs, legal complexity, and community resistance are major obstacles. Grid stability also depends heavily on financial and regulatory support, underscoring the need for infrastructure investment and policy alignment. Access to energy is notably impacted by financial barriers, highlighting affordability as a persistent challenge for many families. These findings reinforce the importance of public policies that provide financing mechanisms, social tariffs, and inclusive cooperative models to support vulnerable populations.

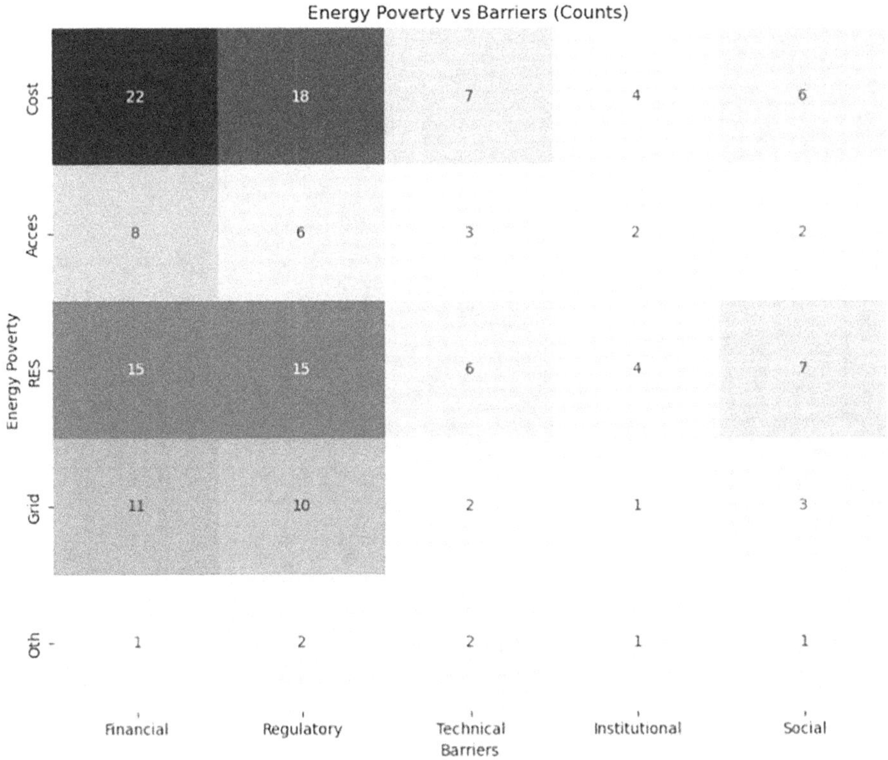

Fig. 4. Exploring energy poverty and key barriers

3.3 Technological Trends in Different National Contexts and Over Time

The initial interpretation of the results of the technology analysis reveals the consistency of "smart grids", energy store, and "digital platforms" as the most addressed solutions over the years. Technologies such as "blockchain", "AI" and "IoT" have shown more recent growth, reflecting emerging trends in the sector. The growing emphasis on "energy storage" may be associated with the greater integration of RES into energy systems (Fig. 5).

An analysis of the performance results of technologies over different periods reveals a progressive increase in interest in technologies applied to energy communities, especially since 2020. This advance coincides with the adoption of more ambitious European policies, such as the European Green Deal, the strengthening of incentives for a fair and decentralized energy transition, and the increase in funding for technological innovation projects. The growing number of studies highlights scientific mobilization around the modernization of the energy sector (Fig. 5).

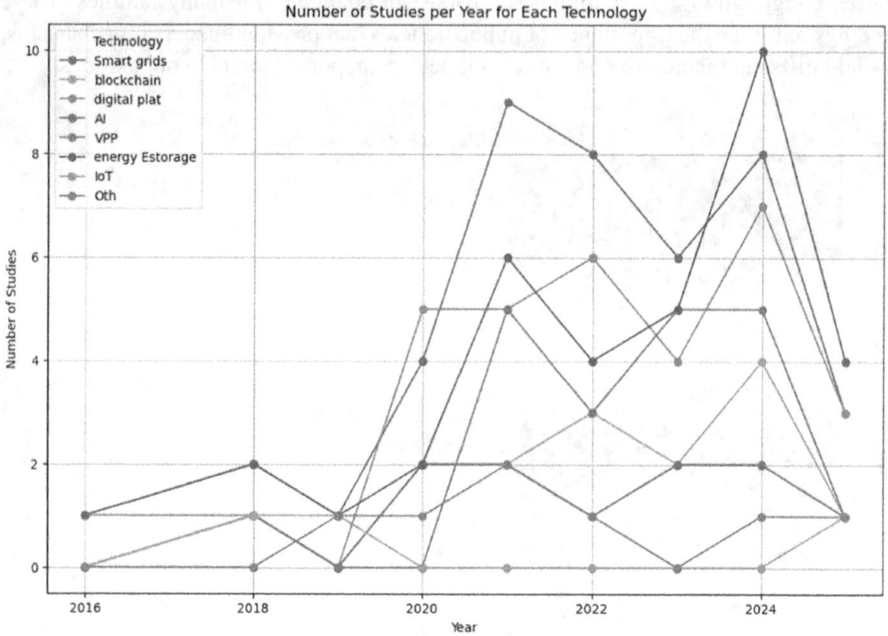

Fig. 5. Technological overview by country and period

Among the most recurrent technologies, "smart grids" stand out, which consistently lead due to their essential role in the integration of renewable sources and the modernization of electricity networks. "Digital platforms" are also gaining prominence, reinforcing the relevance of "digitalization" for management and engagement in energy communities. "Energy storage" has shown more recent growth, reflecting the need to deal with the intermittency of sources such as solar and wind. Emerging technologies, such as "blockchain", "AI" and the "IoT", although still in the early stages of application, show upward curves and potential for expansion. The presence of "VPPs", albeit occasional, signals the beginning of research into the aggregation of distributed energy resources (Fig. 5).

4 Conclusion

To understand how RECs in European countries are adopting technologies, the main barriers to their expansion and how both factors interfere in energy poverty, this research uses an SLR approach. One of the conclusions of the study highlights the crucial role of RECs in the development and democratization of access to energy, especially in view of the challenges of promoting a fair, inclusive and sustainable energy transition in the European space.

The study also reveals a limited interest in the development of energy communities in Europe, as shown by the still modest number of published studies. Although this number has grown in recent years, it remains low considering the benefits of the model for inclusive, low-carbon development. This situation reflects the limited presence of RECs in the region.

Although technologies such as "smart grids" and "digital platforms" are well-associated with energy communities, newer solutions such as "blockchain", "VPP", "IoT" and "battery energy storage" still require greater attention from incentive policies so that they can be used more widely. This is essential to generate efficiency gains, reduce energy waste and boost adoption by communities, promoting significant socioeconomic benefits.

A similar situation occurs regarding barriers, especially financial and technical ones, that hinder the expansion of energy communities. Without specific policies to address the high initial investment costs and limited access to financial resources—especially among the less privileged classes—it will be difficult to promote the accelerated development of RECs and achieve the expected benefits of decarbonization and democratization of the energy sector.

The research and innovation (R&I) area must have structured financial support to foster research, especially those focused on the application of cutting-edge technologies, financial engineering and public-private partnership models. This is essential to strengthen the interest of the scientific community, increase the level of collaboration between the several stakeholders and support the proliferation of RECs in the European space.

Some limitations of this study should be considered, such as the equal treatment of technologies and barriers, without distinction of relative impact, as well as the presence of studies that, although relevant, do not clearly identify the country analyzed. These aspects were treated with caution so as not to compromise the results. However, future work could explore the technological maturity or the scale of implementation of the most recurrent solutions in the development of energy communities.

Disclosure of Interests. The authors have no competing interests to declare that are relevant to the content of this article.

References

1. Sokołowski, M.M.: Renewable and citizen energy communities in the European Union: how (not) to regulate community energy in national laws and policies. J. Energy Nat. Resour. Law **38**(3), 289–304 (2020)

2. Costa, E., Vicente, P., Reis, E.: Driving innovation through cross-sector collaboration with higher education institutions. In: INTED2025 Proceedings, IATED, pp. 2045–2053 (2025)
3. Hanke, F., Guyet, R., Feenstra, M.: Do renewable energy communities deliver energy justice? Exploring insights from 71 European cases. Energy Res. Soc. Sci. **80**, 102244 (2021)
4. Mercier-Laurent, E., Kayakutlu, G.: Artificial Intelligence for Knowledge Management, Energy and Sustainability (2022)
5. Costa, E.: Industry 5.0 and SDG 9: a symbiotic dance towards sustainable transformation. Sustain. Earth Rev. **7**(1), 4 (2024)
6. Mercier-Laurent, E.: Artificial intelligence, sustainability and climate change. In: IFIP International Workshop on Artificial Intelligence for Knowledge Management, pp. 1–10. Springer, Cham (2023)
7. Costa, E., Wells, P., Wang, L., Costa, G.: The electric vehicle and renewable energy: changes in boundary conditions that enhance business model innovations. J. Clean. Prod. **333**, 130034 (2022)
8. Costa, E., Teixeira, A.C.R., Costa, S.C.S., Consoni, F.L.: Influence of public policies on the diffusion of wind and solar PV sources in Brazil and the possible effects of COVID-19. Renew. Sustain. Energy Rev. **162**, 112449 (2022)
9. Mercier-Laurent, E.: The future of artificial intelligence: empowering humanity and help protecting planet. In: Computer Sciences & Mathematics Forum, MDPI, p. 38 (2023)
10. Wu, Y., Wu, Y., Cimen, H., Vasquez, J.C., Guerrero, J.M.: Towards collective energy Community: potential roles of microgrid and blockchain to go beyond P2P energy trading. Appl. Energy **314**, 119003 (2022)
11. Ahmed, S., Măgurean, A.M.: Renewable energy communities: towards a new sustainable model of energy production and sharing. Energ. Strat. Rev. **55**, 101522 (2024)
12. Makridou, G., Matsumoto, K., Doumpos, M.: Evaluating the energy poverty in the EU countries. Energy Econ. **140**, 108020 (2024)
13. Radtke, J., Wurster, S.: Multilevel governance of energy transitions in Europe: addressing wicked problems of coordination, justice, and power in energy policy. Zeitschrift für Politikwissenschaft **33**(2), 139–155 (2023)
14. Vardanyan, Y.: Energy communities in sweden: barriers and driving forces. Curr. Sustain./Renew. Energy Rep. **12**(1), 10 (2025)
15. Becheikh, N., Landry, R., Amara, N.: Lessons from innovation empirical studies in the manufacturing sector: a systematic review of the literature from 1993–2003. Technovation **26**(5–6), 644–664 (2006)

Towards Sustainable, AI-Powered Agriculture: Emerging Technologies, Benefits and Barriers

Marios Paraskevopoulos[1], George Routis[1], and Ioanna Roussaki[1,2](✉)

[1] National Technical University of Athens, Athens, Greece
ioanna.roussaki@cn.ntua.gr
[2] Institute of Communication and Computer Systems, Zografou, Greece

Abstract. Agriculture faces pressures from climate change, resource scarcity, biodiversity loss, and increasing food demand, intensifying the need for sustainable and resilient transformation. Artificial intelligence has emerged as a key part of this transition, offering advanced capabilities for precision agriculture, resource management, supply-chain optimization, and data-driven decision-making. This chapter examines the role of AI in supporting sustainability across the agri-food sector, thoroughly explaining its main benefits, including increased efficiency and productivity, environmental and socio-economic gains, and improved risk assessment and resilience. Following, we critically assess the challenges and barriers that hinder broader AI adoption, including infrastructural, regulatory, economic, and cultural constraints that restrict equitable deployment. Next, the chapter presents a State-of-the-Art review of contemporary AI solutions, encompassing machine learning, remote sensing, IoT-based monitoring, and other emerging multi-modal systems. Finally, the need for ethical AI brings forth some considerations on fairness, inclusivity, transparency, accountability, digital divide, and social conditions. Together, these analyses offer a holistic perspective on how AI can foster sustainable agriculture, provided that there are proper, responsible governance and alignment with broader ecological and societal goals.

Keywords: Artificial Intelligence · Sustainable Agriculture · Digital Transformation · Precision Agriculture · Data Sovereignty · Interoperability · Ethical AI · Food Security · Machine Learning · Remote Sensing · Pest Management · Decision Support Systems · Supply Chain · Climate-Smart Agriculture · Agri-Food Systems · SDGs

1 Introduction

Climate change, global demographic growth, resource scarcity, geopolitical instability, and biodiversity loss are only some of the prominent challenges the world faces. So, what is the common point here? All of them put unprecedented pressure on food systems, severely challenging their resilience capabilities and agriculture stands in the spotlight. Recent insights from the *OECD-FAO Agricultural Outlook* [1] are indicative of the situation, despite only being projections: Population is set to keep growing with an

Published by Springer Nature Switzerland AG 2026
E. Mercier-Laurent et al. (Eds.): AI4S 2025, IFIP AICT 780, pp. 293–306, 2026.
https://doi.org/10.1007/978-3-032-21184-2_23

average rate of 0.8% per year pushing the food demand further. At the same time, global consumption of agri- and aquaculture commodities will increase by 13% and the caloric intake per capita by 6%. Interestingly, developing and middle-income countries make a major contribution to these numbers, while low-income ones still fail to reach the necessary per capita intake and a number of 700 million people is estimated to be suffering from chronic undernourishment. Nevertheless, increased consumption comes at a high environmental cost, as greenhouse gas emissions are projected to increase by 6%. Even though per unit emissions improve, the overall footprint will still be negative. It is estimated that agriculture accounts for about 1/3 of global greenhouse gas emissions. Data calls for urgent measures towards a sustainable agri-food sector.

In this demanding and quite complicated landscape, Artificial intelligence (AI) emerges as one of the -if not the- most prominent game changers. Recent advances have broadened the possibilities and enabled AI-enhanced solutions towards a sustainable, eco-friendly and human-centered agriculture [2]. Technologies such as machine learning, computer vision, robotics, geospatial analytics, and advanced sensing allow efficient use of resources, climate-smart solutions, precise analytics and decision-support and facilitate the adoption of regenerative practices and improved supply-chain coordination. AI enhances comprehensive decision making through the analysis of big, complex datasets, ranging from real-time field -e.g., soil, water, weather and others- measurements to global satellite imagery. Multiple studies demonstrate how these tools can have an impact on every aspect of the sector, including increased production, optimal resource use, environmental resilience and supply-chain transparency [3]. In the following section, we analyze the advantages to an extent.

These benefits are coupled with serious, intertemporal challenges and barriers that hinder widespread and equitable deployment. Many farmers -mostly smallholders- face significant challenges dealing with unbearable investment costs, low digital literacy and other technological barriers connected to inadequate or hard-to-access infrastructure. There are also concerns about data privacy and security, ownership and intellectual property, and fair distribution of profit and mistrust in data-sharing practices. On top of that, the digital landscape is fragmented and interoperability between platforms is limited and non-standard resulting in low and ineffective integration. Beyond these practical barriers, serious ethical considerations arise around inclusivity, explainability, accountability, environmental and social impact, even potential labour displacement. Ensuring that AI fosters sustainability, rather than perpetuating existing inequalities requires careful governance and design [4].

Given this mixture of opportunities, risks, and constraints, there is a clear need to understand both the current capabilities of AI and the conditions required to leverage them. This chapter therefore provides a consolidated state-of-the-art review of the latest AI technologies shaping a sustainable agri-food domain. This section comprises developments in machine learning, deep learning, computer vision, IoT-enabled sensing, and remote monitoring offering a strong technical foundation for efficient evaluation of how these tools can support sustainable, data-sovereign, and interoperable agricultural transformation.

All these elements combined can provide a comprehensive understanding on the current position of AI in relation to sustainable agriculture and the requirements to achieve effective, equitable and socio-ecologically aware deployment.

2 Benefits of Using AI Tools in Agriculture

AI has been introduced in the past few decades as an auxiliary tool to human intelligence and currently enjoys increasing recognition as a transformative power in every sector -sooner or later- including the agri-food sector. Recent and state-of-the-art research shows the main benefits of AI, as it enhances efficiency and productivity by enabling data-driven resource management, improving decision-support, and optimizing input use through precision agriculture techniques. These advancements not only improve yield and economic performance but also contribute to broader sustainability objectives, such as reducing environmental impacts, enhancing resilience to climatic variability, and strengthening supply-chain transparency. AI offers a complete suite of tools capable of addressing multiple sustainability challenges simultaneously, having the additional benefit that similar solutions have worked in other sectors that got digitalized before agriculture.

AI technologies can have a significant impact on improving **efficiency and productivity** in agriculture. By performing real-time data analysis using any kind of sensors and satellite imagery, AI systems optimize resource use and land management, resulting in better yields and optimal land use [5]. We see how all aspects of agricultural operations can be improved using digital tools, from energy and resource usage to improved disease detection and yield predictions [6]. Additionally, AI-powered automation using robotics and AI-enhanced tools (e.g. robotic harvesters) can reduce labor requirements, increase productivity, and lower operational costs [7].

Linked to the previous, but now observed from an **economic point of view**, AI-enabled digital agricultural technologies bring added value to agricultural operations. A comprehensive review of DATs reports economic benefits from increased yields, reduced labor, and lower chemical and fuel use [8]. These cost savings can make farms more profitable and resilient, especially as AI systems optimize fertiliser, pesticide, and water use, effectively reducing crop losses [9]. Moreover, in precision agriculture, predictive analytics help farmers choose optimal planting times, improve crop selection, and reduce resource waste—all of which contribute to a higher return on investment] [10].

AI also offers profound **social and environmental benefits**. Environmentally, digital technologies can reduce gas emissions, lower chemical inputs, and save water. In the aforementioned review, pesticide and fertilizer use reduction by up to 65% and 80% respectively is reported, as well as water savings up to 35% in specific systems [8]. On social level, AI can contribute to food security by improving climate resilience: predictive models can predict drought periods or changing weather patterns, helping farmers adapt more rapidly and reducing associated risks [11].

A major advantage in AI use is the **decision-support** capabilities. AI systems can integrate big data from heterogeneous sources and of different types, e.g. meteorological, soil, crop data, historical yields and deliver timely, actionable insights. These tools help farmers make comprehensive decisions and efficiently schedule planting, irrigation, interventions, even planning for long-term sustainability [12]. Furthermore, "affordable AI" tools can enhance farmers' local knowledge: such systems enable forecasting microclimate conditions, and assist farmers in optimized resource use and risk assessment [13].

AI significantly enables and boosts what we call **precision agriculture**. Advanced digital technologies like IoT and sensor networks, computer vision, and ML solutions for predictive modelling, AI can recognize patterns and prevent or promptly detect diseases, nutrient deficiencies, and infestations in a precise manner [10]. Precision in this context stands for optimized and only when necessary treatments of fertilizers and pesticides. This results again in reducing waste, operational costs, and environmental pollution, while maximizing yield. This positively impacts the universal goal of a green, sustainable AI-enhanced future.

AI-enabled solutions also support the whole **supply chain**. AI improves operational coordination, transparency, and resilience. According to an EU-approved systematic review, AI applications in agri-food supply chains foster logistics, traceability, and decision-making [14]. Another recent study shows that AI-enhanced supply chain systems help organizations predict disruptions, respond faster and more efficiently, and optimize inventory and distribution, improving supply chain resilience and the corresponding performance [15].

Risk management and disaster resilience are also key benefits. AI-powered predictive models can forecast radical climatic changes or incidents, pest outbreaks or other disruptions, warning farmers and supply chain stakeholders to prepare and mitigate the negative impact [16]. Techniques like digital twins facilitate modeling various scenarios, assisting stakeholders in planning more robust contingency strategies.

Finally, AI can facilitate **regulatory compliance** and auditability [17], as already happens in several sectors [18] [19]. Real-time data analytics and monitoring systems can track key environmental indicators (e.g., emissions, water use, chemical application) helping farms meet regulatory requirements and report efficiently. For instance, integrating AI with IoT can support carbon-footprint monitoring architectures to align agriculture with climate regulations and sustainability frameworks.

Observing benefits as a floor plan now, it is evident how AI can act as an accelerator for economic, environmental, and social sustainability in agriculture. By enhancing precision, timeliness, and intelligence of farming and -essentially- the whole supply-chain decisions, AI supports more resource-efficient practices, lowers production risks, and facilitates alignment with regulatory and sustainability frameworks. At the same time, the literature highlights that these are most properly realized when the tools and systems are embedded within the appropriate supportive governance structures, interoperable data spaces and ecosystems, and inclusive, collaborative efforts. While AI facilitates clear and measurable sustainability advantages, long-term impact depends on ensuring equitable access, responsible deployment, and alignment with broader environmental and societal goals.

3 Challenges and Barriers

The coverage of digital transformation in agricultural sites is growing, but evidently the number is still low, especially compared to the adoption of digitalization and AI solutions in other sectors, like health. This can be highlighted by the emphasis that both the EU [20] and worldwide organizations put on the importance of this digital transition. Despite the transformative potential of AI to support sustainability transitions across the domain, several structural, socio-technical, and governance-related barriers continue to limit adoption and subsequently the impact. These challenges arise at multiple levels from technical infrastructure and data governance to social acceptance, regulatory uncertainty, and of course economic constraints.

Significant **initial costs** of technology are one of the most prominent barriers hindering the adoption of AI-enabled sustainability tools. Sensors, drones, decision-support platforms, cloud analytics, and data storage systems require substantial investment, which might be prohibitive especially for smallholders or farmers from low- and middle-income countries (LMIC) [21], where economies are vastly agri-driven. Beyond hardware, the ongoing cost of **data management** -including cleaning, annotation, storage, cybersecurity, and system integration- creates an additional financial burden that often exceeds the capacity of end users, especially smallholders. Additionally, economic concerns also arise from uncertainty about return on investment: many actors are unsure whether those tools can foster tangible financial or sustainability benefits. These economic considerations extend the so-called **digital divide** [22] fostering a two-tier system where those who can afford the initial investment can benefit from advanced digital tools while others risk falling further behind.

Recent advancements have allowed specific technological services or products to become more affordable, but many users lack the necessary **digital skills** to effectively use those AI-based systems [23]. Farmers, local authorities, rural cooperatives, and environmental stakeholders often face steep learning curves, especially using advanced analytics or decision-support platforms that require familiarity with data interpretation, software tools, or automation workflows. Lack of or just limited access to advisory and training services deteriorates the situation, slowing down and hindering the adoption of the technologies and the subsequent added-value they provide or increasing dependency on intermediaries or providers.

Technical constraints comprise a major bottleneck, particularly in rural or remote regions where sustainability challenges are most pressing [24, 25]. Limited broadband connectivity and unreliable network coverage, insufficient edge-computing and other advanced digital capabilities, and lack of access to modern equipment hinder the adoption of AI-based tools. Even when infrastructure exists, interoperability issues, outdated hardware, and incompatibility among devices prevent seamless deployment of digital solutions.

Concerns and uncertainty about **data privacy and security** are an important factor that hinders digital transformation. Many users are unsure and sceptical about how their data will be stored, who can access it, and for what purposes it may be reused [26]. Surveillance, unauthorized data sharing, or data breach incidents further undermine trust in digital systems. In the agricultural and environmental domains, where geolocation, production data, and other sensitive information are collected, the stakes are particularly high, and concerns are justified to an extent. Fear of non-compliance with data protection policies and the uncertain regulatory landscape itself further discourages experimentation with new AI-based solutions.

Speaking of **regulatory uncertainty**, emerging legislation -EU AI Act, data governance frameworks, and sector-specific regulations- create confusion about compliance obligations both for developers and end users. Concerns over liability, system classification, documentation requirements, and risk assessments can delay deployment. Without clear, harmonized policies and predictable governance mechanisms, stakeholders hesitate to integrate AI tools into operational workflows [25].

Ambiguity around **intellectual property rights and data ownership** further impedes data-driven innovation. It is often unclear who owns the data generated by machines, sensors, digital platforms etc. and who gets to benefit from it. This uncertainty emerges especially in multi-actor ecosystems like agri-food or environmental monitoring, where machinery manufacturers, platform providers, and end users often have conflicting interests [27]. Limited or inadequately transparent governance models and fair data-sharing agreements reinforce mistrust and stalls collaborative innovation, even in cases where data sharing facilitates increased accuracy or enables better environmental monitoring. There are efforts that focus on bridging that gap [28], but there is still a lot to do.

AI systems depend heavily on access to high-quality, reliable datasets, yet many actors **hesitate to share data** due to mistrust in governance, unclear or unjust value distribution, or fear of exploitation [29]. Farmers worry that sharing sensitive data, e.g. production or soil information could impair them in negotiations with buyers, while companies may fear losing the competitive advantage of exclusivity in yield predictions, resource use, or production patterns. This **lack of willingness** to participate in data ecosystems reduces the effectiveness of AI models, which rely on diverse data inputs to generate reliable predictions.

Sustainable AI relies on data integration across sensors, platforms, and institutions, yet technical and semantic fragmentation remains pervasive. Different vendors use incompatible data models, proprietary platforms and APIs, and non-standard vocabularies, making it difficult to combine datasets or connect software modules [30]. This **lack of interoperability** increases costs, reduces scalability, and weakens the potential impact of data-sharing ecosystems, the so-called data spaces like CEADS[1]. Standardization efforts are ongoing, but progress remains slow relative to the speed of digital innovation] [31].

Finally, the **social factor** plays a crucial role. Resistance to change, perceived threats to autonomy, concerns about labor displacement, and deeply rooted cultural practices may all hinder AI tools adoption. Users may distrust technologies developed by external actors or feel that digital tools conflict with local knowledge systems or traditional management practices [21, 32]. Addressing these concerns requires community engagement, collaborative and inclusive design, and long-term participatory approaches that respect contextual knowledge.

It is evident that most of the challenges and barriers are interconnected, as they are cross-referred on several occasions throughout the text -and sometimes inevitably before

[1] https://ceads.eu/.

the respective analysis-. Important steps have already been taken and technological advances keep fostering effective approaches, but all stakeholders should do their part to ensure "ethical and compliant use of digitalization and AI systems in the agri-food sector, uphold principles of inclusivity, transparency, fairness, accountability, and data protection, strengthen the profitability and sustainability of farmers and ensure efficiency in production practices" as WFO states in the respective policy paper [33].

4 Related State-of-the-Art

In the current section there is a classification of papers related to the subject analysis of the current survey paper, in order to help the reader have an overview of the state-of-the-art publications. The classes are the following: the crop management, environmental and resource management, pest and disease management, AI application in smart agriculture/Supply chain, market analysis and price prediction and finally the food safety and traceability.

4.1 Crop Management

Five ML models (AdaBoost, GBM, XGBoost, LightGBM, RF) where used for estimating future crop yields, because these models can handle complicated datasets demonstrating increased performance capabilities [34]. The realization of smart computing and ML algorithms (Deep Learning) methods in crop observation that falls in what is known as CIoT (Consumer Internet of Things) rationale, is analyzed [35]. There is a new crop-based smart mechanism that mixes real-time soil monitoring along generative AI in order to provide personalized agricultural knowledge. This novel scheme is called SmartCrop [36]. In another study, there is the introduction of smart crop yield forecasting taking advantage of Random Forest ML algorithm in order to use only the needed levels of water regarding the environmental parameters [37]. A hybrid solution called H-GNNLM-CropField which consists of Graph Neural Networks (GNNs) and LLMs is used for forecasting crop health and autonomous farm control [38].

4.2 Environmental and Resource Management

In order to monitor different cotton sowing dates, when there is a lot of precipitation, ridge regression and structural equation schemes are used for analysis [39]. There is the evaluation of supportable value chain methods in order to market chain growth chances for agricultural outputs in Bangladesh area. Thus, to fight climate swift via community centered adjustment choices [40]. A quite novel model; CFLP-MP, which stands for Capacitated Facility Location Problems incorporating multi-period, is constructed in order to optimize the supply chain model of the different locations as well as transit procedures regarding production and needs for many time periods [41]. IoT, AI, ML algorithms, remote sensing processing and VRT (Variable Rate Technology) can be implemented in forecasting various soil formats, enhance yield performance and identify water stress levels from IoT sensors. Moreover, these means can "perceive" problems in plants (such as diseases or weeds) via the use of image processing [42].

MVF (Multiponics Vertical Farming) mechanism and AI can be used in soil control for various operations, such as classification, detection and prediction. Three famous ML algorithms are used; SVM, DT and NN in the related smart farming area [43].

4.3 Pest and Disease Management

Not only can climate change bring problems to an agricultural system, infecting the availability of food. It can also affect the parasites' management at a macro level, due to the alterations in species compositions. The latter attacks the products that are been cultivated in an area. Eventually there are strategies that can help solving that kind of problems [44]. In another work, there is the presentation and analysis of datasets related to plant diseases, various DL methods including their metrics' scores as well as challenges connected to them [45]. Robotic mechanisms play a significant role in Agriculture 5.0. These systems contain challenges and boundaries, such as possible threats in the agri sector, ranging from climate issues, parasites, diseases and various costs. The idea is how AI and robotics can help towards this direction [46]. A very famous CNN ML model the YOLOv5s is used in UAV-centered realizations, in the area of pest recognition and classification [47].

4.4 AI Application in Smart Agriculture/Supply Chain and Logistics

Regarding climate change and supply break events in potato supply chain, there is a multi-echelon targeted on the chain model via the use of agent-based modeling (ABM) that consists of five types of agents: (a) farmers, (b) shippers, (c) processors, (d) retailers and (e) logistic companies. There is also a ML model to forecast monthly potato at the grower stage including trends and seasonality per year[48]. A proposed novel IoT + blockchain cooperative model is introduced in order to face the new challenges in agriculture regarding Food Supply Chain [49]. A DSS (Decision Support System) is presented for inventory management and allocation, incorporating AI techniques, for instance; Machine Learning schemes, predictive analysis as well as smart optimizations methods [50]. Another study investigates the connections in the following methods (among them); PLS-SEM (Partial Least Squares Structural Equation Modelling), AI (Artificial Intelligence) implementations, IS (Information Sharing) and SCR (Supply Chain Resilience). These 4 act as independent variables, whereas the SCP (Supply Chain Performance) acts as the dependent variable [15].

4.5 Market Analysis and Price Prediction

As it is widely known Phosphorus is very essential material in inorganic fertilizers. For this reason, a model is presented regarding economic and environmental aspects in order to manage a renewable and supportable Phosphorus fertilizer for supplying chain management [51]. A two-level context of agriculture, which consists of a supplier that encounters yield regarding weather unpredictability, as well as retailer market-connected request incertitude. Moreover, an investigation takes place on how the randomness of yield and request affects the earnings and the performance of supply chain parameters

[52]. There is research targeting garlic which is produced in China as the main aspect of the analysis and research and investigates the prices and which parameters affect the price of pork sales as well as consuming garlic needs in China for the last decade [53]. In another work, a realization of research regarding green gram by investigating the weekly wholesale price in Chinese country. The analysis covers a decade, starting from 1st January 2010 up to 3rd January 2020, which demonstrates critical economic effects [54]. An approach regarding current literature regarding Machine Learning applications in agri domain, underlining key aspects and capabilities by demonstrating experimental outcomes, thus, showing the result of changing labels regarding accuracy of the respected data analysis models is depicted in this work [55].

4.6 Food Safety and Traceability

An intelligent Hybrid CNN-Transformer targeting agricultural monitoring using remote sensing datasets. The model is tested in known ML schemes, such as: CNN, RNN, LSTM, GAN, SVM, DT, RF GBM and LR models depicting increased performance regarding the various metrics' outputs, approaching an accuracy of 98.88% [56]. In research, an analysis on data of quality and safety of agricultural products takes place where there is a sorting of the supply chain as well as business approaches. There is investigation of various risk factors regarding heavy metals. Furthermore, evaluation models are produced [57]. A Hyperledger Fabric as well as a Hyperledger Caliper were constructed in order to evaluate algorithms targeting agricultural supply chain control. In their work, they present the efficacy of a system via the use of various metrics, such as throughput, latency and resource usage in order to guarantee food safety [58]. A novel device which can be used for monitoring glyphosate. The proposed device uses a custom-built and automated fluidic operation which via the utilization of MIP (Molecularly Imprinted Polymer) sensors can identify possible glyphosate infection [59].

5 Ethical Considerations

As the adoption of AI technologies in agricultural context escalates, so do the concerns and the subsequent need for proper regulatory action. The World Farmers' Organisation recently discussed and compiled its position on a policy paper -referred earlier in the chapter- over AI solutions and digitalization in the agri-food domain [33]. One of the most prominent considerations is whether AI systems contribute to **inclusivity and fairness**. We need to ensure that both access and the benefits happen in an equitable manner and avoid perpetuating existing imbalance and discriminations. This includes smallholders, marginalized populations and communities that often deal with systemic disadvantages in infrastructure or knowledge access. Equitable design is necessary to address connectivity, accessibility and even language or cultural relevance. Relatively linked to that is the aforementioned barrier of **digital divide**. Bridging this divide is not only a technological challenge but a socio-political one, requiring long-term public investment and multi-stakeholder coordination to ensure no community or individual is left behind in the sustainable digital transition. Another important discussion relates to **bias and imbalance** in AI models. Sustainable development applications often rely on heterogeneous datasets—from geospatial imagery to farm management data—where representation is uneven. Imbalanced training data can lead to skewed predictions, reinforcing environmental or other injustice (e.g., underestimating climate risks for underserved regions). Ethical AI development requires proactive bias detection, transparent documentation of model limitations, and approaches that facilitate validated model outputs by affected communities.

AI systems that facilitate sustainable decision-making must exhibit high levels of **transparency**. Many systems operate in a way that is opaque to all stakeholders. Not only end users but also the developers and providers of such a system or service are often unable to explain how and why a certain outcome occurred. In other words, these systems operate as black boxes [60]. This opacity causes mistrust on the results, since there is unclear or inadequate reasoning behind the decision that was met. Furthermore, it may hide potential deficiencies of the system, e.g. certain bias or discriminations derived by inadequate or imbalance training. Explainability and auditability of decisions is necessary to increase understanding of the logic behind a prediction or decision, which may even result in logical overturn, when certain circumstances have been overlooked by the model. Explainable AI (XAI) techniques are increasingly important to facilitate trust, especially in high-stakes environmental management scenarios. Equally critical is **clear accountability** throughout the AI lifecycle. As sustainability tasks and challenges often involve multiple stakeholders -potentially both public and private-, defining responsibility for errors, impacts, and adaptations is non-trivial and requires legal arrangements. Robust governance frameworks should clarify who is accountable for data quality, model performance, deployment decisions, and even remediation in case things go wrong. Without this, the legitimacy of AI as a sustainability enabler remains fragile.

Ethical AI for sustainability should be undoubtedly aligned with **societal goals**. The most prominent case is the **UN Sustainable Development Goals** (SDGs[2]). This includes ensuring that AI solutions provide a positive environmental and social impact on one or some of the goals without undermining another. For example, solutions for improving yield can affect SDGs 2 (zero hunger) and 8 (economic growth) in a positive manner, but should not compromise responsible production and life on land (SDGs 12 and 15, respectively). SDGs together with life cycle assessments (LCA) are essential tools towards sustainable and ethical AI. Furthermore, the advancements in technology -with AI in the spotlight, of course- should facilitate and promote **biodiverse, ecological and regenerative practices**. This means that the respective solutions must prioritize resilience over short-term productivity, e.g. promoting soil health using biodiverse farming. Integrating agroecological expertise into the models can help ensure that sustainable solutions reinforce natural systems. Finally, an important consideration is potential **labour displacement** from an AI-driven transition. Automation in operational and monitoring tasks can shift the labour market and displace the most vulnerable workers. Ethical deployment requires strategic actions towards protecting and reskilling affected people. AI should facilitate and augment human expertise, not replace it.

---

[2] https://sdgs.un.org/goals.

6 Conclusions

Artificial intelligence steadily becomes a core factor of the sustainable and resilient agri-food transition. While the food security concerns grow and several climate and geopolitical situations put extra pressure, AI offers a complete suite of tools to optimize resource management, enhance environmental performance, facilitate risk assessment and aversion and improve decision-making. Our review demonstrates how machine learning, remote sensing, digital twins and several technologies are not concepts, but can offer operational capabilities and significantly reshape agricultural practices.

Yet, AI comes with a complicated landscape that currently restricts complete, efficient and equitable deployment. Persistent barriers and challenges such as limited or inadequate governance frameworks, data ownership and privacy concerns, digital gap and inequalities, lack of interoperability, and socio-cultural resistance continue to hinder adoption, especially among small and medium-scale farmers. Ethical concerns including fairness, inclusivity, transparency, accountability, and data sovereignty highlight the risk that AI could reinforce this imbalance.

Taking advantage of the full potential requires a lot more than technologic advances themselves. Future progress will be fostered by **human-centered and inclusive AI design**, supported by participatory approaches that include farmers and communities into system development. The field must strengthen **ethical and transparent AI frameworks**, focusing on explainability, bias detection, accountability mechanisms, and environmental auditing across the AI lifecycle. Equally important is the development of **interoperable, sovereign, and trusted data ecosystems** -including standards, ontologies, and decentralized architectures- that enable data sharing while preserving privacy and sovereignty. Emerging work on **energy-efficient, low-carbon AI and context-aware, multi-scale decision-support systems** will be central to aligning digital innovation with long-term sustainability goals. Finally, interdisciplinary research on **socio-economic impacts, adoption behaviour, and governance frameworks** will be vital to ensuring that regulatory, institutional, and skills-related conditions evolve in tandem with technological progress.

If these advancements co-evolve, AI can become the prominent transformative enabler for climate-smart, regenerative, and equitable food systems. Driven by responsible design and supported by robust data infrastructures, AI has the potential to accelerate progress toward the realization of SDGs and contribute to a more resilient agricultural future.

Acknowledgments. The research presented in this paper is partially based on work carried out under the Horizon Europe DIVINE project (Grant agreement ID: 101060884) that is funded by the European Commission under HORIZON.2.6.3 - Agriculture, Forestry and Rural Areas (HORIZON-CL6-2021-GOVERNANCE-01-20).

Disclosure of Interests. The authors have no competing interests to declare that are relevant to the content of this article.

References

1. OECD-FAO Agricultural Outlook 2025–2034 | OECD. https://www.oecd.org/en/publications/2025/07/oecd-fao-agricultural-outlook-2025-2034_3eb15914.html. Accessed 24 Nov 2025
2. Holzinger, A., Fister, I., Kaul, H.-P., Asseng, S.: Human-centered AI in smart farming: toward agriculture 5.0. IEEE Access **12**, 62199–62214 (2024)
3. Latino, M.E., Corallo, A., Menegoli, M., Nuzzo, B.: Agriculture 4.0 as enabler of sustainable agri-food: a proposed taxonomy. IEEE Trans. Eng. Manage. **70**, 3678–3696 (2021)
4. Ehlers, M.-H., Huber, R., Finger, R.: Agricultural policy in the era of digitalisation. Food Policy **100**, 102019 (2021)
5. Chen, X.: The role of modern agricultural technologies in improving agricultural productivity and land use efficiency. Front. Plant Sci. **16**, 1675657 (2025)
6. Raj, M., Prahadeeswaran, M.: Revolutionizing agriculture: a review of smart farming technologies for a sustainable future. Discov. Appl. Sci. **7**, 937 (2025). https://doi.org/10.1007/s42452-025-07561-6
7. Al-sammarraie, M.A.J., Ilbas, A.I.: Harnessing automation techniques for supporting sustainability in agriculture. T **4** (2024). https://doi.org/10.48130/tia-0024-0026
8. Papadopoulos, G., Arduini, S., Uyar, H., et al.: Economic and environmental benefits of digital agricultural technologies in crop production: a review. Smart Agric. Technol. **8**, 100441 (2024)
9. Routis, G., Paraskevopoulos, M., Vetsikas, I.A., et al.: Data-driven and interoperable smart agriculture: an IoT-based use-case for arable crops. In: 2022 IEEE International Conference on Omni-layer Intelligent Systems (COINS), pp. 1–8. IEEE (2022)
10. Gupta, G., Kumar Pal, S.: Applications of AI in precision agriculture. Discov. Agric. **3**, 61 (2025). https://doi.org/10.1007/s44279-025-00220-9
11. Bhairo, P., Maharaj, R.: Harnessing artificial intelligence for food security in a changing climate. Discov Food **5**, 342 (2025). https://doi.org/10.1007/s44187-025-00609-x
12. Mehedi, I.M., Hanif, M.S., Bilal, M., et al.: Remote sensing and decision support system applications in precision agriculture: challenges and possibilities. IEEE Access **12**, 44786–44798 (2024)
13. Kumar, P., Nelson, A., Kapetanovic, Z., Chandra, R.: Affordable artificial intelligence – augmenting farmer knowledge with AI (2021)
14. Reitano, M., Segovia, M.S., Nayga, R.M., Jr.: A systematic review on the impact of artificial intelligence in the agri-food supply chain. Food Policy **137**, 102983 (2025)
15. Yin, F., Lo, M.C., Mohamad, A.A., Sin, K.Y.: The impact of AI applications, information sharing, and supply chain resilience on agricultural supply chain performance. J. Data Inf. Manag. (2025). https://doi.org/10.1007/s42488-025-00155-2
16. Usigbe, M.J., Asem-Hiablie, S., Uyeh, D.D., et al.: Enhancing resilience in agricultural production systems with AI-based technologies. Environ. Dev. Sustain. **26**, 21955–21983 (2023). https://doi.org/10.1007/s10668-023-03588-0
17. Kruse, N., Schöning, J.: Explainable and trustworthy AI compliance for farms. In: 45. GIL-Jahrestagung, Digitale Infrastrukturen für eine nachhaltige Land-, Forst-und Ernährungswirtschaft. Gesellschaft für Informatik eV, pp. 297–302 (2025)
18. Kothandapani, H.P.: Automating financial compliance with AI: a new era in regulatory technology (RegTech). Int. J. Sci. Res. Arch. **11**, 2646–2659 (2024)
19. Emeihe, E., Ajegbile, M., Olaboye, J., Maha, C.: The impact of artificial intelligence on regulatory compliance in the oil and gas industry. Int. J. Life Sci. Res. Arch. **07**, 28–039 (2024). https://doi.org/10.53771/ijlsra.2024.7.1.0059
20. Digitalising the EU agricultural sector | Shaping Europe's digital future. https://digital-strategy.ec.europa.eu/en/policies/digitalisation-agriculture. Accessed 24 Nov 2025

21. Manzoor, F., Wei, L., Siraj, M.: Digital agriculture technology adoption in low and middle-income countries-a review of contemporary literature. Front. Sustain. Food Syst. **9**, 1621851 (2025)
22. Lythreatis, S., Singh, S.K., El-Kassar, A.-N.: The digital divide: a review and future research agenda. Technol. Forecast. Soc. Chang. **175**, 121359 (2022)
23. FAO. The State of Food and Agriculture 2022. Leveraging agricultural automation for transforming agrifood systems. FAO, Rome (2022)
24. Digital trends in Africa 2021: Information and communication technology trends and developments in the Africa region, 2017–2020. In: ITU. https://www.itu.int:443/en/publications/gs/Pages/publications.aspx. Accessed 24 Nov 2025
25. Pretty, J., Benton, T.G., Bharucha, Z.P., et al.: Global assessment of agricultural system redesign for sustainable intensification. Nat. Sustain. **1**, 441–446 (2018). https://doi.org/10.1038/s41893-018-0114-0
26. Dembani, R., Karvelas, I., Akbar, N.A., et al.: Agricultural data privacy and federated learning: a review of challenges and opportunities. Comput. Electron. Agric. **232**, 110048 (2025). https://doi.org/10.1016/j.compag.2025.110048
27. Atik, C.: Towards comprehensive european agricultural data governance: moving beyond the "data ownership" debate. IIC **53**, 701–742 (2022). https://doi.org/10.1007/s40319-022-01191-w
28. Kalatzis, N., Paraskevopoulos, M., Routis, G., Roussaki, I.: Smart farming data and IoT in support of agricultural policy monitoring. In: 2024 IEEE International Conference on Omni-layer Intelligent Systems (COINS), pp. 1–8. IEEE (2024)
29. Wiseman, L., Sanderson, J., Zhang, A., Jakku, E.: Farmers and their data: an examination of farmers' reluctance to share their data through the lens of the laws impacting smart farming. NJAS-Wageningen J. Life Sci. **90**, 100301 (2019)
30. Tzounis, A., Katsoulas, N., Bartzanas, T., Kittas, C.: Internet of things in agriculture, recent advances and future challenges. Biosys. Eng. **164**, 31–48 (2017)
31. Paraskevopoulos, M., Roussaki, I.: Fostering interoperability in agricultural data spaces: a DIVINE use case. In: 2025 IEEE International Conference on Omni-layer Intelligent Systems (COINS), pp 1–6. IEEE (2025)
32. Nxumalo, G.S., Chauke, H.: Challenges and opportunities in smallholder agriculture digitization in South Africa. Front. Sustain. Food Syst. **9**, 1583224 (2025)
33. World Farmers' Organisation. WFO Policy on Digitalization and and Artificial Intelligence. Policy Paper, World Farmers' Organisation, Rome (2025)
34. Gharakhanlou, N.M., Perez, L.: From data to harvest: leveraging ensemble machine learning for enhanced crop yield predictions across Canada amidst climate change. Sci. Total. Environ. **951**, 175764 (2024)
35. Ahmed, I., Ahmad, M., Ghazouani, H., et al.: Intelligent computing for crop monitoring in CIoT : leveraging AI and big data technologies. Expert. Syst. **42**, e13786 (2025). https://doi.org/10.1111/exsy.13786
36. Patel, S.K., Singh, K.: AIoT-enabled crop intelligence: real-time soil sensing and generative AI for smart agriculture. J. Sci. Innov. Adv. Res. (JSIAR) **1**, 163–167 (2025)
37. Dhanaraj, R.K., Maragatharajan, M., Sureshkumar, A., Balakannan, S.P.: On-device AI for climate-resilient farming with intelligent crop yield prediction using lightweight models on smart agricultural devices. Sci. Rep. **15**, 31195 (2025)
38. Patel, P., Patel, R.: AI-driven agricultural robotics: integrating GNNs and LLMs for comprehensive field-based crop health prediction and management. Available at SSRN 5355686 (2025)
39. Wu, F., Guo, S., Huang, W., et al.: Adaptation of cotton production to climate change by sowing date optimization and precision resource management. Ind. Crops Prod. **203**, 117167 (2023)

40. Barua, P., Rahman, S.H., Barua, M.: Sustainable management of agriculture products value chain in responses to climate change for South-Eastern coast of Bangladesh. Mod. Supply Chain Res. Appl. **3**, 98–126 (2021)
41. Jeong, H., Choi, C.: Adaptive supply chain system design for fruit crops under climate change. Systems **11**, 514 (2023)
42. Kumari, K., Mirzakhani Nafchi, A., Mirzaee, S., Abdalla, A.: AI-driven future farming: achieving climate-smart and sustainable agriculture. AgriEngineering **7**, 89 (2025)
43. Chen, Q., Li, L., Chong, C., Wang, X.: AI-enhanced soil management and smart farming. Soil Use Manag. **38**, 7–13 (2022). https://doi.org/10.1111/sum.12771
44. Frontiers | Pest Management in the Postharvest Agricultural Supply Chain Under Climate Change. https://www.frontiersin.org/journals/agronomy/articles/10.3389/fagro.2022.918845/full. Accessed 24 Nov 2025
45. Salman, Z., Muhammad, A., Piran, M.J., Han, D.: Crop-saving with AI: latest trends in deep learning techniques for plant pathology. Front. Plant Sci. **14**, 1224709 (2023)
46. Balaska, V., Adamidou, Z., Vryzas, Z., Gasteratos, A.: Sustainable crop protection via robotics and artificial intelligence solutions. Machines **11**, 774 (2023)
47. Khan, A., Malebary, S.J., Dang, L.M., et al.: AI-enabled crop management framework for pest detection using visual sensor data. Plants **13**, 653 (2024)
48. Rahman, M.M., Nguyen, R., Lu, L.: Multi-level impacts of climate change and supply disruption events on a potato supply chain: an agent-based modeling approach. Agric. Syst. **201**, 103469 (2022)
49. Awan, S., Ahmed, S., Ullah, F., et al.: IoT with blockchain: a futuristic approach in agriculture and food supply chain. Wirel. Commun. Mob. Comput. **2021**, 5580179 (2021). https://doi.org/10.1155/2021/5580179
50. Wang, H.: Agricultural product supply chain inventory control and allocation decision support system integrating artificial intelligence technology. Int. J. Hi Spe Ele Syst. 2540556 (2025). https://doi.org/10.1142/S012915642540556X
51. Shokouhifar, M., Sohrabi, M., Rabbani, M., et al.: Sustainable phosphorus fertilizer supply chain management to improve crop yield and P use efficiency using an ensemble heuristic–metaheuristic optimization algorithm. Agronomy **13**, 565 (2023)
52. Shi, Y., Wang, F.: Agricultural supply chain coordination under weather-related uncertain yield. Sustainability **14**, 5271 (2022)
53. Wang, Y.: Agricultural products price prediction based on improved RBF neural network model. Appl. Artif. Intell. **37**, 2204600 (2023). https://doi.org/10.1080/08839514.2023.2204600
54. Jin, B., Xu, X.: Wholesale price forecasts of green grams using the neural network. Asian J. Econ. Bank. **9**, 463–490 (2025). https://doi.org/10.1108/AJEB-01-2024-0007
55. Elbasi, E., Zaki, C., Topcu, A.E., et al.: Crop prediction model using machine learning algorithms. Appl. Sci. **13**, 9288 (2023)
56. Maguluri, L.P., Geetha, B., Banerjee, S., et al.: Sustainable agriculture and climate change: a deep learning approach to remote sensing for food security monitoring. Remote Sens. Earth Syst. Sci. **7**, 709–721 (2024). https://doi.org/10.1007/s41976-024-00161-9
57. Li, C., Lu, Y., Bian, Y., et al.: Design of safety evaluation and risk traceability system for agricultural product quality. Appl. Sci. **14**, 2980 (2024)
58. Sugandh, U., Nigam, S., Khari, M., Misra, S.: An approach for risk traceability using blockchain technology for tracking, tracing, and authenticating food products. Information **14**, 613 (2023)
59. Radogna, A.V., Latino, M.E., Menegoli, M., et al.: A monitoring framework with integrated sensing technologies for enhanced food safety and traceability. Sensors **22**, 6509 (2022)
60. TechDispatch - Publications Office of the EU. https://op.europa.eu/en/publication-detail/-/publication/59e38fb7-8436-11ee-99ba-01aa75ed71a1/language-en. Accessed 24 Nov 2025

Artificial Intelligence for Global Security

A SWOT Analysis of AI Technology Sovereignty Landscape Assessment

Mariya Evtimova-Gardair[1(✉)], Frédérique Segond[1], Patrick Fabiani[2], Thibault Lahire[2], and Michel Rottmaier[3]

[1] Institut national de recherche en informatique et en automatique, Le Chesnay-Rocquencourt, France
{mariya.evtimova-gardair,frederique.segond}@inria.fr

[2] Dassault Aviation, Saint-Cloud, France
{patrick.fabiani,thibault.lahire}@dassault-aviation.fr

[3] Airbus, Ulm, Germany
michael.rottmaier@airbus.com

https://www.dassault-aviation.com/fr/, https://www.airbus.com/en

Abstract. Due to the current societal and geopolitical environment, the topic of digital sovereignty is increasing its importance in many application domains. Some of the major digital sovereignty issues are related to security, privacy, and the availability of policy instruments concerning government and citizens' digital information. The paper proposes an Artificial Intelligence and machine learning sovereign technology classification for defense applications. A methodology for assessing the sovereignty of an AI pipeline is critical in the development of functionalities for military aircraft due to the high stakes involved in national security, system reliability, and strategic autonomy. In this context, "sovereignty" refers to the degree of control and independence that a nation or defense organization maintains over the entire AI development and deployment process, including data handling, algorithm design, software infrastructure, and model interoperability. The paper defines a SWOT (Strengths, Weaknesses, Opportunities, Threats) analysis of sovereign software technologies that can help to observe technology priorities. It is proposed as a description of the SWOT AI sovereignty framework. As a result of the analysis conducted, the sovereignty technology was identified, the sovereignty degree, and the sovereignty criteria. The sovereignty degree measures the level of sovereignty concerning the description of corresponding sovereignty criteria. Fuzzy logic is used to identify the degree of sovereignty.

Keywords: Sovereignty · SWOT analysis · Defense

1 Introduction

Artificial intelligence (AI) capabilities in software technology integration have grown exponentially. AI is an evolving technology used to solve a wide range

Published by Springer Nature Switzerland AG 2026
E. Mercier-Laurent et al. (Eds.): AI4S 2025, IFIP AICT 780, pp. 309–321, 2026.
https://doi.org/10.1007/978-3-032-21184-2_24

of issues. In this paper is discussed more particulary the sovereignty AI assessment concerning the military aircraft regarding important aspects as safeguarding national security, enhancing system robustness, and autonomous strategic capability. Moreover, national and international standards are in the process of development to achieve technical interoperability, reliability, and transparency [1]. In [2], software quality is defined as the degree to which a system, component, or process meets specified requirements concerning customer needs or expectations.[1] Measurement provides a quantitative value that facilitates interpretation [3]. According to [4], you cannot control what you cannot measure. So [3] defines the measurement as follows: "The measurement is the process by which numbers or symbols are assigned to attributes of entities in the real world to describe such entities according to clearly defined rules". It is important to mention that the metrics are used by the software industry to enhance the quality and reduce the possible risks during software development. Furthermore, the use of metrics can be more productive when organizations choose the appropriate metrics according to their need at the right time. The practical applications of AI and machine learning hold promising prospects, and much research is dedicated to this area. However, the challenges in this area concerning internal and external problems also need to be considered. The EU proposes many regulations, but unique regulations for AI sovereignty are currently not presented. In general, AI has been looking to help humans execute their daily life tasks, starting with the automation of processes via robots, followed by introducing expert systems that could provide high-quality results based on knowledge and examples from human experts.

Co-funded by the European Union

2 Artificial Intelligence and Machine Learning Sovereign Technologies Classification

The European Union is implementing regulations for artificial intelligence (AI) that need to be applied to AI technologies. Also, efforts are being made in Germany to put the industry standards into practice [5]. In [5], three dimensions of AI Applications are introduced: methods, capabilities, and criticality. An AI Expert group position paper is used to create the link between AI methods and AI capabilities [5].

[1] This publication was co-funded by the European Union under the Grant Agreement 101103669-EICACS (European Initiative on Collaborative Air Combat Standardization). Its contents are the sole responsibility of the author and do not necessarily reflect the views of the European Union or the European Commission. Neither the European Union nor the granting authority can be responsible for them.

Artificial intelligence, by definition, is the ability of computer systems to perform tasks that can be commonly associated with intelligent creatures [6]. Another definition is that AI includes software and hardware methods that imitate or reproduce human behavior and thinking. Concerning certain literature sources, AI can be divided into weak AI, strong AI, or general AI [5,7,8]. The purpose of the AI is to solve individual problems, and the level of accuracy of these solutions is satisfactory for practice.

Concerning the classification review proposed in [5,9,10], and the practical experience survey, AI software technologies can be classified as Cognitive AI, Hybrid AI, Data-driven AI, and Model-driven AI. The classification schema is proposed in Fig. 1. Also, this category can be divided into subcategories as follows: Robotics, Machine Learning, Expert Systems, Natural Language Processing, Speech, Vision, and Planning. The machine Learning category can be presented generally as Supervised Learning, Unsupervised Learning, Semi-Supervised Learning, Reinforcement Learning, and Deep Learning. On the other side, Natural Language Processing can be classified into Machine Translation, Content Extraction, Question Answering, Information Retrieval, Sentiment Analysis, Text Generation, and Topic Modeling [9,11].

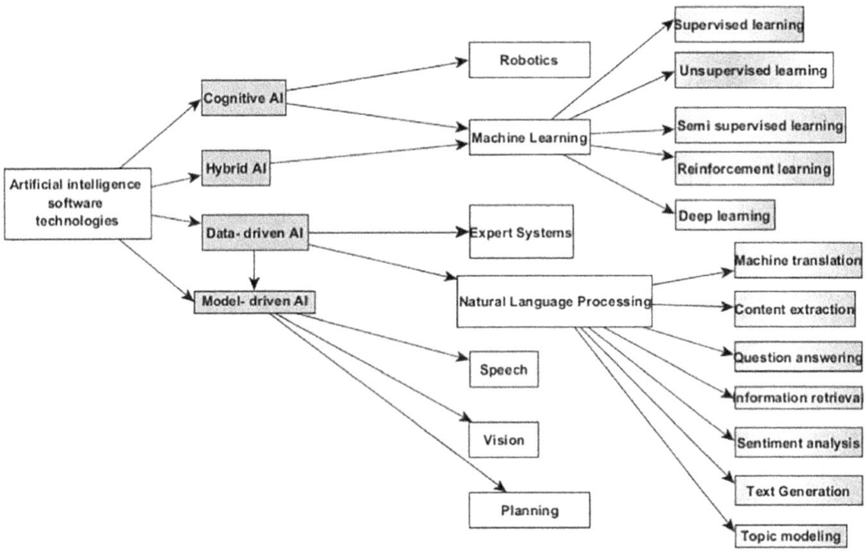

Fig. 1. Classification of AI software technology concerning AI methods.

The advancement of AI-driven methods directly reflects the content of the AI technology's structure and leads to the deviation of the human role in this domain.

3 Related Work

As far as we know, no other study has been conducted to investigate the AI technology sovereignty landscape using SWOT analysis for military airplanes. Therefore, there is no comprehensive explanatory knowledge for conducting an AI sovereignty technology SWOT analysis.

However, several articles apply SWOT analysis to AI technologies.

For example, [12] describes a SWOT analytical framework that classifies factors into four categories: strengths, weaknesses, opportunities, and threats. The SWOT framework proposes that the factors in the strengths and weakness categories to be represented as internal aspects of the organization, and the factors in opportunities and threats categories to be presented as external aspects of the organization.

Another article [1] proposes a study that evaluates software security metrics and their effectiveness. The study analysis applies a SWOT analysis that highlights their strengths, weaknesses, opportunities, and threats.

A paper [13] investigated strategically the possibility of alignment of ontologies to software engineering that is applied to the intelligent software development process. The SWOT analysis is presented, which shows the usage of the ontologies for enriching and enhancing software engineering processes.

Research proposed in [14] performs a SWOT analysis and its application in solving research tasks.

In the research software technology, sovereignty is a relatively young research field, and not a lot of information is available. However, related to data sovereignty, more research information is presented. As the definition of sovereignty is defined in [15] and [16]. The open question exists to ensure the technological sovereignty of AI when sharing data and working with different holders. That is the improvement of interoperability, technology sovereignty, and the possibility of sharing across different legacy systems. All that defines the technical challenges concerning sovereignty. It is important to mention the work with heterogeneous and complex data that is required to be more auditable and transparent concerning the AI technology sovereignty principles [17]. Many systems rely on ensuring sovereignty over the usage of access control mechanisms with authentication. However, this leads to certain constraints as the data cannot be viewed and monitored. Concerning this, there is a defined need to control the usage concerning temporality and to continuously monitor and control the exploitation of resources such as files and services. In distributed systems, the implementation and integration of usage control as a technical implementation of data sovereignty is a challenging but necessary task [18].

Projects that concern questions from the data sovereignty area, including developing technical solutions, are FIWARE, the International Data Spaces (IDS), and Gaia-X have been built up [19].

4 Description of the SWOT Analysis for Sovereignty Software Technologies

Many ways of developing AI tools, more constraints, and more principles will allow for development that functions **Research questions** The classification of the factors based on the SWOT framework can assist practitioners in getting a thorough understanding so that they can address the key knowledge areas for successfully implementing software AI technology in the environment. In our study, the following research questions were developed in order to answer the reported research problem [12].

RQ1: What are the criteria that positively influence the sovereignty of AI technology?
RQ2: What are the gaps that negatively influence the sovereignty of AI technology?
RQ3: What are the most critical sovereignty AI technology criteria that positively impact or sovereignty AI technology that negatively impact including the gaps in the landscape of the EU sovereignty technology representation?
RQ4: How could the identified sovereignty criteria and gaps be mapped with the four categories of the SWOT framework?

5 SWOT AI Sovereignty Framework Description

SWOT analysis of the AI Technologies represents the internal and external elements that can have an impact on the selection of the AI technology.

5.1 Strengths

1. Data Sovereignty and Security: Using sovereignty AI software permits governments or organizations to control their data, ensuring it is not stored or processed by foreign entities. That improves the security performance against data breaches, espionage, or external possibility for surveillance.
2. Customization and Localization: The software can be shaped concerning specific national or local needs, including language processing, regional data sets, and local legal frameworks, ensuring a more relevant AI output.
3. National Regulations Compliance: Sovereignty AI software can be defined to conform with local laws and regulations that reduce issues with foreign software providers. The software needs to meet certain requirements, such as GDPR in Europe or other data protection standards.
4. Low dependence on AI solutions: When using national AI software, governments and organizations can reduce the external dependence that can be linked with the global tech giants, which might have different priorities or potential geopolitical agendas.

5.2 Weaknesses

1. Increased Development and Maintenance Costs: Creating and maintaining national AI software requires significant investment concerning infrastructure, human resources, and continuous development.
2. Limitation concerning innovation:
   (a) External AI providers benefit from a global user base, providing them with massive datasets and feedback loops for innovation. Sovereignty AI software restricted to one country or organization might lack this scope for rapid technological advancement.
   (b) Human resource limitation capacity concerning research and innovation.
3. Interoperability constraints: AI systems developed with national requirements in mind may have difficulty integrating with global systems, standards, or international partners.

5.3 Opportunities

1. Data Privacy Insurance: There is a growing need for privacy and data security. This increases the need to seek AI solutions that guarantee data sovereignty. This provides an opportunity to implement AI software based on sovereignty.
2. Geopolitical situation: Some countries may invest resources in AI sovereignty to ensure national sovereignty and limit foreign influence.
3. National Digital Transformation: Governments can modernize infrastructure and improve public services with sovereign solutions that include artificial intelligence.
4. Application and integration: the use of sovereign AI can include both data processing and end-to-end applied solutions, in industries such as healthcare, defense, and finance.

5.4 Threats

1. Challenges in the use of AI: Sovereign's AI software competes with large, established, and well-funded global players in the AI market such as Google, Microsoft, and Amazon, which have huge data sets and sophisticated R&D resources.
2. Cybersecurity Threats: Artificial intelligence systems can be controlled within national borders, but even so, they are still susceptible to cyberattacks by sophisticated foreign actors, making the software an attractive target.
3. Lack of Global Standardization: As global AI systems advance, countries may find themselves isolated if their AI systems, driven by sovereignty concerns, are not interoperable with global platforms and standards.
4. Delayed innovation: global technology companies are constantly working with innovative AI technologies, offering new software solutions or updates. Sovereign AI software must struggle to keep up with the times to contribute to global systems.

6 Results of the Analysis

6.1 Identification of the Researched Questions

In this section, we discuss every research question using the evidence extracted from the identified resources.

Table 1 defines sovereignty criteria and their description.

RQ1: What are the criteria that positively influence the sovereignty of AI technology?

The degree of each sovereignty criterion is different concerning the evaluation score that provides the necessity to apply the unification algorithm of the output values from each sovereignty criteria scoring (for example, between 0 and 1) using a fuzzy membership function.

$$\mu(x) = \frac{1}{1 + \frac{X^2}{2}} \tag{1}$$

In the formula, the value x is defined as the membership function of the minimum and maximum value and the threshold. The results from this research question analysis are described in Table 1. In the table, are defined sovereignty criteria and their corresponding sovereignty degree.

Index S defines the representation of sovereignty degree. So if the index S has a value more than 0.5 the criteria for certain AI technology positively influence sovereignty as described in Table 1.

RQ2: What are the aspects that negatively influence the sovereignty of AI technology? This question concerns the gaps that can be observed and that have an impact on AI technology. The results from this research question analysis are described in Table 2. In the table are defined sovereignty criteria and their corresponding sovereignty degree when S has a value below 0.5. That is the value that corresponds to the negative influence of the sovereignty AI technology.

RQ3: What are the most critical sovereignty AI technology criteria that impact positively or sovereignty AI technology that impact negatively including the gaps in the landscape of the EU sovereignty technology representation? The results from this question are represented in Table 2. In the table are defined criteria that are considered as most critical concerning the sovereignty of AI technology. The idea of critical factors was initially proposed by [20]. Critical factors represent the most important areas that organizations must focus on. The limited attention given to critical factors can degrade the business performance. In order to test the criticality of the identified factors for AI technology sovereignty assessment.

A Systematic Literature Review (SLR) method was applied to identify and to get a thorough understanding of the critical factors that positively or negatively impact agile practices in the software AI technology context for military aircraft. Systematic literature reviews in general define a manner to synthesize scientific evidence to answer a particular research question in a way that is transparent and reproducible. The synthesis of the scientific evidence investigates all published evidence on the topic and appraises the quality of this evidence [21]. Systematic

Table 1. Mapping sovereignty criteria in a SWOT analysis.

Criteria	Description
Software ownership	The most fundamental aspect of sovereignty is the ownership of the source code. Ensure that the organization or individual developing the software retains complete control over the source code and underlying models
Software country of origin	Country from which the software originates and the country where most of the current maintainers are from or affiliated to
Software license	Type of license the technology is distributed under
Data Ownership	The most fundamental aspect of sovereignty is the ownership of the Data. Ensure that the organization or individual developing the software retains complete control over the Data, especially in applications dealing
Data country of origin	Country from which the Data originates and the country where most of the current maintainers are from or affiliated to
Data License	Type of license the dataset is distributed under
Community and Ecosystem	For open-source projects, evaluate the size and health of the community and ecosystem around the software. A strong community can provide support and reduce dependence on a single entity
Regulatory and Legal Compliance	1. Ensure that the software development adheres to legal and regulatory requirements. 2. Comply with the corresponding regulations
Funding and sustainability	Evaluate the financial sustainability of the project. Dependency on external funding or organizations may affect sovereignty
Interoperability	The ability of the software to work seamlessly with other systems and tools
Development processes	Processes that exist for contribution to open-source projects, and the way that the contributions are evaluated and examined
Vendor Lock-in	To have control over Technology and Infrastructure

Table 2. Description of sovereignty criteria that are most critical

Criteria	Description
Data Governance	Strong data governance frameworks ensure data sovereignty, security, and compliance with EU regulations
Open-source ecosystem	Presence of open-source AI tools and frameworks that reduce dependency on non-EU technologies
Regulatory Alignment	Harmonization of AI regulations across EU member states to facilitate innovation and compliance
Interoperability	Compatibility of AI systems with existing European digital infrastructure and standards

Literature Review studies follow the following workflow: identification of the relevant primary papers, extraction of the required data, analysis, and synthesis of results [22]. The workflow steps can gain further and broader insight into the investigated AI technology domain for military aircraft. A systematic review of the scientific literature is an essential analysis for the AI technology domain for military aircraft, so as to define the research questions and to justify the future research for AI technology for military aircraft. Performing a systematic literature review is a complex activity for young scientific researchers, especially if the skills are not developed for searching and filtering information, and without particular knowledge of databases specific to the domain of study. Furthermore, the systematic reviews aim to identify all scientific research in correlation with a specific question so as to propose a balanced and unbiased summary of the literature [23]. Systematic Literature Review [24,25] uses methods to identify studies for inclusion that have been developed specifically to identify the negative studies that can be published in low-index-class journals or within conference proceedings, which are not indexed in the bibliographic databases. That can correct the results found in the positive studies that are more easily identified.

RQ4: How could the identified sovereignty criteria and gaps be mapped with the four categories of the SWOT framework? The result from the analysis of that research question is presented in Table 3. The table contains the Success Factors and the Barriers that can be identified as internal and external factors, and identifies the Strengths, Weaknesses, Opportunities, and Threats for the sovereignty of AI technologies.

6.2 Description of Particular Use Case

In this section, a particular use case for the Sci-kit learn library is defined. Table 4 evaluates the sovereignty criteria with a particular example of the Sci-kit learn library.

Table 3. Mapping sovereignty criteria in a SWOT analysis

	Success Factors	Barriers
Internal factors	**Strength** 1. Data Sovereignty and Security 2. Customization and Localization 3. National Regulations Compliance 4. Low dependence on AI solutions	**Weakness** 1. Increased development and Maintenance Costs 2. Limitation concerning innovation 3. Interoperability constraints
External factors	**Opportunities** 1. Data Privacy Insurance 2. Geopolitical situation 3. National Digital Transformation 4. Application and Integration	**Threats** 1. Challenges in the use of AI 2. Cybersecurity Threats 3. Lack of Global Standardization 4. Delayed Innovation

Table 4. Evaluation of the sovereignty criteria for Sci-Kit learn library

Criteria	Degree	Result
Software ownership	1	$S > 0.5$
Software country of origin	1	$S > 0.5$
Software license	0,8	$S > 0.5$
Data Ownership	0,6	$S > 0.5$
Data country of origin	1	$S > 0.5$
Data License	0,8	$S > 0.5$
Community and Ecosystem	0,6	$S > 0.5$
Regulatory and Legal Compliance	0,8	$S > 0.5$
Funding and sustainability	0,6	$S > 0.5$
Interoperability	0,8	$S > 0.5$
Development processes	0,7	$S > 0.5$
Vendor Lock-in	0,6	$S > 0.5$

Each criterion has an associated fuzzy value that, in the current case, reflects positively the overall sovereignty assessment of the software. To illustrate, consider an AI-based target recognition function deployed in a military aircraft. This function may leverage computer vision algorithms to identify enemy assets in real-time from aerial imagery. The risks associated with outsourcing any component of this function—such as the model training phase or the use of proprietary black-box software libraries—include untraceable failure modes, potential backdoors, hidden biases, and legal or geopolitical constraints that limit operational freedom. An evaluation methodology for AI pipeline sovereignty would enable defense stakeholders to assess systematically:

- The origin and auditability of software libraries used (e.g., open-source vs proprietary),
- The traceability and control over training data,
- The transparency of model behavior, and
- The reproducibility and portability of the system across platforms.

From this perspective, using a tool like Scikit-learn [26] a widely used, open-source machine learning library developed and maintained in Europe, offers clear advantages [27]. It aligns with European data protection laws, supports full transparency of algorithmic behavior, and fosters reproducibility and traceability—key pillars of technical sovereignty. Furthermore, as Scikit-learn avoids opaque abstraction layers and relies on interpretable algorithms such as logistic regression [28], decision trees [29], or k-nearest neighbors [30], it is particularly suitable for critical defense applications where explainability and verification are paramount [31].

7 Conclusion

In conclusion, a sovereignty-aware methodology does not merely assess technical performance but incorporates geopolitical, legal, and ethical dimensions. It is a foundational element in the trustworthy deployment of AI in military aviation, ensuring that decision-making functions—such as target recognition—remain under full national control, are robust to adversarial risks, and comply with strategic defense imperatives. The paper presents an artificial intelligence and machine learning sovereign technologies classification. A SWOT analysis is performed concerning sovereignty technologies, and strengths, weaknesses, opportunities, and threats are identified. That paper represents as innovation the creation of SWOT analysis for sovereign AI technologies and, more particularly, their representation for military aircraft. The objective of this study is to investigate and classify the sovereignty criteria to identify the needs and, respectively, the possible gaps. Each criteria have a different degree concerning fuzzy numbers that classify the corresponding technology as AI Technology as a strength or weakness. The results are represented concerning the defined research questions. The result of the SWOT analysis contains the success factors and barriers concerning internal and external factors. This analysis aims to support the evaluation of AI Technology sovereignty. More precise analysis could be provided in the future as an extension of this analysis concerning the particular needs and gaps for the military aircraft industry.

References

1. Khalid, A., et al.: A SWOT analysis of software development life cycle security metrics. J. Softw. Evol. Process **37**(1), e2744 (2025)

2. IEEE Computer Society. Software Engineering Technical Committee: IEEE Standard Glossary of Software Engineering Terminology. IEEE **729**(1983) (1983)
3. Rehman, S., Khan, S.U.: Swot analysis of software quality metrics for global software development: a systematic literature review protocol. IOSR J. Comput. Eng. **2**(1), 1–7 (2012)
4. DeMarco, T..: Controlling Software Projects: Management, Measurement, and Estimates. Prentice Hall PTR (1986)
5. Schmid, T., Hildesheim, W., Holoyad, T., et al.: The AI methods, capabilities and criticality grid: a three-dimensional classification scheme for artificial intelligence applications. KI-Künstliche Intell. **35**(3), 425–440 (2021)
6. Mclean, S., Read, G.J.M., Thompson, J., et al.: The risks associated with artificial general intelligence: a systematic review. J. Exp. Theor. Artif. Intell. **35**(5), 649–663 (2023)
7. Tizhoosh, H.R., Pantanowitz, L.: Artificial intelligence and digital pathology: challenges and opportunities. J. Pathol. Inform. **9**(1), 38 (2018)
8. Singh, P., Elmi, Z., Lau, Y., et al.: Blockchain and AI technology convergence: applications in transportation systems. Veh. Commun. **38**, 100521 (2022)
9. Mukhamediev, R.I., Popova, Y., Kuchin, Y., et al.: Review of artificial intelligence and machine learning technologies: classification, restrictions, opportunities and challenges. Mathematics **10**(15), 2552 (2022)
10. Lins, S., Pandl, K.D., Teigeler, H., et al.: Artificial intelligence as a service: classification and research directions. Bus. Inf. Syst. Eng. **63**(4), 441–456 (2021)
11. Mastropaolo, A., Escobar-Velasquez, C., Linares-Vasquez, M.: From triumph to uncertainty: the journey of software engineering in the AI era. ACM Trans. Softw. Eng. Methodol. **34**(5), 1–34 (2025)
12. Sinha, R., Shameem, M., Kumar, C.: SWOT: strength, weaknesses, opportunities, and threats for scaling agile methods in global software development. In: Proceedings of the 13th Innovations in Software Engineering Conference (Formerly Known as India Software Engineering Conference), pp. 1–10 (2020)
13. Bhatia, M., Kumar, A., Beniwal, R.: SWOT analysis of ontology driven software engineering. Indian J. Sci. Technol. **9**(38) (2016)
14. Petricek, P., Klir, R., Kalvsky, P.: Swot analysis and its application in solving research tasks. In: 2020 New Trends in Aviation Development (NTAD), pp. 197–201. IEEE (2020)
15. Hummel, P., Braun, M., Tretter, M., et al.: Data sovereignty: a review. Big Data Soc. **8**(1), 2053951720982012 (2021)
16. Mugge, D.: EU AI sovereignty: For whom, to what end, and to whose benefit? J. Eur. Public Policy **31**(8), 2200–2225 (2024)
17. Munoz-Arcentales, A., Lopez-Pernas, S., Pozo, A., et al.: An architecture for providing data usage and access control in data sharing ecosystems. Procedia Comput. Sci. **160**, 590–597 (2019)
18. Hellmeier, M., Pampus, J., Qarawlus, H., et al.: Implementing data sovereignty: requirements & challenges from practice. In: Proceedings of the 18th International Conference on Availability, Reliability and Security, pp. 1–9 (2023)
19. Khan, A.A., Keung, J., Niazi, M., et al.: Systematic literature review and empirical investigation of barriers to process improvement in global software development: client-vendor perspective. Inf. Softw. Technol. **87**, 180–205 (2017)
20. Altendeitering, M., Pampus, J., Larrinaga, F., et al.: Data sovereignty for AI pipelines: lessons learned from an industrial project at Mondragon corporation. In: Proceedings of the 1st International Conference on AI Engineering: Software Engineering for AI, pp. 193–204 (2022)

21. Torres-Carrion, P.V., Gonzalez-Gonzalez, C.S., Aciar, S., et al.: Methodology for systematic literature review applied to engineering and education. In: 2018 IEEE Global Engineering Education Conference (EDUCON), pp. 1364–1373. IEEE (2018)
22. Dziopa, F., Ahern, K.: A systematic literature review of the applications of Q-technique and its methodology. Methodology (2011)
23. Lame, G.: Systematic literature reviews: an introduction. In : Proceedings of the Design Society: International Conference on Engineering Design, pp. 1633–1642. Cambridge University Press (2019)
24. Kitchenham, B., Brereton, O.P., Budgen, D., et al.: Systematic literature reviews in software engineering-a systematic literature review. Inf. Softw. Technol. **51**(1), 7–15 (2009)
25. Van Dinter, R., Tekinerdogan, B., Catal, C.: Automation of systematic literature reviews: a systematic literature review. Inf. Softw. Technol. **136**, 106589 (2021)
26. Bisong, E.: Introduction to scikit-learn. In: Building Machine Learning and Deep Learning Models on Google Cloud Platform: A Comprehensive Guide for Beginners, pp. 215–229. Apress, Berkeley (2019)
27. Nelli, F.: Machine learning with scikit-learn. In: Python Data Analytics: with Pandas, numPy, and Matplotlib, pp. 259–287. Apress, Berkeley (2023)
28. Suh, Y.S., Shin, S.K., Baang, D., et al.: A performance evaluation method of a machine learning programming of simple logistic regression. In: Transactions of the Korean Nuclear Society Autumn Meeting, Goyang, Korea (2019)
29. Testas, A.: Decision tree regression with pandas, scikit-learn, and PySpark. In: Distributed Machine Learning with PySpark: Migrating Effortlessly from Pandas and Scikit-Learn, pp. 75–113. Apress, Berkeley (2023)
30. Zhang, Z.: Introduction to machine learning: k-nearest neighbors. Ann. Transl. Med. **4**(11), 218 (2016)
31. Hackeling, G.: Mastering Machine Learning with Scikit-Learn. Packt Publishing Ltd. (2017)

Rule-Based Coordination Strategies for Protection in Multi-agent Defense Scenarios

Thibault Lahire(✉)

Dassault Aviation, Saint-Cloud, France
thibault.lahire@dassault-aviation.com

Abstract. Protecting High Value Assets in contested environments remains a critical challenge for autonomous multi-agent systems. This study investigates coordination strategies among allied drones tasked with escorting a strike drone through hostile territory. We introduce a discrete-grid simulation inspired by tactical military settings, where enemy drones spawn periodically and follow a deterministic pursuit policy. Focusing on interpretable and robust rule-based systems, we evaluate coordination protocols under varying agent ratios. Key results show that, when enemy drones are outnumbered, strategies that balance close protection of the strike drone with support to engaged allies yield higher mission success rates. Beyond performance, our work contributes to the standardisation of coordination rules for multi-agent defense tasks, supporting future interoperability across autonomous systems.

Keywords: Multi-Agent Systems · Rule-Based Systems · Collaborative Air Combat

1 Introduction

In complex operational settings, the protection of High Value Assets (HVAs) by autonomous systems represents a central challenge in distributed artificial intelligence and collaborative robotics. Aerial defense scenarios, in particular, demand effective coordination between heterogeneous autonomous agents operating in dynamic, partially observable, and potentially hostile environments. In this work, we focus on a simplified but representative scenario: escorting an air-to-ground strike drone with air-to-air combat drones through a hostile area populated with enemy threats.

Leveraging prior knowledge of enemy generation and behavior, our goal is to design efficient coordination strategies among allied agents. To achieve this, we implement and evaluate a range of strategies using a set of custom-defined performance metrics. Rather than exploring deep reinforcement learning approaches, this study focuses on rule-based systems, motivated by two key factors: first, their interpretability, which is essential in mission-critical contexts; and second, their robustness to out-of-distribution generalization errors, which often affect learned

Published by Springer Nature Switzerland AG 2026
E. Mercier-Laurent et al. (Eds.): AI4S 2025, IFIP AICT 780, pp. 322–336, 2026.
https://doi.org/10.1007/978-3-032-21184-2_25

models. Through this environment, we aim to answer a fundamental question: what level of coordination (and how many agents) is required to guarantee a tactical advantage in the studied setting?

The proposed environment is built on a discrete 20×16 grid, inspired by tactical simulations commonly used in military contexts. It strikes a balance between sufficient complexity to induce emergent behavior and enough simplicity to enable controlled study of coordination protocols. The game proceeds as follows: a strike drone must reach and destroy a dynamic ground target while facing successive waves of enemy drones. The mission immediately fails if the strike drone is destroyed, making protection by escort drones essential.

A key aspect of this environment is the clearly defined set of engagement rules and win/loss conditions, as well as the periodic appearance of enemy drones guided by a simple but effective policy. These enemy drones, although controlled by a deterministic strategy (minimizing distance to the strike drone), spawn at random locations and pose a persistent threat. This forces allied agents to coordinate dynamically to maximize the mission's chances of success.

By designing and comparing various rule-based coordination strategies, we aim to identify organizational principles that enable effective multi-agent cooperation. This work also contributes to the standardisation of coordination protocols for multi-agent protection missions, facilitating interoperability across autonomous systems. The analysis considers both the absolute performance of each strategy and its scalability: how performance evolves as we increase or decrease the number of allied and enemy agents.

In this study, we analyze a specific coordination strategy: when the number of enemy drones is equal to or greater than the number of allied escort drones, each allied drone is assigned to intercept the nearest enemy. This distance-based matching ensures that all threats are directly engaged. However, when the number of enemy drones is lower than the number of allied escorts, some allies remain unassigned. In such cases, we explore two possible reassignment strategies: (i) repositioning unassigned drones closer to the strike drone for direct protection, and (ii) deploying them to assist allied drones already engaged with enemies.

Our findings reveal that the most effective cooperation strategy lies between these two extremes. While assigning unengaged drones to close protection roles is beneficial, not all should follow the same pattern; some should instead support their engaged teammates. The success of this hybrid strategy depends on improved spatial coverage, enabling a more agile and robust tactical response to emerging threats.

The remainder of this article is structured as follows. Section 2 reviews related work on multi-agent coordination in protection missions. Section 3 presents the simulation environment. Section 4 evaluates the balance between coordination strategies, and Sect. 5 discusses the limitations of this study and offers concluding remarks.

2 Background and Related Work

The coordination of multiple autonomous agents in dynamic and adversarial environments has been a longstanding research challenge in artificial intelligence, robotics, and control theory. The problem becomes particularly acute when agents must cooperate to achieve shared objectives under resource constraints and threats from intelligent adversaries. This is especially true in defense applications, where the survivability of key assets may depend critically on the ability of autonomous systems to coordinate effectively in real time.

2.1 Multi-agent Coordination

A central theme in multi-agent systems (MAS) research is *coordination*—how agents synchronize actions, share information, allocate roles, and jointly plan to maximize team performance. Coordination may be *explicit*, involving direct communication, predefined roles, or shared plans, or *implicit*, emerging from decentralized policies or environmental signals. Classical coordination tasks include *formation control* [15], *coverage and task allocation* [8], and *pursuit–evasion* [11]. These paradigms are deeply connected to our aerial escort mission, where agents must coordinate to shield a moving drone while simultaneously targeting adversaries.

Within this broader field, *cooperative defense tasks*—where agents protect a moving high-value asset—are especially relevant. Such scenarios require balancing multiple objectives: intercepting threats, denying adversarial access, and maintaining proximity to the protected asset. Previous studies have explored similar frameworks, particularly in *robot-soccer formations* [6] and *perimeter defense* [13]. Moreover, deep reinforcement learning has been applied to convoy defense scenarios, achieving robust escort behavior though with significant computational complexity [9].

2.2 Rule-Based Vs Learning-Based Approaches

While deep reinforcement learning (DRL) has recently become the dominant paradigm in multi-agent systems (MAS) research—particularly in large-scale, high-dimensional environments such as StarCraft [24] or Google Research Football [12]—there are well-documented limitations to applying DRL in critical operational contexts. First, DRL policies often lack interpretability, complicating efforts to audit decisions or diagnose failures [18]. Second, they may exhibit limited generalization to unseen environments or altered agent configurations, particularly when coordination behaviors become brittle or overfit to training regimes [23].

In contrast, rule-based systems present a compelling alternative in domains where reliability, transparency, and human-in-the-loop validation are essential. Rule-based coordination—whether derived from expert knowledge or structured around domain-specific logic—enables explicit control over agent behavior, simplifies debugging, and can provide formal guarantees under well-specified constraints [4]. This approach is particularly valued in defense and aerospace applications, where mission predictability and adherence to safety standards are paramount.

Rule-based coordination has been extensively explored through frameworks such as finite-state machines, behavior trees [5], and communication protocols [17]. In swarm robotics, a long tradition of using simple local rules to produce emergent global behaviors further validates this approach [2]. The escort scenario proposed in this paper draws on both traditions: it leverages structured, rule-based coordination while navigating the uncertainties of a strategically hostile environment.

2.3 Game-Theoretic and Tactical Models

Theoretical frameworks such as differential games, multi-agent pathfinding, and Stackelberg games have been extensively used to model attacker-defender interactions in both continuous and discrete domains [1,21,22]. These models offer valuable insights into optimal behaviors, often under assumptions of perfect or partial observability and rational play. They also support equilibrium analysis in adversarial settings. However, their practical applicability is frequently constrained by computational intractability and modeling assumptions such as full observability or synchronized movements [25].

More pragmatic approaches, particularly in multi-agent reinforcement learning (MARL), have emerged as data-driven alternatives for strategy learning in complex, high-dimensional environments [10]. Despite significant progress, coordination remains a key bottleneck, especially as agent populations scale and communication becomes limited. MARL systems are prone to non-stationarity, policy oscillations, and failures in achieving consistent coordination under adversarial stress [7,16].

Our work builds upon these foundational lines of research but diverges in one crucial respect: rather than pursuing optimal or learned policies, we systematically evaluate interpretable, rule-based coordination strategies within a fixed threat model. This shifts the focus toward frameworks more aligned with protocol design and tactical decision-making [17], where transparency, repeatability, and robustness across a range of adversarial conditions are of paramount importance.

2.4 Simulation Environments and Evaluation Protocols

There exists a broad spectrum of simulation platforms developed for evaluating multi-agent coordination strategies, from classical environments such as the Multi-Agent Particle Environment (MPE) [14] to more sophisticated frameworks inspired by military or tactical domains, including OpenAI's Gym environments [3], AirSim for aerial robotics [20], and dedicated pursuit-evasion simulators [11]. These environments have enabled the testing of reinforcement learning policies and decentralized strategies at various levels of complexity. However, most are tailored for learning-based agents and offer limited native support for deploying rule-based coordination logic or for systematically varying the threat landscape.

To address this gap, our work introduces a custom environment that explicitly supports the implementation and evaluation of rule-based multi-agent strategies under configurable adversarial pressure. This environment is designed to make the trade-offs between coordination structures visible—revealing, for instance, the tension between decentralization and robustness, or between mobility constraints and defensive effectiveness.

By grounding our experimental protocol in a discrete, interpretable simulation space, we contribute a methodology aimed at filling a persistent gap in the literature: the lack of standardized tools for stress-testing interpretable, rule-based systems in adversarial, time-sensitive coordination scenarios [17,19]. Our goal is to support reproducible testing that emphasizes both agent-level behavior and emergent team dynamics.

3 Environment

We propose a custom-designed aerial combat simulation environment to conduct our studies on optimal strategies in the high-value asset protection scenario. This environment provides a simplified but operationally meaningful abstraction, where a convoy of heterogeneous drones must accomplish a strategic air-to-ground strike mission while defending against adversarial interception. The environment is fully observable and deterministic, offering a controlled space to explore and compare coordination protocols.

The simulation takes place on a discrete two-dimensional grid of size 20×16, where the horizontal axis (columns) spans from 0 to 19, and the vertical axis (rows) from 0 to 15. The allied agents are initialized on the left-hand side of the grid (column 0), and the adversarial region is defined on the right-hand side (columns 17 to 19). The mission objective is to traverse this space and destroy a randomly positioned ground target located in the enemy zone.

There are two types of allied drones in this scenario: a single *air-to-ground* (A2G) drone, whose primary objective is to destroy the ground target, and P *air-to-air* (A2A) drones, whose sole responsibility is to protect the A2G drone from enemy threats. Together, they form a convoy of $P+1$ agents, initialized on the same grid cell in column 0. At each time step, every allied drone may either move to a neighboring cell (North, South, East, or West) or remain stationary (NoOp). All agents are assumed to have perfect situational awareness and unrestricted communication capabilities, which simplifies the coordination dynamics and isolates the contribution of the rule-based decision logic.

The ground target is randomly initialized in one of the cells within columns 17 to 19 and any of the 16 rows. It moves slowly, shifting to a randomly selected adjacent cell every five time steps. A mission is deemed successful if the A2G drone reaches a position within Chebyshev distance 1 of the target (i.e., in any of the eight surrounding cells), resulting in the destruction of the target.

As the convoy progresses eastward, adversarial drones are introduced into the simulation. These enemy drones follow a greedy policy: at each time step, they calculate the shortest path to the A2G drone and move to the adjacent cell that minimizes this distance. The A2G drone is considered destroyed if an enemy drone comes within a Chebyshev distance of 2, representing successful interception.

Combat mechanics are rule-based and proximity-triggered. A2A drones can destroy enemy drones if they are in adjacent cells. However, in scenarios where an enemy drone, an A2A drone, and the A2G drone are all within range of one another, precedence is given to the enemy drone. That is, if both an A2A and an enemy drone are adjacent to the A2G unit, the A2G drone is still destroyed. This prioritization avoids degenerate strategies where A2A drones "shadow" the A2G drone continuously, forming an impenetrable protective shield. Friendly fire is not modeled; A2A drones may occupy the same or adjacent cells as the A2G drone without negative consequences.

Enemy drones spawn dynamically throughout the mission. At time $t = 0$, M enemy drones are randomly generated according to spatial constraints: they must appear beyond a five-cell perimeter around the A2G drone's current position and are restricted to horizontal locations between $x + 2$ and the right boundary of the grid, and vertical locations between $y - 8$ and $y + 8$, where (x, y) is the A2G drone's position. Once all active enemy drones are eliminated, a new wave of M enemy drones is spawned under the same spatial rules. This process continues until the mission ends.

The simulation terminates under one of three conditions: (i) the target is destroyed (success), (ii) the A2G drone is intercepted (failure), or (iii) a maximum of 80 time steps has elapsed. This time limit encourages urgency in mission planning and rewards efficient trajectories.

To assess the performance of the coordination strategies, we define a reward function that reflects both mission outcomes and temporal efficiency. Each time step incurs a penalty of -1, incentivizing swift execution. A reward of $+100$ is given if the target is successfully destroyed, and a penalty of -100 is applied if the A2G drone is eliminated. From this, we derive two evaluation metrics: the average cumulative reward per episode, and the success rate across multiple episodes, defined as the proportion of simulations in which the target was successfully destroyed. Each strategy is tested on 10000 randomly generated tactical situations.

This simulation environment, illustrated on Fig. 1, provides a well-defined, discrete testbed for evaluating the effectiveness of rule-based coordination policies. By abstracting away from low-level control and focusing on high-level tactical reasoning, it serves as a valuable benchmark for exploring the scalability, robustness, and interoperability of coordination protocols under adversarial pressure. Its design also enables future extensions toward more complex communication models, partial observability, and learning-based hybrid strategies.

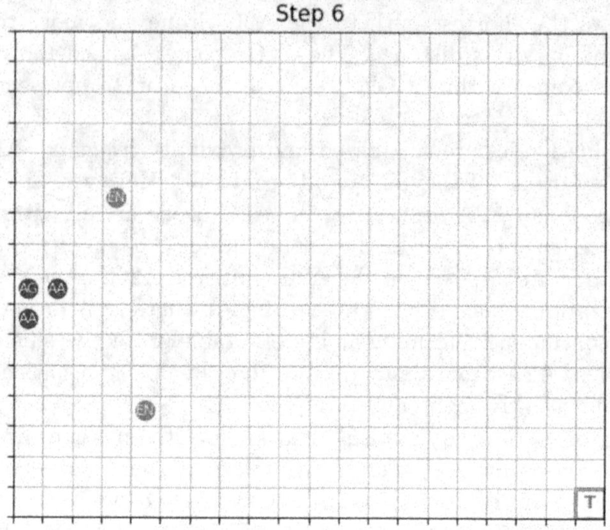

Fig. 1. Illustration of the environment, with 2 air-air (AA) drones facing 2 enemies (EN). The goal of the blue team is to strike the target (T) with the air-to-ground (AG) drone. (Color figure online)

4 Optimal Cooperation

The purpose of this section is to analyze and evaluate cooperation strategies between allied air-air (AA) drones in the defense of a single air-to-ground (A2G) agent against enemy drones. The A2G agent follows a simple evasion policy: whenever an enemy drone enters a Chebyshev distance of 4 cells or less, the A2G agent flees by selecting the action that maximizes its distance from the closest enemy drone.

Our focus is therefore on determining how to allocate the allied AA drones to intercept enemy drones in a manner that maximally protects the A2G agent. The central heuristic we propose is a distance-based assignment rule: each allied drone is allocated to the closest enemy drone, based on the current spatial configuration. However, this problem becomes less straightforward when the number of allied drones exceeds the number of enemies, either due to the nature of the enemy attack or the destruction of one or more enemy drones during the mission. In such scenarios, we investigate alternative coordination strategies that utilize the surplus allied drones.

Two principal approaches are explored. In the first, once each enemy has been assigned a unique allied drone, the remaining unassigned allies are redirected to closely protect the A2G agent. This ensures a layer of direct defense. However, it forgoes redundancy in enemy interception and may lead to suboptimal spatial positioning for handling subsequent waves of enemies. The second approach permits multiple allied drones to be assigned to the same enemy drone, introducing redundancy in the interception process and potentially yielding superior spatial coverage.

The experiments presented in this section evaluate these two strategies across various enemy configurations, with the aim of identifying trade-offs and discovering optimal hybrid policies. The results demonstrate that a balanced approach—allocating some drones to shared interception duties while maintaining partial close protection—achieves superior performance.

4.1 Two Allied Drones

We begin by studying scenarios where the number of allied AA drones is fixed at two, and the number of enemy drones varies. For each enemy configuration, we evaluate two coordination strategies: one that prioritizes diversity in assignments (i.e., each enemy is targeted by a different ally and surplus allies protect the A2G agent), and another that allows redundant assignments (i.e., multiple allies may engage the same enemy). The performance metrics are the overall score (a mission-level performance index) and the number of targets killed by the A2G drone.

Two Enemy Drones. In the configuration where two AA drones face two enemy drones, the assignment rule based on distance naturally yields a one-to-one mapping. Both allied drones are allocated to the nearest enemy. When there is one remaining enemy drone, the closest AA drone is assign with the attack of the enemy, whereas the other AA drone goes close to the A2G drone. This strategy yields a score of 50 ± 36 and an average of 0.96 ± 0.21 enemy targets neutralized per episode. The second strategy, where both allied drones are assign the remaining drone yields a slightly improved outcome with a score of 53 ± 35 and a target elimination rate of 0.96 ± 0.20. Although the difference is marginal, it suggests that in cases of symmetric configurations, redundant targeting does not necessarily degrade performance and may offer resilience.

One Enemy Drone. In cases where a single enemy drone is confronted by two AA allies, the strategic options diverge. The first approach allocates only the nearest ally to intercept the enemy, while the second is redirected toward close protection of the A2G agent. This yields a score of 62 ± 19 and a high target kill rate of 0.99 ± 0.092. The alternative strategy, where both allies converge on the single enemy, produces similar results: a slightly lower score of 61 ± 26 and a comparable target kill rate of 0.98 ± 0.13. This indicates that redundant interception does not compromise mission success in this configuration, but neither does it significantly enhance it.

Three Enemy Drones. When two AA drones face three enemy drones, the assignment problem becomes under-constrained. In this setting, some enemy

Table 1. Results for 3 AA drones versus 3 enemies drones

Rules for 2 drones remaining	Rules for 1 drone remaining	Scores	Target killed (%)
2 E/1 A2G	1 E/2 A2G	49 ± 36	0.95 ± 0.21
2 E/1 A2G	2 E/1 A2G	52 ± 29	0.97 ± 0.16
2 E/1 A2G	3 E	53 ± 32	0.97 ± 0.18
3 E	1 E/2 A2G	51 ± 33	0.96 ± 0.19
3 E	2 E/1 A2G	54 ± 30	0.97 ± 0.16
3 E	3 E	51 ± 38	0.95 ± 0.21

drones remain unengaged by allied drones. The first strategy continues to allocate the nearest allies to the nearest threats, leaving at least one enemy drone free to approach the A2G agent. Under this strategy, the results are considerably degraded, with a score of -0.59 ± 82 and an average of 0.69 ± 0.46 targets eliminated.

The redundant strategy, in which both allies are reassigned to the same enemy when one remain, fares even worse. In this case, the score drops to -12 ± 87, with only 0.62 ± 0.49 targets eliminated on average. This outcome emphasizes the limits of redundant targeting when enemy presence outnumbers allied capacity: unaddressed threats can penetrate defense lines and endanger the A2G agent.

The goal of this subsection was to introduce our methodology to find the best assignment scheme. The findings in this subsection highlight that, when the enemies outnumber the allies, redundant assignments severely degrade results, and close protection is preferable. However, this has to be treated cautiously, since there are only a small number of allied AA drones. In the next subsection, we extend this analysis to configurations involving three allied AA drones.

4.2 Three Allied Drones

In this scenario, three allied air-to-air (AA) drones are deployed against enemy drones, always with the goal of neutralizing threats and protecting a designated air-to-ground (A2G) drone. The engagement logic is governed by rules that adapt to the number of enemy drones remaining, and different strategies are tested to assess their impact on mission performance.

Three Enemy Drones. Initially, when all three enemy drones are still active, the standard behavior across all tested configurations is to assign each allied AA drone to engage the nearest enemy drone. Once again, this distance-based allocation ensures a fast response and maximizes the likelihood of interception before the enemy can threaten the A2G asset.

When the number of enemy drones decreases to two, several decision rules are evaluated. In some strategies, two of the allied AA drones are assigned to the remaining enemy drones based on proximity, while the third AA drone is redirected to protect the A2G drone in close formation. This behavior is denoted as the "2 E/1 A2G" configuration in the results table. The other behavior, denoted as "3 E", assigns two allied drones to the two remaining enemies, and the third drone is assign to its closest enemy.

When only a single enemy drone remains, strategies diverge further. One approach dispatches the closest allied drone to intercept the enemy while the remaining two provide close-range escort to the A2G drone, a configuration labeled "1 E/2 A2G". Another variation maintains offensive pressure by assigning two or all three AA drones to engage the last remaining threat ("2 E/1 A2G" or "3 E", respectively), based on distance, emphasizing rapid neutralization over protective coverage.

The results, summarized in Table 1, reveal that the "3 E then 2 E/1 A2G" strategy combination yields the highest score (54 ± 30) and target kill rate (0.97 ± 0.16), highlighting the effectiveness of sustained offensive pressure when supported by adaptive engagement policies. On the other hand, configurations favoring early reallocation of drones for protection, such as "2 E/1 A2G" followed by "1 E/2 A2G tend to result in lower scores and slightly reduced target kill ratios, indicating a trade-off between direct engagement and protective behavior.

Two Enemy Drones. In this scenario, three allied AA drones are deployed against only two enemy drones. This simplified tactical context provides an opportunity to analyze how different allocation rules affect performance in a favorable engagement.

As in the previous scenario, initial targeting when both enemy drones are present follows the defined rule for the two-drone phase. Configurations vary between assigning two AA drones to intercept the enemy while the third escorts the A2G drone ("2 E/1 A2G"), and committing all three AA drones to offensive action ("3 E").

When the number of enemy drones is reduced to one, multiple reallocation strategies are explored. These include sending only one AA drone to neutralize the final threat while the remaining two protect the A2G asset ("1 E/2 A2G"), or maintaining offensive pressure with two or all three drones continuing the engagement ("2 E/1 A2G" and "3 E", respectively).

The results, shown in Table 2, demonstrate an overall improvement in both scores and kill rates compared to the 3 vs. 3 scenario. The best-performing configuration assigns all three AA drones to the enemy in the two-drone phase and retains two of them for the final engagement ("3 E then 2 E/1 A2G"), achieving the highest score (61 ± 15) and a perfect or near-perfect kill rate (1.00 ± 0.07). Notably, even strategies that reallocate one or two drones to protective roles at later stages show relatively strong performance, suggesting that in this more favorable setting, flexibility in defensive coverage does not significantly compromise mission outcomes.

Table 2. Results for 3 AA drones versus 2 enemies drones

Rules for 2 drones remaining	Rules for 1 drone remaining	Scores	Target killed (%)
2 E/1 A2G	1 E/2 A2G	53 ± 30	0.97 ± 0.17
2 E/1 A2G	2 E/1 A2G	58 ± 20	0.99 ± 0.10
2 E/1 A2G	3 E	59 ± 14	1.00 ± 0.07
3 E	1 E/2 A2G	53 ± 36	0.97 ± 0.18
3 E	2 E/1 A2G	61 ± 15	1.00 ± 0.07
3 E	3 E	59 ± 25	0.98 ± 0.12

Three Allied Drones Versus Four Enemy Drones. This scenario increases the difficulty by pitting three allied AA drones against four enemy drones. The numerical disadvantage poses a significant challenge, especially in terms of protecting the A2G asset while managing offensive pressure.

As before, allocation decisions differ across configurations. During the two-enemy-drone phase, we test the strategy reallocating one AA drone to escort duties ("2 E/1 A2G"), while others continue assigning all drones to offensive engagement ("3 E"). Once only one enemy drone remains, once again, we test the "1 E/2 A2G", "2 E/1 A2G", and "3 E" strategies.

Table 3 reports the results, which reflect the increased difficulty of the scenario. Overall scores and kill rates are significantly lower than in the previous configurations. The best result is observed when all three AA drones are used offensively until only one enemy remains, at which point two are reassigned for final engagement ("3 E then 2 E/1 A2G. This configuration achieves a score of 33 ± 63 with a target kill rate of 0.87 ± 0.34. This shows that the trade-off between close protection of the A2G asset and the redundancy during attacks is beneficial under unfavorable conditions.

Interestingly, the configuration that commits all drones offensively until the end ("3 E/3 E") performs poorly in this setting, with the lowest score (22 ± 72) and kill rate (0.80 ± 0.40). This result underscores the importance of balancing attack and defense when facing a superior number of threats.

General Insights on Rule-Based Allocation Strategies. The analysis of rule-based allocation strategies across different scenarios—ranging from balanced (3 vs. 3) to advantageous (3 vs. 2) and finally disadvantageous (3 vs. 4) engagements—reveals consistent patterns in the relationship between tactical decisions and mission outcomes.

First, configurations that prioritize offensive engagement by assigning all three AA drones to enemy targets tend to perform best when the threat level is moderate or low. In the 3 vs. 2 scenario, such aggressive strategies consistently yield high scores and nearly perfect kill rates, confirming the effectiveness of overwhelming force when the enemy is outnumbered or matched evenly.

Table 3. Results for 3 AA drones versus 4 enemies drones

Rules for 2 drones remaining	Rules for 1 drone remaining	Scores	Target killed (%)
2 E/1 A2G	1 E/2 A2G	28 ± 64	0.85 ± 0.35
2 E/1 A2G	2 E/1 A2G	36 ± 57	0.89 ± 0.31
2 E/1 A2G	3 E	29 ± 65	0.84 ± 0.36
3 E	1 E/2 A2G	29 ± 64	0.85 ± 0.35
3 E	2 E/1 A2G	33 ± 63	0.87 ± 0.34
3 E	3 E	22 ± 72	0.80 ± 0.40

Second, as the threat level increases, a more nuanced allocation becomes necessary. In the 3 vs. 3 scenario, the best performances are obtained when drones are reallocated dynamically—intensifying the offensive effort when only one enemy remains while preserving enough defensive coverage to protect the A2G asset. This balance appears crucial in situations where the outcome hinges on small tactical advantages.

Third, in the most adverse condition (3 vs. 4), the overall degradation of performance metrics suggests the limits of rule-based strategies under numerical inferiority. However, even in this setting, a conservative reallocation—preserving some offensive pressure while increasing protection—offers better outcomes than purely offensive postures. Notably, the configuration that retains two AA drones on the final enemy while one protects the A2G drone ("3 E then 2 E/1 A2G") consistently outperforms the purely offensive counterpart ("3 E then 3 E"), both in terms of score and target survival rate.

These findings highlight the importance of adaptive rule design: strategies must not only consider the current number of enemy drones, but also anticipate the evolving tactical situation and adjust the balance between offensive and defensive priorities accordingly. While aggressive rules can be effective in favorable scenarios, their limitations become apparent as the threat level increases—calling for more strategic flexibility.

5 Limitations, Discussion, and Conclusion

While the rule-based system developed in this study exhibits strong performance in a variety of tactical situations, it is not without its limitations. One of the most salient weaknesses lies in its occasional failure even under scenarios of tactical superiority. This phenomenon can be attributed to unfavorable initial configurations not accounted for by the predefined rules.

To illustrate this point, let us consider a specific configuration, as illustrated in Fig. 2. An A2G drone is positioned centrally within the domain, while two AA drones are located slightly further east and north. Now, suppose an enemy drone appears directly east of the A2G unit, moving westward in its direction. In accordance with its self-protection protocol, the A2G drone will begin to flee

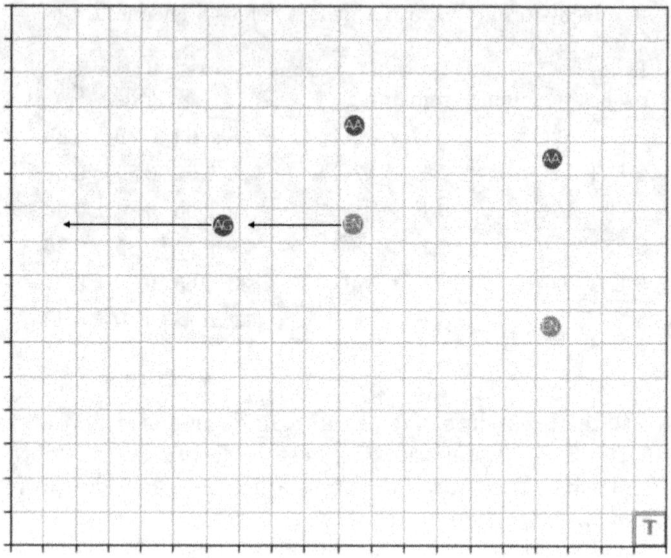

Fig. 2. The A2G drone flees west the enemy, and the AA drones are too far away to neutralize the threat. This leads to a mission failure, while a simple north move could potentially brings the enemy closer to the AA drone...

westward, away from the threat. Simultaneously, the AA drones, whose behavior is governed by a rule minimizing their distance to the enemy they are assigned to, will also move westward, rather than descending southward to intercept. This leads to a situation in which the A2G drone may find itself cornered at the western boundary of the domain, potentially leading to mission failure. Ironically, a simple vertical maneuver to the north would have forced the enemy to follow suit and potentially led them directly into the engagement range of the AA units.

This example highlights a fundamental limitation of rule-based systems: their inherent rigidity. While such systems enable standardization of agent behavior and are appealing for reasons of simplicity and interoperability, this standardization constrains the system's ability to adapt to unanticipated situations. In scenarios where nuance or situational awareness is critical, the fixed nature of rule-based strategies can become a liability.

To overcome the ceiling imposed by static rules, we propose the adoption of a multi-agent reinforcement learning (MARL) approach. To the best of our knowledge, MARL represents the most promising path toward improving the coordination and adaptability of such systems. Nevertheless, training a multi-agent system from scratch poses a formidable challenge, both in terms of computational cost and the intrinsic instability of the learning process. Consequently, we envision a staged learning strategy in which the current rule-based system is used to generate expert demonstrations. These demonstrations can then be leveraged to bootstrap a multi-agent imitation learning phase, serving as a stable initialization prior to deploying reinforcement learning. The implementation of this two-phase learning process is left for future work.

Another pertinent issue concerns the predictability of the rule-based system proposed. If the behavior of our agents can be anticipated by adversaries, then the effectiveness of the system may be compromised. This brings forth the broader challenge of adversarial robustness. In future studies, we intend to explore the optimization of enemy generation, specifically aiming to identify and exploit vulnerabilities in the learned behaviors of our agents. Such adversarial testing will be crucial to ensure that the system maintains robust performance under more sophisticated threats.

This discussion naturally leads to a broader reflection on military superiority in cooperative autonomous systems. While finely-tuned cooperation protocols can indeed be anticipated by adversaries, they are only part of the equation. In many scenarios, numerical superiority plays a dominant role. As observed in our experiments, increasing the number of allied drones by even one unit can lead to significantly improved performance, sometimes reaching 100% success rates. This emphasizes the importance of quantitative dominance in the broader tactical landscape.

In conclusion, we have proposed a rule-based cooperation protocol for the protection of a high-value asset (HVA) in a simplified grid-based environment. Despite its inherent limitations, the proposed system provides a solid and operational foundation for standardization. This opens up the possibility of ensuring interoperability among heterogeneous platforms—for instance, AA drones from different manufacturers or countries. While future improvements will likely involve adaptive learning-based systems, the rule-based approach remains a valuable and pragmatic step toward coordinated autonomy in defense scenarios.

Acknowledgments. This publication was co-funded by the European Union under the Grant Agreement 101103669-EICACS (European Initiative on Collaborative Air Combat Standardization). Its contents are the sole responsibility of the author and do not necessarily reflect the views of the European Union or the European Commission. Neither the European Union nor the granting authority can be responsible for them.

References

1. Basar, T., Olsder, G.J.: Dynamic Noncooperative Game Theory. SIAM (1999)
2. Brambilla, M., Ferrante, E., Birattari, M., Dorigo, M.: Swarm robotics: a review from the swarm engineering perspective. Swarm Intell. **7**(1), 1–41 (2013)
3. Brockman, G., et al.: OpenAI gym. arXiv preprint arXiv:1606.01540 (2016)
4. Brooks, R.A.: A robust layered control system for a mobile robot. IEEE J. Robot. Autom. **2**(1), 14–23 (1986)
5. Colledanchise, M., Ögren, P.: Behavior Trees in Robotics and AI: An Introduction. CRC Press (2018)
6. Duan, Y., Cui, B.X., Xu, X.H.: A multi-agent reinforcement learning approach to robot soccer. Artif. Intell. Rev. **38**, 193–211 (2012)

7. Foerster, J., Nardelli, N., Farquhar, G., et al.: Stabilising experience replay for deep multi-agent reinforcement learning. In: International Conference on Machine Learning, pp. 1146–1155 (2017)
8. Gerkey, B.P., Mataric, M.J.: A formal analysis and taxonomy of task allocation in multi-robot systems. Int. J. Robot. Res. **23**(9), 939–954 (2004)
9. Hasan, Y.A., Garg, A., Sugaya, S., Tapia, L.: Defensive escort teams for navigation in crowds via multi-agent deep reinforcement learning. IEEE Robot. Autom. Lett. **5**(4), 5645–5652 (2020)
10. Hernandez-Leal, P., Kartal, B., Taylor, M.E.: A survey and critique of multi-agent reinforcement learning. ACM Comput. Surv. (CSUR) **51**(6), 1–35 (2019)
11. Isaacs, R.: Differential Games: A Mathematical Theory with Applications to Warfare and Pursuit, Control and Optimization. Courier Corporation (1965)
12. Kurach, K., Raichuk, A., Stanczyk, P., et al.: Google research football: a novel reinforcement learning environment. arXiv preprint arXiv:1907.11180 (2020)
13. Lee, E.S., Zhou, L., Ribeiro, A., Kumar, V.: Graph neural networks for decentralized multi-agent perimeter defense. Front. Control Eng. **4**, 1104745 (2023)
14. Lowe, R., Wu, Y., Tamar, A., Harb, J., Abbeel, P., Mordatch, I.: Multi-agent actor-critic for mixed cooperative-competitive environments. In: Advances in Neural Information Processing Systems, vol. 30 (2017)
15. Olfati-Saber, R.: Flocking for multi-agent dynamic systems: algorithms and theory. IEEE Trans. Autom. Control **51**(3), 401–420 (2006)
16. Papoudakis, G., Christianos, F., Schramowski, P., Albrecht, S.V.: Benchmarking multi-agent deep reinforcement learning algorithms in cooperative tasks. arXiv preprint arXiv:2006.07869 (2021)
17. Poslad, S.: Specifying protocols for multi-agent systems interaction. ACM Trans. Autonomous Adapt. Syst. (TAAS) **2**(4), 15–es (2007)
18. Puiutta, E., Veith, T.: Explainable reinforcement learning: a survey. Proceedings of the 19th International Conference on Autonomous Agents and MultiAgent Systems, pp. 1114–1122 (2020)
19. Raboin, E., Kuter, U., Nau, D.: Generating strategies for multi-agent pursuit-evasion games in partially observable Euclidean space. In: Proceedings of the 11th International Conference on Autonomous Agents and Multiagent Systems, vol. 3, pp. 1201–1202. Citeseer (2012)
20. Shah, S., Dey, D., Lovett, C., Kapoor, A.: AirSim: high-fidelity visual and physical simulation for autonomous vehicles. In: Field and Service Robotics, pp. 621–635. Springer (2018)
21. von Stackelberg, H.: Marktform und gleichgewicht. Springer (1934)
22. Stern, R., Atzmon, D., Boyarski, E., et al.: Multi-agent pathfinding: definitions, variants, and benchmarks. AI Mag. **43**(1), 32–42 (2022)
23. Strouse, D., McKee, K., Botvinick, M., Hughes, E., Everett, R.: Collaborating with humans without human data. Adv. Neural. Inf. Process. Syst. **34**, 14502–14515 (2021)
24. Vinyals, O., Babuschkin, I., Czarnecki, W., et al.: Grandmaster level in StarCraft II using multi-agent reinforcement learning. Nature **575**(7782), 350–354 (2019)
25. Zhang, K., Yang, Z., Başar, T.: Multi-agent reinforcement learning: a selective overview of theories and algorithms. In: Handbook of Reinforcement Learning and Control, pp. 321–384 (2021)

RedTeamLLM: An Agentic AI Framework for Offensive Security

Brian Challita[1(✉)] and Pierre Parrend[1,2]

[1] Laboratoire de Recherche de l'EPITA, 14–16 Rue Voltaire, 94270 Le Kremlin-Bicêtre, France
{brian.challita,pierre.parrend}@epita.fr
[2] ICube, UMR 7357, Université de Strasbourg, CNRS, 300 bd Sébastien Brant, CS 10413, 67412 Illkirch Cedex, France

Abstract. From automated intrusion testing to discovery of zero-day attacks before software launch, agentic AI calls for great promises in security engineering. This strong capability is bound with a similar threat: the security and research community must build up its models before the approach is leveraged by malicious actors for cybercrime. We therefore propose and evaluate RedTeamLLM, an integrated architecture with a comprehensive security model for automatization of pentest tasks. RedTeamLLM follows three key steps: summarizing, reasoning and act, which embed its operational capacity. This novel framework addresses four open challenges: plan correction, memory management, context window constraint, and generality vs. specialization. Evaluation is performed through the automated resolution of a range of entry-level, but not trivial, CTF challenges. The contribution of the reasoning capability of our agentic AI framework is specifically evaluated.

Keywords: Cyberdefense · AI for cybersecurity · generative AI · Agentic AI · offensive security

1 Introduction

The recent strengthening of Agentic AI [10] approaches poses major challenges in the domains of cyberwarfare and geopolitics [23]. LLMs are already commonly used for cyber operations for augmenting human capabilities and automating common tasks [3,38]. They already pose significant ethical and societal challenges [21], and a great threat of proliferation of cyberdefence and -attack capabilities, which were so far only available for nation-state level actors. Whereas there current recognized capabilities are still bound to the rapid analysis of malicious code or rapid decision taking in alert triage, and they pose significant trust issues [28], there expressivity and knowledge-base are rapidly ramping up. In this context, Agentic AI, *i.e.* autonomous AI systems that are capable of performing a set of complex tasks that span over long periods of time without human supervision [1], is opening a brand new type of cyberthreat. They follow two

Published by Springer Nature Switzerland AG 2026
E. Mercier-Laurent et al. (Eds.): AI4S 2025, IFIP AICT 780, pp. 337–354, 2026.
https://doi.org/10.1007/978-3-032-21184-2_26

complementary strategies: goal orientation, and reinforcement learning, which have the capability to dramatically accelerate the execution of highly technical operations, such as cybersecurity actions, while supporting a diversification of supported tasks.

In the defense landscape, cyberwarfare takes a singular position, and targets espionage, disruption, and degradation of information and operational systems of the adversary. More than in traditional arms, skill is a strong limiting factor, especially since targeting critical defense systems heavily relies on the exploitation of rare, unknown vulnerabilities, which are most often than not 0-days threats. Actually, whereas financial criminality aims at money extorsion and thus targets a broad range of potential victims to exploit the weakest ones, defense operations aim at entering and disrupting highly exposed, and highly protected, technical environments, where known vulnerabilities are closed very quickly. In this context, operational capability relies so far in talented analysts capable of discovering novel vulnerabilities. This high-skill, high-mean game could face a brutal end with the advent of tools capable of discovering new exploitable flows at the heart of the software, thus enabling smaller actors to exhibit a highly asymetric threats capable of disrupting critical infrastructures, or launching large-scale disinformation campaigns. Agentic AI has the capability to provide such a tool, and LLMs themselves in their stand-alone versions, have already proved capable of detecting these famous 0-day vulnerabilities: Microsoft has published, with the help of its Copilot tools, no less that 20 (!!) vulnerabilities in the Grub2, U-Boot and barebox bootloaders since late 2024[1].

This is the public side of the medal, by a company who seeks to advertise its software development environment, and create some noise on vulnerabilities on competing operating systems. No doubt malicious actors have not waited to take the same tool at their advantage to unleash novel capabilities to their arsenal, beyond the malicious generative tools analyzed by the community: WormGPT[2], DarkBERT [13], FraudGPT [5]. In the domain of autonomous offensive cybersecurity operations, the probability and likely impact of proliferation of agentic AI frameworks are high. Understanding their mechanism to leverage these tools for defensive operations, and for being able to anticipate their malicious exploitation, is therefore an urgent requirement for the community.

We therefore propose the RedTeamLLM model to the community, as a proof-of-concept of the offensive capabilities of Agentic AI. The model encompasses automation, genericity and memory support. It also defines the principles of dynamic plan correction and context window contraint mitigation, as well as a strict security model to avoid abuse of the system. The evaluations demonstrate the strong competitivity of the model wrt. state-of-the-art competitors, as well as the necessary contribution of its summarizer, reasoning and act components. In particular, RedTeamLLM exhibit a significant improvement in automation capability against PenTestGPT [4], which still show restricted capacity.

---

[1] https://www.microsoft.com/en-us/security/blog/2025/03/31/analyzing-open-source-bootloaders-finding-vulnerabilities-faster-with-ai/.

[2] https://flowgpt.com/p/wormgpt-6.

The remainder of this paper is organised as follows: Sect. 2 presents the state of the art. Section 3 defines the requirements, and Sect. 4 present the RedTeam-LLM model for agentic-AI based offensive cybersecurity operations. Section 5 presents the implementation and Sect. 6 the evaluation of the model. Section 7 concludes this work.

2 State of the Art

The advent, under the form of LLMs, of computing processes capable of generating structured output beyond existing text, is a key driver for a renewed development of agent-based models, with so-called 'agentic AI' models [26], which are able both to devise technical processes and technically correct pieces of code. These novel kind of agents support multiple, complex and dynamic goals and can operated in dynamic environments while taking a rich context into account [1]. They thus open novel challenges and opportunity, both as generic problem-solving agents and for highly complex and technical environment like cybersecurity operations.

2.1 Research Challenges for Agentic AI

The four main challenges in Agentic AI are: analysis, reliability, human factor, and production. These challenges can be mapped to the taxonomy of prompt engineering techniques by [25]: *Analysis:* Reasoning and Logic, knowledge-based reasoning and generation, meta-cognition and self-reflection; *Reliability:* reduce hallucination, fine-tuning and optimisation, improving consistency and coherence, efficiency; *Human factor:* user interaction, understanding user intent, managing emotion and tones; *production:* code generation and execution.

The first issue for supporting reasoning and logic is the capability to address complex tasks, to decompose them and to handle each individual step. The first such model, chain-of-thought (CoT), is capable of structured reasoning through step-by-step processing and proves to be competitive for math benchmarks and common sense reasoning benchmarks [34]. Automatic chain-of-thought (Auto-CoT) automatize the generation of CoTs by generating alternative questions and multiple alternative reasoning for each to consolidate a final set of demonstrations [39]. Trees-of-thought (ToT) handles a tree structure of intermediate analysis steps, and performs evaluation of the progress towards the solution [36] through breadth-first or depth-first tree search strategies. This approach enables to revert to previous nodes when an intermediate analysis is erroneous. Self consistency is an approach for the evaluation of reasoning chains for supporting more complex problems through the sampling and comparative 1evaluation of alternative solutions [31].

Text generated by LLM is intrinsically a statistical approximation of a possible answer: as such, it requires 1) a rigorous process to reduce the approximation error below usability threshold, and 2) systematic control by a human operator. The usability threshold can be expressed in term of veracity, for instance in the domain of news[3]. For code generation, it matches code that is both correct and effective, *i.e.* that compiles and run, and that perform expected operation. Usable technical processes, like in red team operations, are defined by reasoning and logic capability. reducing hallucination: Retrieval Augmented Generation (RAG) for enriching prompt context with external, up-to-date knowledge [17]; REact prompting for concurrent actions and updatable action plans, with reasoning traces [37].

One key issue for red teaming tasks is the capability to produce fine-tuned, system-specific code for highly precise task. Whereas the capability of LLMs to generate basic code in a broad scope of languages is well recognized [20], the support of complex algorithms and target-dependent scripts is still in its infancy. In particular, the articulation between textual, unprecise and informal reasoning and lines of code must solve the conceptual gap between the textual analysis and the executable levels. Structured Chain-of-Thought [19] closes this gap by enforcing a strong control loop structure (if-then; while; for) at the textual level, which can then be implemented through focused code generation. Programatically handling numeric and symbolic reasoning, as well as equation resolution, requires a binding with external tools, such as specified by Program-of-Thought (PoT) [2] or Chain-of-Code (CoC) [18] prompting models. However, these features are not required in the case of read teaming tasks.

2.2 Cognitive Architectures

Three main architectures implement the Agentic AI approach: ReAct (Reason and Act), ADaPT (As needed Decomposition and Planning) and P&E (Plan and Execute).

ReAct [37] first reasons about the analysis strategy, then rolls out this strategy. It performs multiple rounds of reasoning and acting, executing one action at each round then collecting observation. This enables a strong reduction of the error margin. As shown in Fig. 1, ReAct input is built with an explicit objective and an optional context. Reasoning then summarizes the goal and context and plan next action, each through a call to an LLM agent. The selected action is then executed, again based on an LLM call. If the analysis is not completed, the pipeline returns to the goal definition step, with a given subgoal. If the goal is achieved, the pipeline terminates. The main limits of this architecture, whether it is used with prompting or with complex pipelines, is the absence of memory, which requires each prompt to embed all context and knowledge about previous analysis steps. Since the context windows of current LLMs are strongly limited, information start being ignored as the context and history start exceeding the context window's limit, which can lead to reduced performance and inaccurate outputs.

ADaPT [24] takes a greedy approach to decomposition: it keeps decomposing the task until it reaches subtasks that can be executed, through recursive decomposition which avoids a saturation of agent capability. The decomposition stops

[3] https://www.cjr.org/tow_center/we-compared-eight-ai-search-engines-theyre-all-bad-at-citing-news.php.

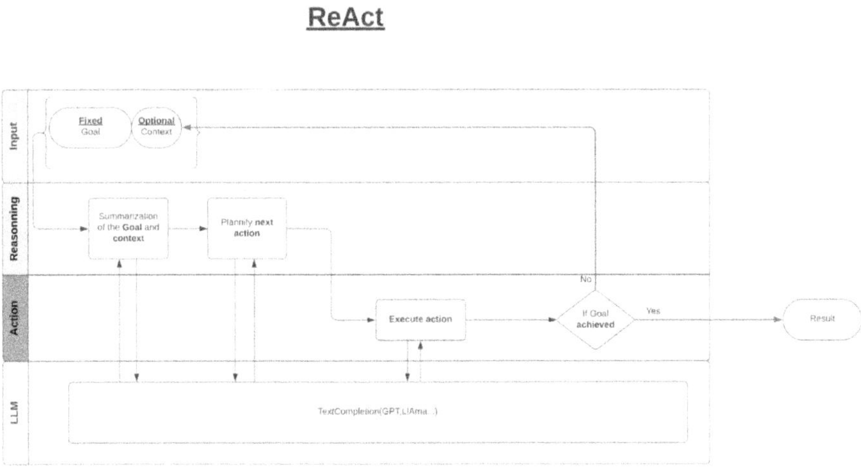

Fig. 1. Process diagram of ReAct

either when a task can be executed directly, or when a max depth is reached. Unlike ReAct and P&E, ADaPT can't be a prompting method as it is based on recursion. ADaPT completely solves the problem of context window size restriction, by decomposing as much as needed. Execution of leaves are then be carried out independently. However, many complications come along the way: plan correction (if a task fails completely, how can we correct the rest of the plan?) and new discoveries (the agent might stumble upon information that can lead to a complete change of plan), in particular, are not supported.

P&E [29] aims to decompose a task into multiple subtasks that are executed independently from one another. This architecture defines first solutions to ReAct's weak points, by decomposing a task and isolating the subtask's execution. Prompt's length is thus minimized, which slows down the consumption of the context window capacity. Task execution becomes more efficient. However, one key issue remains: context window is eventually reached; and a new one is introduced: error handling, since, on a subtask's failure, the whole execution fails.

2.3 Agentic AI and Cybersecurity

Recent offensive-security agents all converge on a narrow design spectrum: a frontier LLM in a ReAct-style loop that plans, executes a single tool call, observes, then repeats [8]—yet none of them store or revise a global plan the way ADaPT or other deliberative-memory systems do. AutoAttacker couples ReAct with an episodic "Experience Manager," but that memory is consulted only to validate the current action rather than to update or back-track the plan itself [35]. LLM-Directed Agent preserves the classic four-stage ReAct chain (NLTG → CFG → CG → NLTP) and likewise discards alternative branches once the CFG selects one [16]. One-Day Vulnerabilities' Exploit [6] and Hack-Websites [7] expose different toolsets to the same ReAct controller, and performance collapses as soon as GPT-4 is replaced by weaker models. CyberSecEval 3 uses an even leaner single-prompt ReAct wrapper to probe Llama-3 and contemporaries, finding that all models stall long before complex exploitation [30]. HackSynth strips the pattern down to just a Planner and a Summariser—still a think-then-act loop—and shows that temperature and context-window size, not architectural novelty, dominate success rates [22]. The sole departure from ReAct is PenTestAgent, which hard-codes a pentesting workflow (Reconnaissance → Search → Planning → Execution) without agentic recursion [27], and PenTestGPT, whose Plan-and-Execute modules shuffle intermediate results between Reasoning, Generation and Parsing stages but never revisit earlier strategies once execution starts [4]. Although defensive models exhibit promising properties [11], the exploitation of Agentic AI for malicious operations is a key concern to the community [21]. Across current systems, memory is used only as a scratch-pad for latest observations; none implement hierarchical plan refinement, long-horizon memory, or roll-back of faulty plans.

3 Requirements

In this section we explicit the specific challenges of agentic AI offensive cybersecurity operations. We address context window's limit, continuous improvement, genericity and automation. One major issue of LLM agent-based systems is their limited context window. Complex tasks usually require many iterations between the agent and a changing environment especially using ReAct, so tracking what has happened is essential for high-quality results. A common way to address this challenge is recursive planning [24], in which a task is broken down into many subtasks that are executed individually; each subtask then passes the key points of its outcome to the next ones. A difficulty arises when a subtask fails, potentially blocking the subtasks that follow. To prevent this, a plan-correction mechanism [32] is applied: whenever a subtask fails, the overall plan is adjusted so execution can proceed smoothly. These two techniques are crucial for building a high-performance agent, but further refinements are still possible. Repeating the same mistakes on every run wastes time, money, and computation. Introducing a memory manager during task planning lets the agent avoid exploratory paths that have already failed. Moreover, genericity is essential. Allowing the agent full freedom to choose its own tools and techniques fosters creativity and broadens its capabilities beyond a fixed toolset. In our case, the agent has unrestricted execution privileges through root access to a terminal. Finally a key part to consider is automation; refining an agent system is important but not useful unless the whole process is automated, not requiring human interaction during the process. Thus, integrating a tool call of an interactive terminal access within this context is rudimentary.

Consolidated requirements for our penetration-testing agent are thus:

1. **Dynamic Plan Correction**—Handling subtask or action failures without halting the entire workflow [33].
2. **Memory Management**—Managing large amounts of contextual data in long-running tasks, which enables continuous self-improvement.
3. **Context Window Constraints mitigation**—Preventing critical information loss due to an LLM's limited prompt size [37].
4. **Generality vs. Specialization**—Balancing the need for specialized pentesting tools with broader adaptability.
5. **Automation**—Automating the interaction of the agent with its designated environment; in our case a terminal.

4 RedTeamLLM

In this section, we propose a novel architecture, supported features and related memory management mechanism for an offensive cybersecurity agentic model. Given the high capability and autonomy of the RedTeamLLM model, a robust security model is also required.

4.1 The Architecture

The architecture of RedTeamLLM is composed of seven components 2: **Launcher**, **RedTeamAgent**, **Memory Manager**, **ADaPT Enhanced**, **Plan Corrector**, **ReAct**, and **Planner**. On a run, the Launcher retrieves the input task and gives it to the **RedTeamAgent** while acting as the user interface (showing number of tasks running, memory access, failed and successful tasks, and allowing intervention in a task's operation, e.g., stopping it or modifying its plan). Upon receiving the task, the **RedTeamAgent** has two objectives: pass it to **ADaPT Enhanced** and await a tree structure representing the full agent execution, then save that structure to the **Memory Manager**. **The Memory Manager**, which is the storage area for operation's knowledge, embeds and stores each node's description from a task tree in a database, thus providing full access to previous task structures and dependencies. **ADaPT Enhanced** then takes that task, passes it to the **Planner** (which returns a tree of subtasks) and traverses it to execute leaves and pass results to siblings. The **Plan Corrector** can then adjust the plan and resume execution on any failure. All leaf executions are performed by the **ReAct** component, which carries out multiple rounds of reasoning, execution, and observations with terminal access.

4.2 Features

To support autonomous offensive operations, the proposed model must address many challenges and effectively meet essential requirements. Thus here are the principal features:

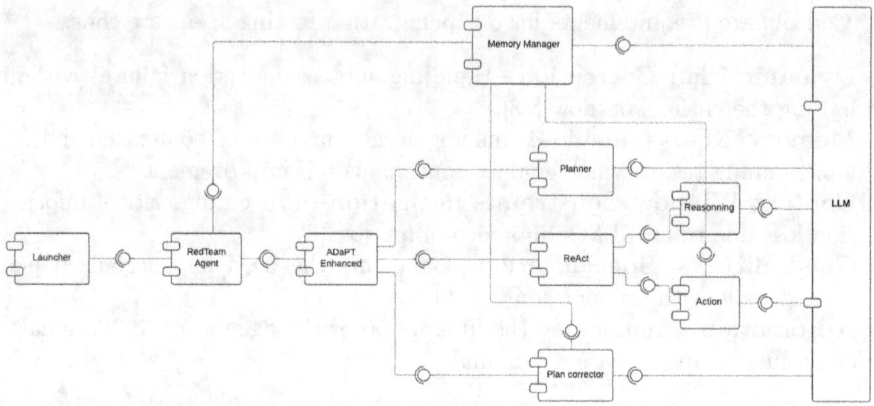

Fig. 2. Software Architecture for Red Team LLM Model

- To address **context window's limit** the model needs to decompose a task recursively as much as needed. This is accomplished by the **ADaPT** component.
- With subtasks comes dependencies that need monitoring to avoid fatal execution failure on error. Here comes the **plan corrector** that has the ability to modify a task's accordingly to lattest outcomes.
- In order to support continuous improvement of the model capabilities, the **Memory Management** comes to improve planning over time by storing past execution in a tree-like way 3.
- Finally the model needs to be generic and avoid restriction to cover a wider range of tasks and not be limited to a set of tools. Here comes **ReAct** with full terminal access.This allows full automation, having full control and autonomy on the task it is executing.

4.3 Memory Management

Memory management is an essential part of the model to be implemented. In all other competing models, memory is just used at the execution part to retrieve already executed commands for a similar task. In our case, memory is used at a higher phase in which the agent decides how to create the execution plan. In fact, at the end of each execution the traces of the whole process are stored in form of a tree that is saved using task's description embedding. Thus, at every decomposition, the planner queries the saved node's hence having access to their success/failure reason, sub-tasks and detailed execution. This technique helps the agent improve over time and especially when re executing a task where he eventually narrows all the possibilities to the right path. This way the **RedTeam-LLM model**, improves over time and has more chances to complete a task over multiple rounds of execution.

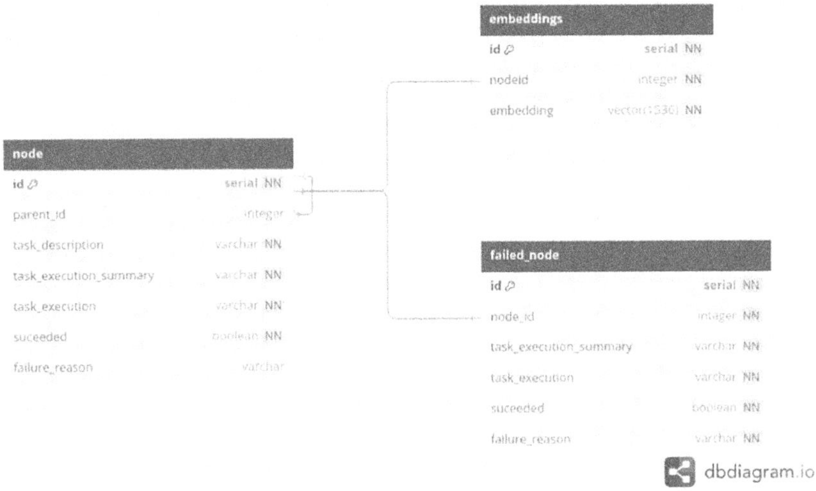

Fig. 3. Database schema for Memory management Model

4.4 The Security Model

The Red-Team LLM architecture supports a powerful autonomous process for pen-testing, including error recovery when the process meets dead-ends, and automation of offensive action. The architecture is thus exposed to two main threat families: hijacking of the execution process, on the one hand, and inversion of dependency from the LLM agents towards the framework, on the other hand. A strong security model is thus required to address the key vulnerabilities of agentic AI models: attack surface expansion, data manipulation and prompt injection, API usage and sensitive data exposure [14]. Its five key components, shown in Fig. 4 are: 1) a dedicated authentication, authorization and session management module, 2) network and system isolation of the runtime environment, 3) systematic command validation by the user before any offensive action, 4) logging in append-only mode for a posteriori analysis and 5) a kill switch to shut the platform down. The threats related to containment and inversion of dependency are shown in Fig. 5. Isolation prevents unauthorized access to network entities or configurations, and to system capabilities. Command validation by the user ensures the alignment between the ongoing security task and performed operations, and prevent accidental calls to unwanted or dangerous tools upon proposal by the agent. Following and, when necessary, reconstructing the execution track is supported by the logging facility. To enhance the reaction capability and to pave the way to greater autonomy of the framework, a kill switch is set up to immediately halt any agent over which the supervision, or the control over actual operations, would have been weakened or lost.

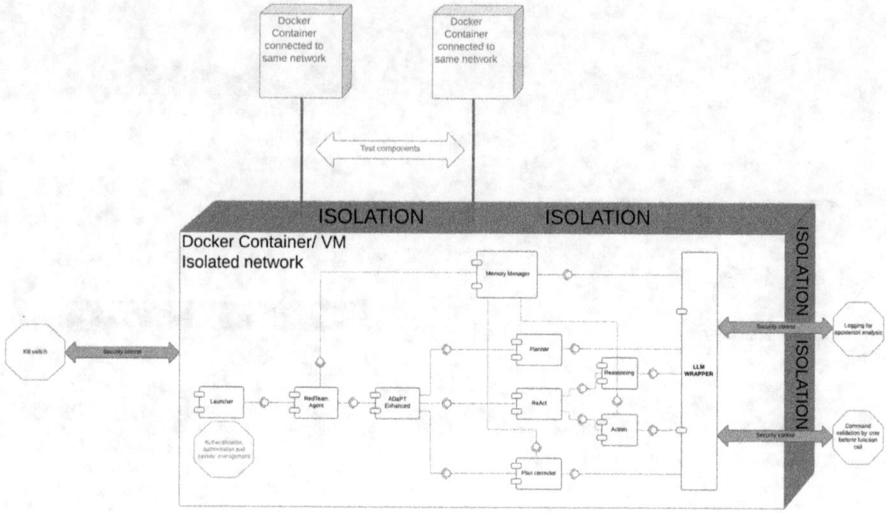

Fig. 4. Security layers wrapping the LLM agent

The LLM itself is used in its default configuration, and with a benevolent user that have not intend to abuse it. Consequently, typical threats like prompt injection attacks [15] or app store abuses [9] are not relevant to RefTeam LLM.

5 Implementation

The proof of concept for the RedTeamLLM model, that we evaluate in following section, entails the **ReAct** component for task execution. The current state of the implementation also covers **ADaPT** for recursive planning, **Memory management** for continuous improvement, and **Plan correction** to support operation continuity after task failure. However, these are less mature, and not evaluated here. RedTeamLLM and related tests are available for the community[4].

The evaluation is tested on a docker container over a Thinkpad e14 gen 5 with 16GB of RAM ddr4 /I513420h processor, and uses OpenAI's API with GPT4-o.

5.1 Three-Step Pipeline

The ReadTeamLLM implementation uses a three-step pipeline, each step handled by a separate LLM session:

1. Reasoning. Before executing any action, the agent reasons about the next steps. Reasoning occurs in an isolated LLM session which elicits an explicit output of its process, detailed steps, and a plan. When the user provides the

[4] https://github.com/lre-security-systems-team/redteamllm.

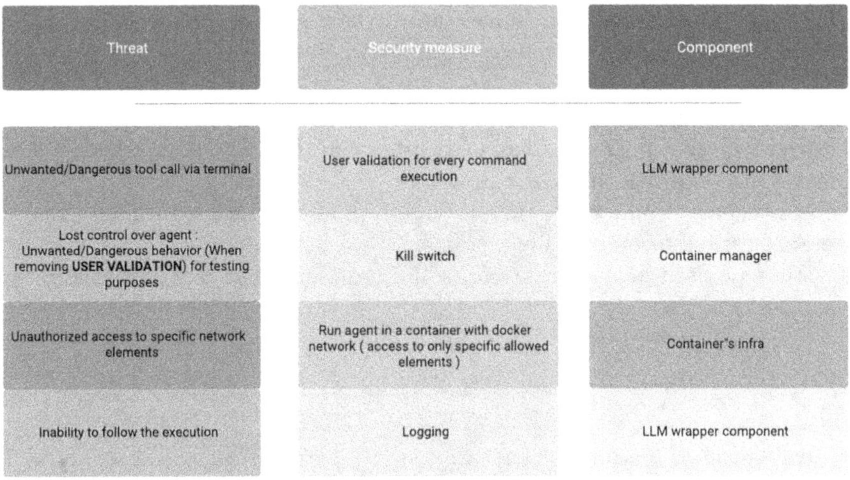

Fig. 5. Security challenges and how RedTeamLLM address them

task definition to the model, it is forwarded to the reasoning component; its output is then passed to the Act component. After each tool call, the executed command and its output are fed back to the reasoning component to generate further analysis.

2. Act. The output of Reasoning is treated as an assistant message by the Act session, which enforces adherence to the plan and reduces the model's inclination to interrupt execution with additional reasoning or safety checks. This setup allows the LLM to focus solely on executing the recommended action. For tool execution, the LLM session has full access to a quasi-interactive, root-privileged Linux terminal. A current challenge is determining when a process requires input; we address this using `strace`, but it is not perfectly precise because some processes read from multiple file descriptors, not only `stdin`. After each tool execution, if the output is too long, it is passed to a summarizer to avoid exceeding the context window.

3. Summarizer. The summarizer is a stateless LLM session: for each request, it summarizes the given command's output. Because this session does not maintain context about the agent's overall goal, it sometimes omits important information. We plan to address this limitation in future work.

5.2 Sample Run

A sample run proceeds as follows:

1. A task is given to the agent (e.g., Obtain root access to the machine with IP `x.x.x.x`").

2. The task is forwarded to the reasoning session as a user message.
3. The reasoner generates a result, which is provided as an assistant message to the acting session.
4. The act session recommends a tool call (e.g., `nmap` or `sqlmap`).
5. After execution, if the command output is lengthy, it is summarized and sent back to the reasoner as a user message.
6. The reasoner produces further thoughts, and the loop continues until the reasoner stops recommending actions.
7. At that point, the system prompts the user for input (e.g., Continue" or a new task).

6 Evaluation

The evaluation is performed in three steps: a qualitative evaluation of the RedTeamLLM capability to autonomously perform offensive operations; a comparative study between the cognitive mechanisms involved in these operations; an ablation study focused on the evaluation of the impact of the presence, or absence, of the reasoning capability.

6.1 Use Cases

The choice of the benchmark to evaluate the RedTeamAgent is based on two factor: reproducibility and variability. We therefore selected 5 use cases: Sar, CewiKid, Victim1, WestWild, CTF4, from VULNHUB repository, which cover a broad range of technical difficulties and various security techniques, are easily deployable, and support reproducible executions. The objective of this work is focused on creating a proof on concept for ReadTeam LLM model, with the evaluation of cognitive operations: summarize, reason, act; and with a processing engine restricted to REaCT component. The 5 selected use cases are embedded in virtual machines from the easy category. This selections also allows us to compare our results since TAPT Benchmark [12] tested PentestGPT [4] on the same target VMs.

RedTeamLLM proves to be competitive for the target use cases, and surpasses PentestGPT on almost all the VM's when using GPT4-o. The decisive factors for these performance are the following ones. First is reasoning, the difference without this step is really important. The agent used block more on same thoughts and doesn't keep a stable execution plan. Launching the agent multiple times, he sometimes completely changes strategy. Having an important amount of tokens dedicated to strategy, output analysis and reasoning help the agent to stay on track. Regularly, without reasoning the agent stops what he's doing to ask for permission. Additionally, giving complete control over a terminal not giving a limited set of tools to the agent, helps with his creativity; being able to chose whatever path to take in order to achieve his goal. Sometime a specific version of a program isn't sufficient so he installs another one, sometime he launches scripts, sometimes he save operation information in a file. Moreover,

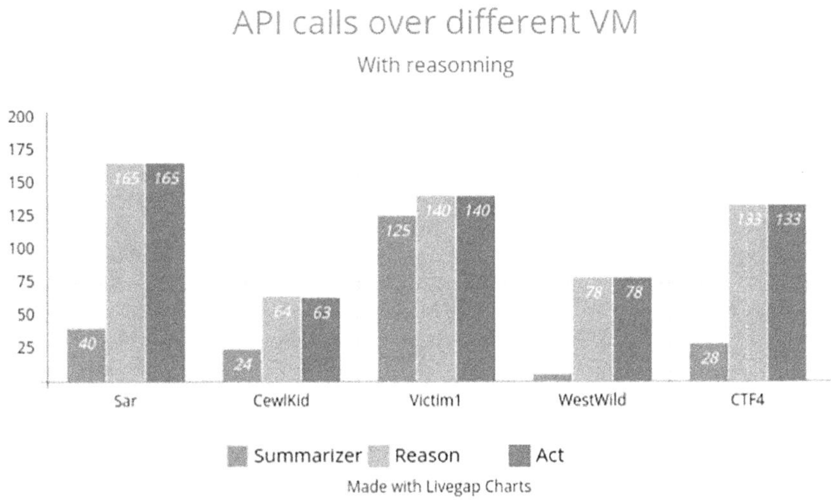

Fig. 6. Number of API calls in Summarizer, Reason, Act steps for the 5 use cases

the fact that he is directly executing the commands himself saves token on other topics. Finally automation is a key part of the agent, which enables longer and more complex automation without the need for manual supervision.

6.2 Cognitive Steps

The RedTeamLLM implementation evaluated in this work is built around the ReACT analysis component. It entails 3 LLM session, *i.e.* 3 interaction dialogs built by assistant and user messages: 3) the summarizer that summarizes command outputs; 2) the reasoning component that reasons over tasks and their outputs, and 3) the Act component that execute the tasks. Figure 6 shows the total number of API calls for each component, over the different use cases after 10 tests on each VM. The Summarizer typically consumes between 9,5% (CTF4) and 15,9% (Cewlkid) of API call tokens, with a low at 3,1% for the WestWild use case and a peek at 30,9% for the Victim1 use case. This peek enables a strong reduction of the required tool calls (See Fig. 7). Reason and Act processes perform a very similar number of API calls.

RedTeamLLM outperforms PenTestGPT in 3 use cases out of 5: *wrt.* the use case write-up, it completes 33% more steps than PentestGPT-Llama (4 successful CTF levels vs. 3) and 300% more than PentestGPT4-o (4 vs. 1) for Victim1 use case, 33% more steps than PentestGPT4-o or PentestGPT-Llama (4 vs. 3) for WestWild use case, 75% more than PentestGPT4-o (3.5 vs. 2) and 250% than PentestGPT-Llama (3.5 vs. 1) for CTF4. PenttestGPT-Llama outperforms RedTeamLLM for Sar by 17% (7 vs. 6) and by 100% (4 vs. 2) for CewiKid use case, while PentestGPT4-o is similar or weaker that RedteamLLM for these 2 test cases.

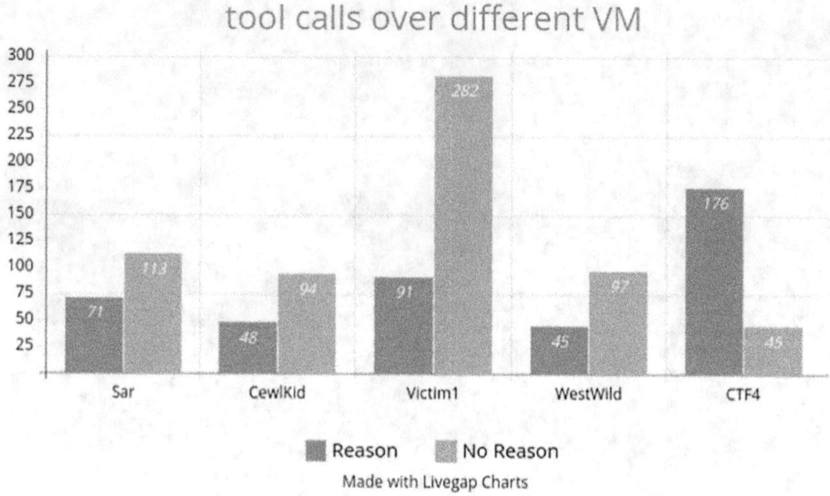

Fig. 7. Number of tool calls without and with reasoning for the 5 use cases

6.3 Reasoning: a Strong Optimization Lever

The ablation study aims to evaluate the contribution of reasoning to the RedTeamLLM framework. Figure 7 shows the number of tool calls without and with reasoning for the 5 use cases. Every LLM session can have tool calls. A tool calls is a specific API response from an LLM session that triggers the use of provided tools (in our case a terminal). For example: when the agent executes a terminal command `ls`, that is a tool call response suggested by the LLM. The total tool calls over the 5 vms with 10 tests on each VM is sumed up: 5 with reasoning, 5 without reasoning. These are only the tool calls with the Act components only because this is where execution is performed. We can clearly see that the agent consumes significantly less tool calls with reasoning in 4 out of 5 use cases: the drop is tool calls range from 37% (Sar) to 68% (Victim1). Only for CTF4, the use of reasoning is bound with an increase of 291% of tool calls, to support a slightly better achievement of the target operation (see Fig. 8). In short, the agent performs more analysis before performing actions, and thus chooses better strategies to perform.

The degree of completion is computed for each use cases, using the write-up, which contains the listing of correct steps to complete the security challenge, as reference. Figure 8 shows these results. The write-up bar shows the total numbers of steps required to achieve the CTF (total of recon, general technique, exploit and privilege escalation). The Reason and No Reason bars show how many steps the agent has completed for each use case with and without reasoning respectively. The test process is similar to previous evaluation: RedTeamLLM handles 5 tests with reasoning and 5 tests without reasoning for every use case. The maximum number of steps achieved over the 5 runs is considered, *i.e.* the better execution. Reasoning improves the results in 4 cases out of 5. In two of

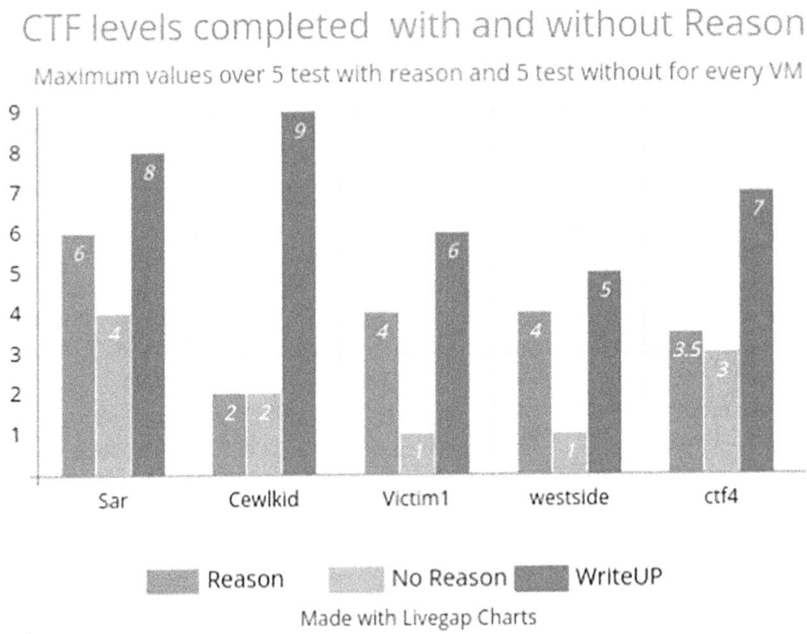

Fig. 8. CTF level completed by the RedTeamLLM framework without and with reasoning for the 5 use cases

these cases, the number of steps mastered pass from 1 to 4. A significant result of our experiments is that this improvement is coupled or with a strong gain in efficiency wrt. to tool calls (see Fig. 7). In one case (Cewlkid), reasoning does not improve the offensive capability.

These results highlight the contribution of the reasoning step to security operation by RedTeamLLM model.

7 Conclusions and Perspectives

Beyond generative AI and now wide-spread Large Language Models (LLMs), Agentic AI is opening wide novel opportunities and threat to global security, and cybersecurity in particular. The objective of this work is to specify a reference model for agentic AI as applied to offensive cyber operations, so that the community can better understand these tools and their capability, leverage them for securing their information systems, and control this novel attack vector.

In this work, we define the key requirements for offensive agentic AI, propose a reference architecture model, and make a proof-of-concept of this architecture focused on iterative task analysis and execution through the ReACT component. The evaluation demonstrates that, though partial, our implementation beats state-of-the art competitors like PentestGPT in 60% of the use cases. It also validate our hypothesis that reasoning is a key feature for agentic AI, since it enables a strong reduction of the necessary tool calls in 80% of the use cases while improving offensive capabilities in 80 % of the use cases. Interestingly enough, in 20% of the use cases, it only supports reduction of tool calls, and thus process costs, and in 20%, the gain in offensive capability requires a 4 times increase in tool calls. This proves that while RedTeamLLM improves both parameters in 60% of cases, it is also efficient in dropping operation costs OR increasing operational capabilities in more complex tasks.

The key insight of this study is that leveraging the dual capability of LLMs to analyze and decompose processes, on the one hand, and to generate code for well-defined tasks, on the other, brings a radical improvement to automation and genericity of ReACT-based offensive cybersecurity frameworks. These first promising results pave the way to structuring the research effort in agentic AI for global security, in particular *wrt.* methodologies for evaluation of cost and automation capabilities of these models. The evaluation of recursive planning, memory management and plan correction is also a necessity to better understand the underlying mechanics and capabilities of agentic models.

References

1. Acharya, D.B., Kuppan, K., Divya, B.: Agentic AI: autonomous intelligence for complex goals–a comprehensive survey. IEEE Access (2025)
2. Bi, Z., Zhang, N., Jiang, Y., Deng, S., Zheng, G., Chen, H.: When do program-of-thought works for reasoning? In: Proceedings of the AAAI Conference on Artificial Intelligence, vol. 38, pp. 17691–17699 (2024)
3. Chowdhury, A.G., Islam, M.M., Kumar, V., Shezan, F.H., Jain, V., Chadha, A.: Breaking down the defenses: a comparative survey of attacks on large language models. arXiv preprint arXiv:2403.04786 (2024)
4. Deng, G., et al.: {PentestGPT}: evaluating and harnessing large language models for automated penetration testing. In: 33rd USENIX Security Symposium (USENIX Security 2024), pp. 847–864 (2024)
5. Falade, P.V.: Decoding the threat landscape: ChatGPT, fraudGPT, and wormGPT in social engineering attacks. arXiv preprint arXiv:2310.05595 (2023)
6. Fang, R., Bindu, R., Gupta, A., Kang, D.: LLM agents can autonomously exploit one-day vulnerabilities. arXiv preprint arXiv:2404.08144 (2024)
7. Fang, R., Bindu, R., Gupta, A., Zhan, Q., Kang, D.: LLM agents can autonomously hack websites. arXiv preprint arXiv:2402.06664 (2024)
8. Heckel, K.M., Weller, A.: Countering autonomous cyber threats. arXiv preprint arXiv:2410.18312 (2024)
9. Hou, X., Zhao, Y., Wang, H.: On the (in) security of LLM app stores. arXiv preprint arXiv:2407.08422 (2024)
10. Hughes, L., et al.: Ai agents and agentic systems: a multi-expert analysis. J. Comput. Inf. Syst. 1–29 (2025)
11. Ismail, I., et al.: Toward robust security orchestration and automated response in security operations centers with a hyper-automation approach using agentic AI (2025)
12. Isozaki, I., Shrestha, M., Console, R., Kim, E.: Towards automated penetration testing: introducing LLM benchmark, analysis, and improvements. arXiv preprint arXiv:2410.17141 (2024)

13. Jin, Y., Jang, E., Cui, J., Chung, J.W., Lee, Y., Shin, S.: DarkBERT: a language model for the dark side of the internet. arXiv preprint arXiv:2305.08596 (2023)
14. Khan, R., Sarkar, S., Mahata, S.K., Jose, E.: Security threats in agentic ai system. arXiv preprint arXiv:2410.14728 (2024)
15. Labunets, A., Pandya, N.V., Hooda, A., Fu, X., Fernandes, E.: Fun-tuning: characterizing the vulnerability of proprietary LLMs to optimization-based prompt injection attacks via the fine-tuning interface (2025)
16. Laney, S.P.: LLM-directed agent models in cyberspace. Ph.D. thesis, Massachusetts Institute of Technology (2024)
17. Lewis, P., et al.: Retrieval-augmented generation for knowledge-intensive NLP tasks. Adv. Neural. Inf. Process. Syst. **33**, 9459–9474 (2020)
18. Li, C., et al.: Chain of code: reasoning with a language model-augmented code emulator. arXiv preprint arXiv:2312.04474 (2023)
19. Li, J., Li, G., Li, Y., Jin, Z.: Structured chain-of-thought prompting for code generation. ACM Trans. Softw. Eng. Methodol. **34**(2), 1–23 (2025)
20. Li, J., et al.: EvoCodeBench: an evolving code generation benchmark with domain-specific evaluations. Adv. Neural. Inf. Process. Syst. **37**, 57619–57641 (2024)
21. Malatji, M., Tolah, A.: Artificial intelligence (AI) cybersecurity dimensions: a comprehensive framework for understanding adversarial and offensive ai. AI Ethics 1–28 (2024)
22. Muzsai, L., Imolai, D., Lukács, A.: HackSynth: LLM agent and evaluation framework for autonomous penetration testing. arXiv preprint arXiv:2412.01778 (2024)
23. Oesch, S., Hutchins, J., Austria, P., Chaulagain, A.: Agentic ai and the cyber arms race. Computer **58**(5), 82–85 (2025)
24. Prasad, A., et al.: Adapt: as-needed decomposition and planning with language models. arXiv preprint arXiv:2311.05772 (2023)
25. Sahoo, P., Singh, A.K., Saha, S., Jain, V., Mondal, S., Chadha, A.: A systematic survey of prompt engineering in large language models: Techniques and applications. arXiv preprint arXiv:2402.07927 (2024)
26. Shavit, Y., et al.: Practices for governing agentic AI systems. Research Paper, OpenAI (2023)
27. Shen, X., et al.: PentestAgent: incorporating LLM agents to automated penetration testing. arXiv preprint arXiv:2411.05185 (2024)
28. Sun, L., et al.: TrustLLM: trustworthiness in large language models. arXiv preprint arXiv:2401.05561 (2024)
29. Sun, S., Liu, Y., Wang, S., Zhu, C., Iyyer, M.: PEARL: prompting large language models to plan and execute actions over long documents. arXiv preprint arXiv:2305.14564 (2023)
30. Wan, S., et al.: CYBERSECEVAL 3: advancing the evaluation of cybersecurity risks and capabilities in large language models. arXiv preprint arXiv:2408.01605 (2024)
31. Wang, X., et al.: Self-consistency improves chain of thought reasoning in language models. arXiv preprint arXiv:2203.11171 (2022)
32. Wang, Y., Wu, Z., Yao, J., Su, J.: TDAG: a multi-agent framework based on dynamic task decomposition and agent generation. arXiv preprint arXiv:2402.10178 (2024)
33. Wang, Y., Wu, Z., Yao, J., Su, J.: TDAG: a multi-agent framework based on dynamic task decomposition and agent generation. Neural Netw. 107200 (2025)
34. Wei, J., et al.: Chain-of-thought prompting elicits reasoning in large language models. Adv. Neural. Inf. Process. Syst. **35**, 24824–24837 (2022)

35. Xu, J., et al.: AutoAttacker: a large language model guided system to implement automatic cyber-attacks. arXiv preprint arXiv:2403.01038 (2024)
36. Yao, S., et al.: Tree of thoughts: deliberate problem solving with large language models. Adv. Neural. Inf. Process. Syst. **36**, 11809–11822 (2023)
37. Yao, S., et al.: ReAct: synergizing reasoning and acting in language models. In: International Conference on Learning Representations (ICLR) (2023)
38. Yao, Y., Duan, J., Xu, K., Cai, Y., Sun, Z., Zhang, Y.: A survey on large language model (LLM) security and privacy: the good, the bad, and the ugly. High-Confidence Comput. 100211 (2024)
39. Zhang, Z., Zhang, A., Li, M., Smola, A.: Automatic chain of thought prompting in large language models. arXiv preprint arXiv:2210.03493 (2022)

Author Index

Published by Springer Nature Switzerland AG 2026
E. Mercier-Laurent et al. (Eds.): AI4S 2025, IFIP AICT 780, pp. 355–356, 2026.
https://doi.org/10.1007/978-3-032-21184-2